Why Allies Rebel

Why do powerful intervening militaries have such difficulty managing comparatively weak local partners in counterinsurgency wars? Set within the context of costly, large-scale military interventions such as the US war in Afghanistan, this book explains the conditions by which local allies comply with (or defy) the policy demands of larger security partners. Analysing nine large-scale post-colonial counterinsurgency interventions including Vietnam, Afghanistan, Iraq, Sri Lanka, Yemen, Lebanon, Cambodia, and Angola, this book utilizes thousands of primary source documents to identify and examine over 450 policy requests proposed by intervening forces to local allies. By dissecting these problematic partnerships, this book exposes a critical political dynamic in military interventions. It will appeal to academics and policymakers addressing counterinsurgency issues in foreign policy, security studies and political science.

BARBARA ELIAS is Assistant Professor of Government at Bowdoin College specializing in international relations, counterinsurgency warfare, national security, Islam and politics and US foreign policy. She was the Director of the Afghanistan/Pakistan/Taliban Documentation Project at The National Security Archive in Washington DC.

Why Allies Rebel

Defiant Local Partners in Counterinsurgency Wars

Barbara Elias

Bowdoin College

CAMBRIDGE
UNIVERSITY PRESS

CAMBRIDGE
UNIVERSITY PRESS

University Printing House, Cambridge CB2 8BS, United Kingdom

One Liberty Plaza, 20th Floor, New York, NY 10006, USA

477 Williamstown Road, Port Melbourne, VIC 3207, Australia

314–321, 3rd Floor, Plot 3, Splendor Forum, Jasola District Centre, New Delhi – 110025, India

79 Anson Road, #06–04/06, Singapore 079906

Cambridge University Press is part of the University of Cambridge.

It furthers the University's mission by disseminating knowledge in the pursuit of education, learning, and research at the highest international levels of excellence.

www.cambridge.org
Information on this title: www.cambridge.org/9781108490108
DOI: 10.1017/9781108784979

First published 2020

A catalogue record for this publication is available from the British Library.

Library of Congress Cataloging-in-Publication Data
Names: Elias, Barbara, author.
Title: Why allies rebel : defiant local partners in counterinsurgency wars / Barbara Elias, Bowdoin College, Maine.
Other titles: Defiant local partners in counterinsurgency wars
Description: Cambridge, United Kingdom ; New York : Cambridge University Press, 2020. | Includes bibliographical references and index.
Identifiers: LCCN 2020004488 (print) | LCCN 2020004489 (ebook) | ISBN 9781108490108 (hardback) | ISBN 9781108748063 (paperback) | ISBN 9781108784979 (epub)
Subjects: LCSH: Counterinsurgency–History–20th century. | Counterinsurgency–History–21st century. | Alliances. | Intervention (International law)–Case studies.
Classification: LCC U241 .E55 2020 (print) | LCC U241 (ebook) | DDC 355.02/18–dc23
LC record available at https://lccn.loc.gov/2020004488
LC ebook record available at https://lccn.loc.gov/2020004489

ISBN 978-1-108-49010-8 Hardback

Contents

Figures

Tables

Acknowledgments

This book exists due to the insights, generosity, and patience of the scholars who taught me – Avery Goldstein, Alex Weisiger, Michael Horowitz, Edward Mansfield, Jim Blight, janet Lang, Ian Lustick, Rudy Sil, Bob Vitalis, Jennifer Amyx, Ellen Kennedy, Matthew Levendusky, Julia Lynch, and the determined experts at The National Security Archive – Tom Blanton, Malcolm Byrne, Michael Evans, Sajit Gandhi, Sue Bechtel, Maria Lorena Martinez, Svetlana Savranskaya, Joyce Battle, John Prados, and Mary Curry.

I am grateful for my superb mentors and kindhearted friends in the Department of Government & Legal Studies at Bowdoin: Ericka Albaugh, Laura Henry, Allen Springer, Henry Laurence, Chris Potholm, Mike Franz, Andrew Rudalevige, Jeff Selinger, Paul Franco, Jean Yarborough, Shana Starobin, Chris Heurlin, Janet Martin, Chryl Laird, Maron Sorenson, and Lynne Atkinson. Thank you as well to Barbara Levergood, and those generous enough to help proofread, copyedit, and check citations, specifically Penny Harper, Lorenzo Meigs, Apekshya Prasai, Paloma Tisaire, Kien Pham, Cathy Sunshine, and William Minter. At Cambridge University Press thank you to John Haslam and the anonymous reviewers whose insightful comments inspired essential improvements to the book. It is also important to me to acknowledge the phenomenal caregivers and teachers at the Bowdoin Children's Center and the Williams-Cone School, whose thoughtful dedication to their jobs enabled me to do mine.

Several institutions provided support at various stages of the project, including Bowdoin College, the University of Pennsylvania, the Wharton Risk Management Center, the Christopher Browne Center, and the Smith Richardson Foundation. Earlier versions of select material were previously published in "The Big Problem of Small Allies: New Data and Theory on Defiant Local Counterinsurgency Partners in Afghanistan and Iraq," *Security Studies* 27, no. 2 (April 3, 2018): 233–62 © Taylor & Francis 2018, all rights reserved. I thank the publishers for their permission.

I am also thankful for my amazing mom, fiercely generous dad, my smart nieces Aimee and Mary, wonderful nephew Cameron, and extraordinary sister Anna for their relentless support as I took a decade to put together this book. Similarly, thank you to my friends who offered me encouragement, cooked me amazing meals, and helped in every way – thank you Kim Turner, Johanna Lacoe, Camille Bryan, Rosella Cappella, Allison Evans, Ryan Grauer, Kaija Schilde, Murad Idris, Karin Hoepker, Michael Kolster, Calvin Kolster, Christy Shake, Lucy, Chris, Egan, and Zach Dawson, Maggie Solberg, Morten Hansen, Ingrid Nelson, Allyson Casey, and Miles and Lydia Casey-Nelson. I must also thank my yia-yia and poppou, Andreas and Anna Elias, who I carry with me in life and through these pages.

A heartfelt thank you to my children, Nathaniel and Gabriel, for sharing their beautiful, earnest, generous, and joyful hearts with me every day.

Last, but really first, this book is dedicated to Jens, my kindhearted and unfailing partner and love. Thank you.

1   Introduction

> In war it is not always possible to have everything go exactly as one likes.
> In working with allies it sometimes happens that they develop opinions
> of their own.   —Winston Churchill

In 2011 the United States seemed victorious in Iraq. The hard-won yet
precarious achievement cost the USA over $2 billion, leaving 4,497
Americans and almost half a million Iraqis dead.[1] With palpable relief,
in December President Barack Obama announced, "We're leaving
behind a sovereign, stable and self-reliant Iraq, with a representative
government."[2] With the grueling intervention over, Iraq seemed like an
Iraqi problem, no longer a divisive quagmire for the United States.

Even as the president's words lauding Iraqi sovereignty hung in the air,
US allies in Baghdad were using their autonomy to overturn the anti-
sectarian policies Washington considered critical to Iraq's stability.
Baghdad's return to divisive anti-Sunni policies fueled the rapid rise of
the Islamic State (ISIS), a devastating outcome that threatened Iraqi and
US security. As General David Petraeus, former commander of US
forces in Iraq, commented in 2014, Baghdad's reversion to sectarianism
was "inexplicable," creating "fertile fields for the replanting of the seeds
of insurgency and rejectionism. And it made ISIS's task much easier than
it should have been."[3] As the situation deteriorated, Washington
lamented its lack of influence over its Iraqi allies, who were reversing
the counterinsurgency gains made just a few years earlier. Petraeus
confessed that despite copious classified briefings on Iraq's unraveling
and its disastrous consequences for US security, "at the end of the day, if
[an allied] government decides on a certain course, there are limits to
what a [supporting] country's influence can achieve."[4] This was not the
first time US defense experts had arrived at this conclusion after a costly
intervention. Forty years earlier, war-weary officials returning from
Vietnam reflected, "We cannot fight a counterinsurgency war as a surro-
gate of a threatened ally; this was true even after we had introduced large
numbers of American combat forces into Vietnam. We have had to return
again and again to the hard fact that it was basically our ally's war."[5]

1

Washington's inability to sway the very same partners that US forces had empowered and sustained in both Iraq and Vietnam can seem puzzling. After all, the United States is a superpower, so why would US officials have such trouble influencing smaller, allied regimes that critics deride as "puppets"?[6] To explore this question, we must begin by recognizing that the difficulties in coercing ostensibly weak partners are not limited to US interventions. Indeed, the difficulty that large states encounter in trying to manage smaller but nonetheless crucial security allies has been a persistent challenge in interventions, fittingly dubbed the "big influence of small allies" by Robert Keohane.[7] The Soviets in Afghanistan, Indians in Sri Lanka, and Egyptians in Yemen all similarly expressed dissatisfaction with their small, local counterinsurgency partners, persistently remarking on the mediocre performance of these local allies and on their own lack of leverage to pressure these allies to adopt better strategic policy.[8] As Robert Komer lamented about US partners in Saigon, "We became their prisoners rather than they ours – the classic trap into which great powers have so often fallen in their relationships with weak allies. The GVN [Government of Vietnam] used its weakness as leverage on us far more effectively than we used our strength to lever it."[9]

Scrutinizing dynamics between local and intervening counterinsurgency partners in military interventions, this book seeks to explain what makes local allies more or less likely to comply with the policy requests of intervening partners in order to better understand a key component of counterinsurgency (COIN) warfare. Doing so requires tackling several big questions: When do intervening forces have coercive leverage over local partners? When can local partners coerce their powerful patrons? What makes these partnerships so notoriously problematic? To address these questions, I examine nine interventions: the United States in Vietnam, Afghanistan, and Iraq; the USSR in Afghanistan; India in Sri Lanka; Egypt in Yemen; Vietnam in Cambodia; Cuba in Angola; and Syria in Lebanon. Drawing on these case studies, I present a theoretical framework arguing that the key to understanding patterns in local compliance is the interaction between the respective interests and dependencies of both partners.

For instance, when the interests of local and intervening allies converge over a policy requested by intervening forces and intervening forces cannot implement the policy proposed without the participation of the local partner, compliance is likely since local allies must contribute in order to fulfill the policy and secure the payoff. However, local ally compliance with a request from intervening partners is less likely when the interests of both allies converge and intervening forces can implement the request unilaterally, if necessary, due to incentives for local counterinsurgents to evade responsibility under these conditions and free ride off the efforts of wealthy partners.

Conversely, I propose that when the interests of intervening and local counterinsurgency partners diverge and the intervening forces are unable to implement the request independently if necessary, local compliance is unlikely. Under these circumstances, interveners have few opportunities for leverage against smaller clients. However, when the interests of intervening and local allies diverge and, if necessary, intervening forces have the capacity to execute the policy unilaterally, compliance by local partners is likely. In this case local regimes have an incentive to participate in order to avoid being isolated or undermined by unilateral action undertaken by the foreign ally.

The proposed theory has implications for the process of managing local partners, better contextualizing coercive diplomatic strategies, and crafting strategies based on a realistic understanding of alliance relations in counterinsurgency. The focus here is not on the prudence or effectiveness of counterinsurgency interventions more broadly. Judgments about the consequences of particular approaches to defeating insurgents are set aside in order to focus on alliance politics as an often-overlooked yet critical component of counterinsurgency warfare.

Why Local Allies Matter in Counterinsurgency

Alliance politics are decisive in counterinsurgency. The nature of the partnership between foreign and local forces can influence essential features of an intervention, including its duration, costs, and outcome. The ability of intervening forces to effectively motivate local COIN allies to accept increased responsibility for providing public services and combating insurgents can help determine how long an intervention will last. Additionally, the length of an intervention and how much assistance foreign forces are expected to provide affects wartime expenditures, which can be enormous. In 2018 the Costs of War project at Brown University calculated that the USA had spent over $5.93 billion on post–September 11, 2001 military efforts, including the counterinsurgency interventions in Iraq and Afghanistan.[10]

Furthermore, the effectiveness of alliances between local and intervening forces can affect war outcome. The 2006 US COIN field manual, *FM 3-24*, instructs US forces to adopt an approach to counterinsurgency that relies on a coherent and effective partnership between US and local forces; indeed, throughout most of the manual the US and its local allies are more often than not conceived as a unitary force.[11] Moreover, according to *FM 3-24*, victory ultimately hinges on the success of local partners, who are expected to win over the local population, provide legitimacy, and eventually take over responsibility for governance and security, summarizing their primacy by claiming "the primary objective of any COIN operation is to foster development of effective governance by a legitimate government."[12] Conrad Crane, a key contributor to the field manual, explained, "The first principle of counterinsurgency emphasized that unless the United States could leave behind a legitimate indigenous authority, COIN would fail."[13] Nonetheless, although the role of local allies can make the difference between winning and losing, the complex political process of working with local partners has been severely undertheorized.

In response to criticisms regarding the oversimplification of alliance politics in *FM 3-24*, an updated 2014 edition of the field manual acknowledges that local COIN allies often operate under a different set of priorities than American forces. This is problematic, the manual states, since US forces "can never fully compensate for lack of will, incapacity, or counterproductive behavior on the part of the supported government."[14] Yet despite noting the often divergent preferences of local and intervening counterinsurgents, the manual does not say much about the effect of those divisions. It suggests that coercing local partners is a relatively easy endeavor, claiming that "the U.S. has many tools at its disposal to influence a host nation in addressing the root causes of an insurgency. For example, economic and military aid can be tied to certain actions or standards of behavior by a host nation."[15] No further guidance is provided about how to tie aid or set conditions that will make coercive tactics more successful. This logic also presumes that the resources of wealthy patrons can be easily translated into influence over local clients, despite extensive evidence that leveraging aid to induce political reform is a complex process, in particular with politically vital partners such as local COIN partners that are essential for long-term victory.[16]

In sum, prevailing doctrine holds that in order to succeed, intervening forces must assist local partners in gaining capabilities, legitimacy, and sovereignty, yet it fails to specify exactly how intervening forces should go about inspiring or inducing local partners to reform. Little consideration is given to the characteristics of the regimes that typically need US support or to the political and military conditions that may make reforms more or less likely. As Stephen Biddle noted in his 2008 review of *FM 3-24*, "The existence of an insurgency in the first place is often a signal of an illegitimate government with strong leadership interests in an unrepresentative distribution of wealth and power. In many cases, leaders will see U.S.-sponsored reforms as a greater threat to their personal well-being – or even survival – than the insurgency."[17] In Vietnam, El Salvador, Iraq, and Afghanistan, the tendencies of local partners to engage in corrupt, discriminatory, and self-protective behaviors (perhaps to prevent a coup) were part of what enabled insurgents to flourish in the first place.[18] Insurgencies take root in worst-case scenarios, typically in environments where the state lacks capacity, legitimacy, and inclusive institutions.[19] These problematic characteristics also color the types of policies these regimes are inclined to adopt to battle insurgents, which may include repressive and discriminatory approaches.[20] Therefore, understanding when it is possible to pressure local allies, and what means of persuasion or coercion have been most successful, is critical to any larger agenda to win over the population or suppress enemy insurgents.

In light of the pivotal role of local allies in COIN interventions, it is puzzling that most work on counterinsurgency, such as *FM 3-24*, glosses over the issue of managing local partners, instead largely presuming there will be earnest local participation.[21] According to Daniel Byman, despite their "central importance, thinking and scholarship on counterinsurgency tends to ignore the role of allies. Analyses are typically divided into two kinds: those that focus on the insurgents, and those that examine COIN forces, with the latter including both the United States and the host government and thus wrongly assuming they both share the same interests."[22] The curious omission of alliance politics from COIN analysis has traditionally been true not only for counterinsurgency theory broadly, but for particular wars as well. Writing on Vietnam, Robert J. McMahon commented, "Although the nature of the South Vietnamese regime inspired a flood of polemical tracts in the United States during the 1960s and early 1970s, relatively few scholars have probed deeply into the underlying structure[s]" motivating Saigon's behavior.[23]

There are several reasons why alliance relations have been typically ignored. First, as Walter Ladwig has observed, current scholarship on counterinsurgency has "drawn a significant number of its insights from the colonial era, particularly the experiences of the British in Malaya and the French in Algeria, which can blind scholars to the challenge of working through an autonomous local government, because in those cases the European power *was* the government."[24] By building on a counterinsurgency canon produced by colonial powers aiming to conquer foreign territory, as opposed to bolstering a foreign regime, the postcolonial scholarship on counterinsurgency can falsely presume unity of forces. The propensity to look to colonial wars as strategic models reflects the tendency to focus on the best-known cases of counterinsurgency success – success that, as David Kilcullen has observed, tends to be "limited to cases where a colonial or postimperial government was fighting on the territory of its dependent (ex)colonies."[25]

Second, intervening forces often presume that since local and intervening allies share key overarching goals, including eliminating the insurgency and strengthening the local regime, their interests and those of the local regime will inevitably align for the most part. However, like all allies, counterinsurgency partners will sometimes disagree about certain key factors, since all alliances are a combination of overlapping and divergent interests.[26] Yet, it sometimes seems surprising to intervening forces that their local clients have distinctive visions and divergent priorities. In a 2006 memo, US National Security Adviser Stephen J. Hadley asked the National Security Council, "Do we and Prime Minister Maliki share the same vision for Iraq?"[27] Torn between Maliki's "reassuring words" about curbing sectarianism and "repeated reports from our commanders on the ground" regarding the Iraqi government's anti-Sunni policies, Hadley struggled to explain Maliki's sectarian actions, offering suggestions for how the Iraqi leader could "demonstrate his intentions to build an Iraq for all Iraqis and increase his capabilities."[28] Three years into the war, high-level American policymakers were questioning whether Washington and the Shi'a-dominated regime in Baghdad had different visions for achieving stability in Iraq, even though such structural tensions between allies are inevitable and have afflicted other critical American partnerships time and again.[29]

The naiveté regarding the divergent priorities of local partners is likely rooted in the omission of alliance politics in the counterinsurgency canon, as well as assumptions about the inherent power and influence of intervening forces. Because foreign patrons control a preponderance of the resources, they often assume they will exert a preponderance of influence in alliances with local regimes. Such assumptions make it easy to gloss over the complexities of managing structurally weak but politically essential partners.[30] Just as assumptions that technologically superior militaries would easily win wars against penurious insurgents had to be revised, assumptions that wealthy patrons will inevitably triumph in bargaining encounters with resource-poor local allies similarly merits reconsideration based on empirical evidence that wealth does not necessarily translate readily into influence in these conditions. This book explores these critical yet typically overlooked dynamics in counterinsurgency wars.

Colonial versus Noncolonial Counterinsurgencies: The Consequences of Legal Sovereignty for Local Partners

Part of the surprising leverage of local partners over wealthy patrons in counterinsurgency interventions relates to dynamics in the postcolonial era. Under colonialism, imperial powers such as Britain or France provided local governance through local satellite administrations as extensions of their empires. While colonial officials employed a variety of techniques to augment their ranks with indigenous forces like Ulster Protestants, Turkish Cypriots, Punjabi Indians, or Alawite Syrians, it was critical to the colonial project to maintain the "thin white line" of administrative command to distinguish between Europeans and non-Europeans.[31] Providing autonomy or sovereignty, even nominally, would defeat the purpose of colonial rule.

However, counterinsurgents in the postcolonial era, such as the USA in Iraq and Afghanistan, are bound by another set of policies vis-à-vis local governance. Required by domestic and international norms to eschew relationships redolent of colonialism, contemporary counterinsurgents cannot simply conquer and occupy as a long-term strategy, but instead must bolster an ostensibly sovereign local administration as a proxy partner to battle the insurgents.[32] As a result, external intervening forces (sometimes called patrons, principals, or interveners) depend on local allied governments (sometimes referred to as proxies, clients, agents, or host nations) to offer a variety of important services for the war effort. First, intervening forces look to local counterinsurgents for knowledge of the local physical, social, and political terrain, hoping to counteract the significant advantage local insurgents often hold in this regard. While intervening forces enjoy a preponderance of resources, local insurgents have the distinct advantage of local knowledge. Intervening forces hope that by engaging a local counterinsurgent regime as a strategic partner, they can neutralize the insurgents' substantial home turf political advantage.

Furthermore, intervening forces expect local allies to provide legitimacy for the collective counterinsurgent cause.[33] The "foreignness" of external counterinsurgents is a significant political liability for interveners. Afghan commander Abdul Haq maintained that this local-foreign divide would be the definitive political dynamic for the USA in Afghanistan. According to Haq, Afghans "will fight. Despite everything the Taliban has done to destroy human rights, to destroy people's health, education, their livelihoods, they will fight. Afghans will always unite in the face of what they see as a foreign enemy and this will help strengthen the Taliban."[34] In the postcolonial era, promoting the idea of local legal sovereignty is vital for legitimizing the counterinsurgency mission as a long-term solution to civil war and instability. As David Lake has detailed, sovereignty is critical to legitimizing state-building, because local regimes that are considered loyal to foreign patrons are viewed as less legitimate and are therefore less likely to solidify control in the long run.[35]

Policymakers in democracies such the USA are also keen to promote the sovereign character of their local COIN partners in order to mollify public opinion at home and internationally.[36] As Max Boot asserted regarding the US war in Vietnam, General William Westmoreland's "'attrition' strategy worked, but in the wrong country. It broke the will not of North Vietnam, but of America."[37] While promoting the sovereignty of local partners would have been anathema to colonialists such as the French in Vietnam or Algeria, democratic intervening states in the postcolonial era do not have the political option to deny local sovereignty as a long-term strategy.[38] "Operation Iraqi Freedom," the given title of US operations in Iraq, explicitly emphasizes liberation and Iraqi national sovereignty in order to avoid politically risky questions of colonial conquest, oil wealth, and exploitation. As detailed in Chapter 4, "Operation Iraqi Freedom" became an increasingly awkward phrase as Americans staffing the Coalition Provisional Authority served as the government of Iraq for fourteen months in 2003–4.

Like so much in counterinsurgency, employing local regimes to provide legitimacy and to serve as counterinsurgency proxies is much easier said than done. Legitimacy can be a difficult mandate for besieged regimes that required large-scale support from external powers in the first place due to their chronic political and military weaknesses. As Daniel Byman observed regarding Iraq and Afghanistan,

The United States is on the horns of a dilemma when working with allies to fight insurgents. Allies experience insurgencies because of the weakness of the state, as well as other factors such as discrimination and corruption. These problems create tremendous difficulties when the United States expects allied militaries to fight on its behalf – the structural problems that cause the insurgencies also shape how well allies fight them.[39]

Employing regimes suffering from chronic political weaknesses to provide legitimacy is a problematic and perhaps a somewhat fanciful proposition, especially since the influx of cash from wealthy patrons can exacerbate rather than curb corrupt practices.

Regimes that require large-scale intervention by foreign military forces to battle significant insurgent threats to their territorial integrity do not have complete sovereignty in a traditional sense. Yet, as Stephen Krasner has noted, there are a variety of concepts of sovereignty in international politics, some emphasizing legal authority, others stressing territorial control.[40] In large-scale counterinsurgency interventions local regimes do not have territorial control, but they often have legal political authority in regard to the policies directly promulgated by their government. This has significant political consequences. Surrendering legal authority to local regimes instead of maintaining a hierarchical chain of command for policy implementation, from patrons down to local clients, means that local regimes have a good deal of opportunity to refute or negotiate proposed policies, since their buy-in is often necessary.

Furthermore, in the postcolonial world local sovereignty has proved to be a durable norm. The perceived benefits of local legal sovereignty, in terms of legitimizing intervention locally and internationally, are viewed as outweighing the costs of surrendering legal authority to local partners. As detailed in Chapter 8 on the Soviet intervention in Afghanistan, denying local sovereignty in the postcolonial era poses substantial risks for interveners. After the Soviet invasion ushered in a large-scale KGB infiltration of the Afghan regime in Kabul, there was a massive desertion of Afghan civilian and military personnel.[41] Former state employees often defected to the opposition, providing intelligence and strengthening the bureaucratic and political capacity of the insurgency. The exodus of Afghan professionals deepened Kabul's dependency on Soviet staff and weakened the local ally's ability to withstand the withdrawal of intervening forces. Such institutional dependencies are not readily undone.

Thus comprehensively denying local sovereignty is typically considered politically untenable. As US Secretary of Defense Donald Rumsfeld wrote in 2003, a transfer of authority from the US-run Coalition Provisional Authority to an Iraqi-led regime was key for several reasons. According to Rumsfeld,

All agree that we should give Iraqis more authority quickly ... Moving too slowly with respect to passing sovereignty to the Iraqis risks having the center or gravity of the Iraqi population move against the Coalition, their cooperation decline, Iraqis become afraid of joining the police, the Governing Council, etc. and be more likely to work with our enemies. This in turn risks a security deterioration that could cause a loss of support from the American people, the Congress and/or the international community.[42]

The emphasis on local legal sovereignty for strategic reasons provides local allies with leverage, since intervening allies are required to consult, not command local partners on local governance policies. Dick Camp described how the US decision to approach Baghdad as a sovereign regime meant Iraqi leaders had significant leverage over their powerful American patrons. "With the turnover of control from the CPA [Coalitional Provisional Authority] to the new Iraqi government, the command relationship changed. United States forces were no longer pulling the strings. General George Casey, who relieved Sanchez, commanded the coalition forces, but he was militarily handicapped because under the turnover agreement, Prime Minister Allawi was supposed to call the shots."[43] Thus the legal sovereignty of local allies, however manufactured or nominal, is nonetheless politically consequential as it prevents intervening allies from dictating governance policies, even if those policies are deemed mission critical. Along these lines, in 2009 Stanley McChrystal, the commanding US general in Afghanistan, emphasized that "increas[ed] transparency within the Afghan government" was a critical political precondition for winning the war.[44] McChrystal also considered Afghan President Hamid Karzai, a US ally, the "chief stumbling block to meaningful reform,"[45] yet when asked whether Karzai was impeding the US mission and failing to provide transparency, McChrystal retorted, "President Karzai is the commander. I work for President Karzai."[46] McChrystal demonstrates the catch-22: due to its mission of supporting the government in Kabul, the United States was unable to press for reforms considered fundamental to the mission of ensuring Kabul's success.

While the dependency of local proxies on foreign patrons for security assistance is well recognized, the dependency of intervening forces on local partners to implement certain key policies – due to the local regime's legal sovereignty – is frequently overlooked or dismissed as unimportant. There is a faulty assumption that such dependence is readily neutralized by resource asymmetries favoring patrons (it is not). By insisting on the sovereign authority of local regimes, intervening partners turn policy discussions into negotiations between legally sovereign partners, providing local partners with opportunities to defy or alter policies to better suit their interests. It should not be surprising that even a nominal degree of sovereign authority has political consequences, as emphasizing local authority provides legitimacy for the local regime and the intervention writ large. Local sovereignty has the unintended consequence, however, of surrendering leverage to local proxies in inter-alliance bargaining encounters. This can have significant strategic consequences, especially when local allies pursue independent agendas that impose costs on intervening forces.

The diplomatic record demonstrates significant variation in compliance outcomes. At times, local counterinsurgents defy requests from their foreign patrons. The Karzai administration, for example, made no effort to ensure that US funding did not fall into the hands of drug traffickers, despite Washington formally specifying that such action be undertaken as part of a $26.6 million aid package addressing police and justice programs in 2002.[47] Yet local allies are not always defiant, but may cooperate in some cases, even when it is costly. For instance, though it took three years, Saigon eventually bowed to American pressure to lower the draft age in South Vietnam from 20 to 18 to augment allied manpower despite elite resistance to the measure.[48] Outcomes can also fall anywhere along a wide spectrum of partial compliance as local partners implement some but not all aspects of a proposed policy, which effectively imposes compromises on patrons through half-measures and partial cooperation. When the Indian High Commissioner J. N. Dixit requested the Sri Lankan government to extend official recognition to Tamil political parties, for example, Sri Lankan officials responded that they would provide recognition when Tamil groups applied to the Elections Commissioner, giving Colombo some oversight over Tamil groups in exchange for recognition.[49] This variation in compliance presents a formidable challenge for intervening forces in crafting a collective security policy, as there are no guarantees that key strategies involving local participation will actually progress according to design. The risk is that critical counterinsurgency initiatives will be doomed before they can even be attempted, without any effort on the part of the enemy.

The Argument

What makes local allies compliant or defiant in counterinsurgency interventions? In popular discourse, policymakers tend to blame the foibles of the particular local regime or its uncooperative head of state for uncooperative behaviors. Kai Eide, United Nations Special Representative to Afghanistan, for example, claimed that "to President Obama and many Americans, the conflict in Afghanistan is the right war that went wrong, largely due to an 'irrational' and 'unpredictable' Afghan president."[50] Taking a different tack, I instead build on the work of political scientists who argue that policymakers associated with intervening forces have not fully recognized key structural factors that motivate local compliance and defiance, and thus have failed to tailor their approaches accordingly.[51]

This study builds on a tradition of modeling partnerships between local and intervening forces according to principal-agent theory,[52] which explains how one actor (the principal or patron) can provide incentives to another (the agent or client) to undertake tasks on the former's behalf.[53] While principal-agent models do not perfectly describe the counterinsurgency alliances analyzed here, given that local allies in the postcolonial era are strategic partners, not employees of intervening powers, agency theory sheds light on certain key problems that emerge when powerful actors rely on smaller but well-positioned actors to accomplish a task.[54] Principal-agent theory primarily refers to actors with formally or legally defined obligations to each other, but it is also often grouped with patron-client approaches, which tend to include actors with less formal obligations as well. Studies on counterinsurgencies often use the terms in tandem, since both involve asymmetric power relationships and describe common political dynamics. Thus, as Walter Ladwig has noted regarding asymmetric counterinsurgency alliances, "Although the kind of patron–client interaction we are examining … is not a pure example of the principal-agent problem in the sense that the agents (the client state) are not hired by the principal (the patron), agency theory can still apply in situations that lack an overt or official delegation of responsibility."[55]

However, while principal–agent models aptly describe key political dynamics produced by significant resource asymmetries in counterinsurgency partnerships, they can neglect other factors that also affect interalliance bargaining dynamics. For example, principal–agent approaches have overlooked variation in how dependent intervening forces are on local partners for policy implementation. In some cases, due to the nominal legal sovereignty of local regimes, intervening forces are dependent on local partners to implement policies, while in other cases, intervening forces may be able to implement desired policies unilaterally if necessary. This variation in how indispensable (or not) local allies are for policy fulfillment has implications for the likelihood of local compliance and the ability of intervening forces to issue credible threats to pressure compliance.

Thus, in order to take a comprehensive approach, I draw insights from both principal–agent approaches and alliance politics more broadly, framing partnerships here primarily as binary relationships between two actors, typically one wealthy and the other resource-deficient. I contend that local regimes are strategic actors that make cost-benefit calculations when faced with policy demands from intervening forces. Naturally, however, each of these allies comprises a wide array of competing and heterogeneous bureaucracies and personalities. Neither ally is dependably rational or unified.[56] In order to control for the role of bureaucratic politics, albeit imperfectly, I examine high-level negotiations between local and intervening allies, such as meetings between ambassadors and prime ministers, which filters out some of the detail of bureaucratic policy-making processes, as detailed in Chapter 3 on methods.

The proposed model explores patterns in the dependent variable, local compliance, defined as the act or process of conforming, submitting, or adapting to a demand, proposal, or regimen, or to coercion. In particular, I examine the effect of four primary independent variables on the likelihood of compliance with policies proposed by intervening forces, namely (1) the capacity of the local partner to implement the requested policy, (2) whether the respective interests of the local and intervening forces converge or diverge over the policy, (3) the dependency of the intervening ally on the local regime to implement the requested policy, and (4) acute external threats from enemy forces. As detailed in the following chapters, I argue that these four variables are key to understanding the seemingly curious behavior of local COIN partners, who at times seem to undermine the strength of a joint counterinsurgency effort by remaining obstinate against key reforms promoted by intervening patrons.

Instead of simply presuming that local allies comply with such requests when it is in their interest to do so, and refuse when their interests diverge, I argue there is a specific pattern of interaction between interests, on one hand, and the reliance of foreign intervening forces on local actors to implement policy, on the other hand, that affects the likelihood of compliance by local partners with policy demands. The dependency of patrons on their local clients to implement a proposed policy, due to the legal sovereignty of local regimes, will affect how interests play out in terms of compliance outcomes. Even when interests are aligned, if local allies can free ride they will tend to do so, and thus fail to comply. If they are unable to free ride, they are more likely to accept their share of the burden and implement the policy as requested by intervening allies.

The interaction between interests and dependencies plays out in several alternative scenarios that can be traced through a variety of conflicts. When interests of the larger ally and the local partner diverge, and the larger ally is unable to implement its desired policy independently, compliance from the local ally is unlikely. Under these circumstances, intervening forces have few opportunities to influence smaller partners. However, when the interests of local and intervening allies diverge *and* intervening forces are able to execute the policy unilaterally if necessary, compliance from small partners is likely; under such circumstances local allies have an incentive to participate in order to avoid being isolated or undermined when intervening patrons take unilateral action. In Iraq, for instance, the USA was able to use these factors to coerce Iraqi Prime Minister Maliki to engage with Sunni groups in 2010. Maliki accommodated US anti-sectarianism demands in order to avoid being undermined by American efforts to unilaterally support counterinsurgent Sunni militias. Once American forces withdrew in 2011, however, the Americans could no longer unilaterally support Sunni groups, and Washington lost its ability to use threats of US-armed Sunnis to curb Baghdad's anti-Sunni sectarian tendencies. Conversely, local compliance is likely when the interests of both partners converge, and intervening forces cannot implement the request without the participation of local partners. Under these circumstances, local allies must contribute in order to profit. Lastly, when the interests of allies converge and the larger ally can implement the request independently, compliance is less likely, because the local ally can free ride.

Importantly, in the theory proposed, I do not assume that intervening forces always propose smart policies.[57] Equipped with local knowledge, in-country allies can be better positioned to foresee the potential local repercussions of a particular approach, and can fail to comply because they consider the request strategically unwise. Or they may dislike a requested policy for another reason, perhaps because they seek to retain power through coup-proofing, corrupt, or nepotistic strategies that run against requests from a foreign patron for reform and legitimacy. As documented in the cases offered in Chapters 4–9, there are a variety of reasons why local allies may define a policy request as in their interest or not. However, the focus of this study is not those reasons or the merits of any given policy, but rather certain structural conditions under which the local ally is likely to comply or not.

To summarize, assuming local partners have sufficient capacity, I contend that compliance with a proposed policy is likely when (1) the interests of local and intervening allies converge over a request and intervening forces cannot implement the policy without local participation, or (2) the interests of local and intervening allies diverge over a request and intervening forces can implement the policy unilaterally if necessary, or (3) there is a substantial enemy offensive that motivates local clients to focus on providing security above other priorities. Conversely, local compliance is unlikely, even if powerful patrons issue threats, when (1) local allies have insufficient capacity to implement the request, or (2) the interests of allies diverge over the proposed policy and intervening forces do not have unilateral ability to implement the request unilaterally. At times intervening forces can coerce local proxies; at other times they are coerced. Structural variables such as interests and dependencies determine who has the advantage in bargaining.

Methods Summary

In order to test this theory, I consider nine large-scale counterinsurgency interventions, conducting an in-depth methodical archival research for five of them. These five are the US interventions in Vietnam (1964–73), Afghanistan (2001–ongoing), and Iraq (2003–11); India in Sri Lanka (1987–90); and the USSR in Afghanistan (1979–89). Utilizing tens of thousands of primary source documents, for these five cases I first identify and analyze specific policy demands from intervening forces to local partners, such as the June 11, 1965, US request that the government in Saigon increase pay for rural teachers.[58] Second, I measure the dependent variable, local compliance or defiance, by recording the outcome of each policy demand, testing my analysis on 460 policy requests from intervening forces to local regimes as individual observations. For example, in the aforementioned case Saigon did not increase teacher salaries, instead leaving it to an office of the US Agency for International Development (USAID) to augment the salaries of local teachers. Unfortunately, this did not afford the quality-of-life benefits intended by the program, mainly because of soaring inflation in Vietnam caused largely by the US invasion, which outpaced the raises provided by USAID.[59]

The document-based approach enables me to present a model that clarifies the effect of interests on local compliance instead of assuming that interest convergence will always lead to local compliance and divergence will always inspire defiance. Four additional interventions, namely Egypt in Yemen (1962–70), Syria in Lebanon (1975–90), Cuba in Angola (1975–91), and Vietnam in Cambodia (1979–91), are also analyzed in Chapter 9, relying on secondary sources due to the unavailability of primary source materials. This chapter is included in order to ensure that case selection is not biased by only considering the better-known and more often studied wars, which in the counterinsurgency canon typically involve American or British forces. This study offers one of the first glimpses into the history of alliance relations in these four frequently overlooked, but costly and important conflicts.

The Plan of the Book

The book has ten chapters. Chapter 2 clarifies the primary arguments and theoretical framework, while Chapter 3 summarizes case selection, methodology, and data. The first case study, in Chapter 4, analyzes the US intervention in Iraq (2003–11), examining 106 US demands to local partners. The findings from Iraq support the theory proposed in Chapter 2, specifically the interaction of interests and dependencies. One difference evident in Iraq however was Baghdad's response to US requests regarding economic policy. Iraq's historically statist economy conflicted with US demands for free-market solutions, and this created a tension around economic policy unobserved in other interventions. Furthermore, Americans directly governed Iraq under the Coalition Provisional Authority (March 2003–June 2004), which provides an interesting parallel to the Soviet takeover of the Afghan government (1980–9), detailed in Chapter 8.

Chapter 5 explores the US war in Afghanistan (2001–ongoing), examining 148 US demands to allies in Kabul.[60] The hypothesized interaction effect between interests and dependency was significant, with one interesting caveat: local allies in Afghanistan were surprisingly less likely to free ride on US efforts under the conditions theorized when compared to other counterinsurgency partners. Though this might seem to be a positive trend for the counterinsurgency, the causes of this low rate of free riding are tied to Afghanistan's notorious issues with corruption and political fragmentation. Rather than free ride on unilateral American counterinsurgency efforts, it appears that Afghan allies preferred to participate, at least to some extent, in US-designed programs in order to access potential funding opportunities to support patronage networks.

Chapter 6 examines the US war in Vietnam (1964–73), analyzing 105 demands by the USA to its allies in Saigon. The theorized interaction effect was also significant in the Vietnam War, while the Tet Offensive in 1968 provides a powerful example of how a significant external threat can motivate local compliance. Chapter 7 discusses the Indian intervention in Sri Lanka (1987–90) as the first of several non-US interventions examined. In less than three years, over 1,150 Indian troops were killed attempting to implement an Indian-brokered solution to Sri Lanka's so-called Tamil problem. Here also, the interaction effect was found to be significant in an analysis of seventy-nine demands from New Delhi to Colombo. This case study is compelling in part because of the way India shifted its alliances between insurgents and counterinsurgents. New Delhi sought to coerce its Sri Lankan allies into a compromise with Tamil leaders by funding the Tamil Tiger insurgents, the Liberation Tigers of Tamil Eelam (LTTE), while also attempting to coerce the LTTE by sending troops to uphold a primarily Indian-designed agreement seeking to disarm it.

Chapter 8 addresses the Soviet intervention in Afghanistan (1978–89). Only twenty-two requests from the USSR to Afghan allies were identified, all dating back to the first thirteen months of the nine-year intervention. By 1980 the USSR had assumed direct control over Afghanistan and ceased making demands, instead directly implementing policy through either Soviet or Afghan bureaucracies. Taking over local decision-making bodies rather than dealing with the difficulties of negotiating with local allies was an approach to counterinsurgency intervention occasionally adopted by nondemocratic intervening states, including the USSR, Egypt, and Vietnam. Denying local allies legal sovereignty offers certain benefits, such as heightened efficiency in implementing a particular counterinsurgency strategy, but it also imposes substantial costs, such as inhibiting local institutional growth, issues with legitimacy, and, from the cases examined, does not increase the likelihood of success in intervention.

Chapter 9 explores four large-scale counterinsurgency interventions undertaken by small states, including Vietnam in Cambodia (1977–91), Egypt in Yemen (1962–70), Syria in Lebanon (1976–2005), and Cuba in

Angola (1976–91). Smaller interveners often bring an inside perspective to COIN interventions, drawing from their own experiences as insurgents or combating colonial occupation. Alliance dynamics between local and foreign counterinsurgents can shift when the asymmetries in capabilities between allies are less marked and the intervening power is a relatively small state in international affairs. Cuba, for example, stood with its Angolan allies in attempting to maximize Soviet assistance while minimizing Soviet influence, since both Havana and Luanda depended on Soviet arms and financing but sought to maintain their autonomy from Moscow. Furthermore, Havana received partial compensation from its local allies in Angola for certain forms of civil assistance, such as education. The unique dynamics produced by Cuba's smaller size and constrained budget relative to other intervening powers affected the nature of the partnership.

In a concluding summary, Chapter 10 explores the significance of the findings. Intervening foreign forces must anticipate that local partners will have a divergent set of priorities motivating behaviors that are likely to impose costs on foreign patrons and recognize their leverage over local allies is often limited. Furthermore, emphasizing the conditions that are likely to foster compliance, while, wherever possible, minimizing circumstances that promote local defiance, would be a more efficient approach for intervening forces and would minimize these tensions that often characterize these alliances. When we understand why local allies often behave in particular patterns, we can gain a clearer picture of why counterinsurgency interventions are so difficult and how to best manage these tricky, but critical partnerships.

Notes

1 "Iraq Study Estimates War-Related Deaths at 461,000," BBC News, October 16, 2013, www.bbc.com/news/world-middle-east-24547256; Neta C. Crawford, "U.S. Costs of Wars through 2014: $4.4 Trillion and Counting," The Watson Institute for International and Public Affairs, June 25, 2014, http://watson.brown.edu/costsofwar/files/cow/imce/figures/2014/Costs%20of%20War%20Summary%20Crawford%20June%202014.pdf; US Department of Defense, "U.S. Department of Defense Casualty Status," January 19, 2016, www.defense.gov/casualty.pdf.

2 Barack Obama, "Remarks by the President and First Lady on the End of the War in Iraq," whitehouse.gov, December 14, 2011, www.whitehouse.gov/the-press-office/2011/12/14/remarks-president-and-first-lady-end-war-iraq.

3 Priyanka Boghani, "David Petraeus: ISIS's Rise in Iraq Isn't a Surprise," PBS Frontline, July 29, 2014, www.pbs.org/wgbh/frontline/article/david-petraeus-isiss-rise-in-iraq-isnt-a-surprise/.

4 Ibid.

5 Chester L. Cooper et al., "The American Experience with Pacification in Vietnam – Volume I – An Overview of Pacification" (Institute for Defense Analyses, March 1972), IDA Log No. HQ 72-14046 CY # 94.

6 Andrew Hubbard, "Plague and Paradox: Militias in Iraq," *Small Wars & Insurgencies* 18, no. 3 (September 2007): 345–62; Ramzy Mardini, "Iraqi Leaders React to the U.S. Withdrawal" (Institute for the Study of War, November 10, 2011), www.understandingwar.org/sites/default/files/Back grounder_IraqLeadersReacttoWithdrawal.pdf; Joel Rayburn, *Iraq after America: Strongmen, Sectarians, Resistance* (Stanford, CA: Hoover Institution Press, 2014); Paul M. Kattenburg, *The Vietnam Trauma in American Foreign Policy: 1945–75* (New Brunswick, NJ: Transaction Publishers, 1980); Frances Fitz-Gerald, *Fire in the Lake: The Vietnamese and the Americans in Vietnam*, reprint ed. (Boston: Back Bay Books, 2002), 303–5.

7 Robert O. Keohane, "The Big Influence of Small Allies," *Foreign Policy*, no. 2 (Spring 1971): 161–82.

8 Central Intelligence Agency, "Memorandum for: Director of Central Intelligence, From: Acting NIO for Latin America, Subject: Cuban Involvement in Angola, NI-1589-77," June 23, 1977, Declassified January 3, 2006, CREST Database, National Archives and Records Administration; A. I. Dawisha, "Intervention in the Yemen: An Analysis of Egyptian Perceptions and Policies," *Middle East Journal* 29, no. 1 (Winter 1975), 48; Rohan Gunaratna, *Indian Intervention in Sri Lanka: The Role of India's Intelligence Agencies* (Colombo: South Asian Network on Conflict Research, 1993), 178; Vasiliy Mitrokhin, "The KGB in Afghanistan, English Edition" (The Cold War International History Project – Working Paper Series, February 2002), www.wilsoncenter.org/sites/default/files/WP40-english.pdf; Eric Schmitt, "U.S. Envoy's Cables Show Worries on Afghan Plans," *The New York Times*, January 26, 2010, sec. International / Asia Pacific, www.nytimes.com/2010/01/26/world/asia/26strategy.html; "Text of U.S. Security Adviser's Iraq Memo," *The New York Times*, November 29, 2006, sec. International / Middle East, www.nytimes.com/2006/11/29/world/middleeast/29mtext.html.

9 Robert Komer, "Bureaucracy Does Its Thing: Institutional Constraints on U.S.–GVN Performance in Vietnam" (Santa Monica, CA: RAND Corporation), August 1972, www.rand.org/pubs/reports/R967/, 35.

10 Neta C. Crawford, "United States Budgetary Costs of the Post–9/11 Wars through FY2019: $5.9 Trillion Spent and Obligated," The Watson Institute for International and Public Affairs, November 14, 2018, https://watson .brown.edu/costsofwar/files/cow/imce/papers/2018/Crawford_Costs%20of% 20War%20Estimates%20Through%20FY2019.pdf.

11 Stephen Biddle, "Review of the New U.S. Army/Marine Corps Counterinsurgency Field Manual," *Perspectives on Politics* 6, no. 2 (2008): 347–50.

12 US Army and Marine Corps, *Counterinsurgency, FM 3-24* (Washington, DC: Headquarters, Department of the Army, 2006), 1–21.

13 Conrad Crane, "Military Strategy in Afghanistan and Iraq: Learning and Adapting under Fire at Home and in the Field," in *Understanding the U.S. Wars in Iraq and Afghanistan*, ed. Beth Bailey and Richard H. Immerman (New York: New York University Press, 2015), 136.

14 US Army and Marine Corps, *Field Manual FM 3-24 MCWP 3-33.5 Insurgencies and Countering Insurgencies* (Washington, DC: Headquarters, Department of the Army, 2014), chapter 1, 8, www.hqmc.marines.mil/Portals/135/JAO/FM%203_24%20May%202014.pdf.

15 Ibid., chapter 10, 5.

16 Stephen M. Walt, *The Origins of Alliances* (Ithaca, NY: Cornell University Press, 1987), 43–4; Walter C. Ladwig III, *The Forgotten Front: Patron-Client Relationships in Counter Insurgency* (Cambridge University Press, 2017), 57–60, 72–84; R. D. McKinlay and R. Little, "A Foreign Policy Model of U.S. Bilateral Aid Allocation," *World Politics* 30, no. 1 (October 1977): 58–86; Gordon Crawford, *Foreign Aid and Political Reform: A Comparative Analysis of Democracy Assistance and Political Conditionality* (Dordrecht: Springer, 2000); Douglas J. Macdonald, *Adventures in Chaos: American Intervention for Reform in the Third World* (Cambridge, MA: Harvard University Press, 1992), 6–8, 139–59; D. Michael Shafer, *Deadly Paradigms: The Failure of U.S. Counterinsurgency Policy* (Princeton University Press, 1988), 251–4; Andrew Mold, "Policy Ownership and Aid Conditionality in the Light of the Financial Crisis" (Paris: Organization for Economic Cooperation and Development, September 1, 2009), 55–61, www.oecd-ilibrary.org/fr/development/policy-ownership-and-aid-conditionality-in-the-light-of-the-financial-crisis_9789264075528-en.

17 Biddle, "Review of the New U.S. Army/Marine Corps Counterinsurgency Field Manual," 348.

18 Daniel L. Byman, "Friends Like These: Counterinsurgency and the War on Terrorism," *International Security* 31, no. 2 (October 1, 2006): 81.

19 Stephen Watts et al., *Countering Others' Insurgencies: Understanding U.S. Small-Footprint Interventions in Local Context* (Santa Monica, CA: RAND Corporation, 2014), xv.

20 Byman, "Friends Like These," 82.

21 Ladwig, *The Forgotten Front*, 4, 22; Byman, "Friends Like These," 80; Biddle, "Review of the New U.S. Army/Marine Corps Counterinsurgency Field Manual," 348.

22 Byman, "Friends Like These," 80.

23 Robert J. McMahon, *Major Problems in the History of the Vietnam War: Documents and Essays*, 2nd ed. (Lexington, MA: D.C. Heath, 1995), 389. More recent scholarship has addressed the politics of South Vietnam, most focusing on the Diem regime that predated the deployment of significant numbers of US troops. See, for example, Edward Miller, *Misalliance: Ngo Dinh Diem, the United States, and the Fate of South Vietnam* (Cambridge, MA: Harvard University Press, 2013); Geoffrey C. Stewart, *Vietnam's Lost Revolution: Ngô Dình Diem's Failure to Build an Independent Nation, 1955–1963* (New York: Cambridge University Press, 2017).

24 Ladwig, *The Forgotten Front*, 5. For examples of works drawing lessons from twentieth-century colonial wars for modern postcolonial operations see John A. Nagl, *Learning to Eat Soup with a Knife: Counterinsurgency Lessons from Malaya and Vietnam*, 1st ed. (University of Chicago Press, 2005); John A. McConnell, "The British in Kenya (1952–1960): Analysis of a Successful Counterinsurgency Campaign," thesis, Naval Postgraduate School, June 2005, www.dtic.mil/docs/citations/ADA435532.

25 Sebastian Gorka and David Kilcullen, "An Actor-Centric Theory of War: Understanding the Difference between COIN and Counterinsurgency," *Joint Force Quarterly*, no. 60 (January 1, 2011): 16.

26 Mancur Olson and Richard Zeckhauser, "An Economic Theory of Alliances," *The Review of Economics and Statistics* 48, no. 3 (August 1966): 266–79; John R. Oneal and Paul F. Diehl, "The Theory of Collective Action and NATO Defense Burdens: New Empirical Tests," *Political Research Quarterly* 47, no. 2 (June 1, 1994): 373–96; Andrew Bennett, Joseph Lepgold, and Danny Unger, "Burden-Sharing in the Persian Gulf War," *International Organization* 48, no. 1 (1994): 39–75; Glenn H. Snyder, *Alliance Politics* (Ithaca, NY: Cornell University Press, 2007), 165.

27 "Text of U.S. Security Adviser's Iraq Memo."

28 Ibid.

29 For a description of what to expect from partner nations in a counterterrorism context see Stephen Tankel, *With Us and against Us: How America's Partners Help and Hinder the War on Terror* (New York: Columbia University Press, 2018), 19–24.

30 US Army and Marine Corps, *Field Manual FM 3-24 MCWP 3-33.5 Insurgencies and Countering Insurgencies*, chapter 10, 5.

31 A. H. M. Kirk-Greene, "The Thin White Line: The Size of the British Colonial Service in Africa," *African Affairs* 79, no. 314 (January 1, 1980): 25–44; Benjamin N. Lawrance, Emily Lynn Osborn, and Richard L. Roberts, *Intermediaries, Interpreters, and Clerks: African Employees in the Making of Colonial Africa* (Madison: University of Wisconsin Press, 2006); Bruno C. Reis, "The Myth of British Minimum Force in Counterinsurgency Campaigns during Decolonisation (1945–1970)," *Journal of Strategic Studies* 34, no. 2 (April 1, 2011): 245–79; Paul K. MacDonald, " 'Retribution Must Succeed Rebellion:' The Colonial Origins of Counterinsurgency Failure," *International Organization* 67, no. 2 (2013): 253–86.

32 MacDonald, "Retribution Must Succeed Rebellion."

33 David A. Lake, *The Statebuilder's Dilemma: On the Limits of Foreign Intervention* (Ithaca, NY: Cornell University Press, 2016).

34 Keith Dovkants, "Rebel Chief Begs: Don't Bomb Now; Taliban Will Be Gone in a Month," *Evening Standard*, October 5, 2001; Peter Tomsen, *The Wars of Afghanistan: Messianic Terrorism, Tribal Conflicts, and the Failures of Great Powers* (New York: PublicAffairs, 2011), 587.

35 Lake, *The Statebuilder's Dilemma*, 1–8, 93–8.

36 Gil Merom, *How Democracies Lose Small Wars: State, Society, and the Failures of France in Algeria, Israel in Lebanon, and the United States in Vietnam* (Cambridge University Press, 2003); Dan Reiter and Allan C. Stam, *Democracies at War* (Princeton University Press, 2002). There is an extensive literature on audience costs, international crisis, and domestic public opinion debating if and how the perceptions of multiple populations affect a military effort. See, for example, James D. Fearon, "Domestic Political Audiences and the Escalation of International Disputes," *American Political Science Review* 88, no. 3 (September 1994): 577–92; Alastair Smith, "International Crises and Domestic Politics," *American Political Science Review* 92, no. 3 (September

1998): 623–38; Robin Brown, "Spinning the War: Political Communications, Information Operations and Public Diplomacy in the War on Terrorism," in *War and the Media: Reporting Conflict 24/7*, ed. Daya Kishan Thussu and Des Freedman (London: Sage Publications, 2003), 87–100; Michael Tomz, "Domestic Audience Costs in International Relations: An Experimental Approach," *International Organization* 61, no. 4 (October 2007): 821–40. Also note that international opinion can also matter in sustaining a counterinsurgency mission, especially COIN operations led by a coalition like the North Atlantic Treaty Organization (NATO)-led International Security Assistance Force (0049SAF) in Afghanistan. See Takamichi Takahashi, "Japan: A New Self-Defense Force Role … or Not?," in *Coalition Challenges in Afghanistan: The Politics of Alliance*, ed. Gale Mattox and Stephen Grenier (Stanford University Press, 2015), 217; Andrew M. Dorman, "The United Kingdom: Innocence Lost in the War in Afghanistan?," in *Coalition Challenges in Afghanistan: The Politics of Alliance*, ed. Gale A. Mattox and Stephen M. Grenier (Stanford University Press, 2015), 119–20; Timo Behr, "Germany and Regional Command – North," in *Statebuilding in Afghanistan: Multinational Contributions to Reconstruction*, ed. Nik Hynek and Péter Marton (London; New York: Routledge, 2012), 56–8.

37 Max Boot, *The Savage Wars of Peace: Small Wars and the Rise of American Power* (New York: Basic Books, 2014), 304.

38 As detailed in Chapter 8 on the Soviet intervention in Afghanistan, intervening forces from authoritarian regimes can more readily dominate local partners, and thus more closely adhere to hierarchical colonial models in this regard. But it is unclear if such an approach is ultimately strategically sound for counterinsurgents seeking to pacify a population due the institutional dependencies this perpetuates and the difficulty building local legitimacy under such a model. See also Lake, *The Statebuilder's Dilemma*.

39 Byman, "Friends Like These," 82.

40 Stephen D. Krasner, *Sovereignty: Organized Hypocrisy* (Princeton University Press, 1999), 3–6. See also J. L. Holzgrefe and Robert O. Keohane, *Humanitarian Intervention: Ethical, Legal and Political Dilemmas* (Cambridge University Press, 2003), 283–92; Martha Finnemore and Judith Goldstein, *Back to Basics: State Power in a Contemporary World* (Oxford University Press, 2013), 23–5.

41 Edward Girardet, *Afghanistan: The Soviet War* (London: Palgrave Macmillan, 1986), 136.

42 Donald Rumsfeld, "Risk in the Way ahead in Iraq," October 28, 2003, Secret, Declassified September 2007, http://library.rumsfeld.com/doclib/sp/353/re%20Risk%20in%20the%20Way%20Ahead%20in%20Iraq%2010-28-2003.pdf.

43 Dick Camp, *Operation Phantom Fury: The Assault and Capture of Fallujah, Iraq* (Minneapolis, MN: Voyageur Press, 2009), 118. Similarly, when responding to a question regarding applying pressure to Prime Minister Maliki, General David Petraeus commented that Maliki remains "Prime Minister of a sovereign country" and thus there are limits to US influence in Iraq. See Charlie Rose, A Conversation with General Petraeus, *The Charlie Rose Show*, April

26, 2007, www.nytimes.com/2007/04/30/world/americas/30iht-30petraeus-charlie-rose.5499787.html.

44 Rudra Chaudhuri and Theo Farrell, "Campaign Disconnect: Operational Progress and Strategic Obstacles in Afghanistan, 2009–2011," *International Affairs* 87, no. 2 (March 1, 2011): 272.

45 Ibid.

46 International Security Assistance Force Afghanistan, "Afghan Media Round-table with General McChrystal," transcript, April 28, 2010, www.isaf.nato.int/article/transcripts/transcript-afghan-media-roundtable-with-gen.-mcchrystal.html [accessed April 19, 2013]. McChrystal was not alone in voicing such statements. When pressed on US policy in Afghanistan, US Central Command Commander General David Petraeus similarly commented, "President Karzai is the commander in chief. He is the president of a sovereign country." See Remarks to the Press at Conclusion of Rehearsal of Concept (ROC) Drill, US Department of State, Special Representative for Afghanistan and Pakistan, transcript, April 11, 2010, https://2009-2017.state.gov/s/special_rep_afghanistan_pakistan/2010/140010.htm.

47 US Department of State, US Embassy Kabul, "Letter of Agreement on Police and Justice Projects," November 29, 2002.

48 Chester L. Cooper, "Memorandum for Mr. Bundy, Subject: Status Report on Various Actions in Vietnam [The 41-Point Program]" (LBJ Library, June 11, 1965), NLJ 84–130; Neil Sheehan, *A Bright Shining Lie: John Paul Vann and America in Vietnam* (New York: Random House, 1988), 731–2.

49 J. N. Dixit, "Aide Memoire Containing Points Conveyed by the Indian High Commissioner J.N. Dixit to the Sri Lankan President Recalling Certain Actions on the Part of the Sri Lankan Government," June 14, 1988, in *India–Sri Lanka: Relations and Sri Lanka's Ethnic Conflict Documents – 1947–2000*, ed. Avtar Singh Bhasin, Vol. IV, Document 832 (New Delhi: Indian Research Press, 2001), 2257–9.

50 John A. Nagl, "Foreword to the University of Chicago Press Edition: The Evolution and Importance of Field Manual 3-24, Counterinsurgency," in *The U.S. Army/Marine Corps Counterinsurgency Field Manual*, US Army and Marine Corps (University of Chicago Press, 2007), 26–9; David Kilcullen, *Counterinsurgency* (Oxford University Press, 2010), 3.

51 See Walter C. Ladwig III, "Influencing Clients in Counterinsurgency: U.S. Involvement in El Salvador's Civil War, 1979–92," *International Security* 41, no. 1 (July 1, 2016): 99–146; Ladwig, *The Forgotten Front*; Eli Berman and David A. Lake (eds.), *Proxy Wars: Suppressing Transnational Violence through Local Agents* (Ithaca, NY: Cornell University Press, 2019).

52 Ladwig, "Influencing Clients in Counterinsurgency"; Ladwig, *The Forgotten Front*; Stephen Biddle, "Building Security Forces & Stabilizing Nations: The Problem of Agency," *Daedalus* 146, no. 4 (September 21, 2017): 126–38; Berman and Lake (eds.), *Proxy Wars*.

53 Jean-Jacques Laffont and David Martimort, *The Theory of Incentives: The Principal-Agent Model* (Princeton University Press, 2009), 2–4, 146–7; Gary J. Miller, "The Political Evolution of Principal-Agent Models," *Annual Review of Political Science* 8, no. 1 (2005): 203–4; Ladwig, "Influencing Clients in Counterinsurgency," 5.

54 Ladwig, *The Forgotten Front*, 28.

55 Ibid.

56 Speaking on the limits of assuming local regimes are unitary actors, US General David Petraeus commented that Iraqi Prime Minister Malki was

> chosen by a number of different parties, each of them from different sects and ethnic groups are represented *(sic)* in the ministries in the ministries in various leadership positions. Sometimes, in fact, what comes out of all of them is a bit discordant. So it's not enough to pressure Maliki. It's not to say, Prime Minister, you must produce this legislation. You really have to help him encourage all the different factions and political parties to get something done in Iraq. And it's a challenging environment, and I laid that out, I think, to the press corps today.

See Charlie Rose, A Conversation with General Petraeus.

57 Komer, "Bureaucracy Does Its Thing," chapter vii.

58 Cooper, "Memorandum for Mr. Bundy, Subject: Status Report on Various Actions in Vietnam [The 41-Point Program]."

59 Takashi Oka, *Newsletters about the Vietnamese War* (Washington, DC: Institute of Current World Affairs, 1964); Douglas C. Dacy, *Foreign Aid, War, and Economic Development: South Vietnam, 1955–1975* (Cambridge University Press, 1986); United States Senate Foreign Relations, *Vietnam: Policy and Prospects, 1970: Hearings before the Committee on Foreign Relations, U.S. Senate, on Civil Operations and Rural Development Support Program, Feb. 17, 18, 19, and 20, and March 3, 14, 17, and 19, 1970* (Washington, DC: US Government Printing Office, 1970), 588.

60 The 148 US requests to Afghan partners date from January 10, 2002 to February 26, 2010, providing coverage of the war from its beginning to early 2010. It was not possible to extend analysis beyond 2010 due to classification of those materials.

2 Why Local Allies Defy or Comply with Requests from Intervening Allies

> The present GVN [Government of South Vietnam] continued, as they had so often before, to agree readily in conversations with us to the principle of national reconciliation; yet any concrete implementation remained elusive even through another top level meeting with the President.
> —The Pentagon Papers

The introduction to the 2006 US operational field manual for waging COIN warfare labels counterinsurgencies as "the most complex and maddening type of war."[1] This pessimistic assessment stands in stark contrast with once commonly held expectations that technologically advanced, conventionally formidable militaries would readily triumph over impoverished rebels. As one journalist remarked while touring a US aircraft carrier in 1965, "They just ought to show this ship to the Vietcong – that would make them give up."[2] Yet despite the overwhelming US advantage in resources, victory eluded the Americans in Vietnam – an unsuccessful outcome that has become increasingly common for third-party counterinsurgency interventions.[3]

The difficulties of counterinsurgency have been attributed to a multitude of shortcomings: ill-fitting or flawed strategy,[4] poor implementation of strategy on the battlefield,[5] an overreliance on technology.[6] It has also been suggested that the mission of state-building while fighting entrenched insurgents is so inherently problematic that success is virtually impossible.[7]

Similarly, blame for losing long, irregular counterinsurgency wars has been assigned to a variety of factors, including problematic local partners. Inevitably, arguments are offered that the inability or unwillingness of local allies to reform or defend themselves contributed to failure, while intervening partners were trapped in a quagmire supporting these ineffectual regimes.[8] Yet the two assumptions here – that local allies are uncooperative and obstructive, and that their unreasonableness contributed to defeat – are typically glossed over in the counterinsurgency literature. Indeed, for all the ink spilled on unconventional warfare, only

24

a few studies have systematically considered alliance politics, despite the importance of these partnerships to the success of the mission.[9] This book aims to address these complex partnerships to help further our understanding of counterinsurgency interventions in the process.

In this chapter I offer a theory to explain variation in local compliance with the policy prescriptions of intervening forces.[10] This is vital because without understanding the politics of these partnerships, especially the sources of leverage and limitations of foreign influence on local regimes, policymakers tasked with winning interventions by propping up local allies are left with dangerously underspecified mandates. As David Kilcullen has asserted, when it comes down to winning and losing, "the long history of counterinsurgency emphasizes that foreigners can't fix all these issues. It has to be the locals."[11] Therefore, understanding how to influence local partners is a critical component of COIN warfare.

There is a variety of competing ideas about the key elements that sustain an insurgency, and about what counterinsurgents should target in order to win. Enemy-centric counterinsurgency strategies, for instance, underscore the importance of destroying enemy forces, essentially viewing COIN as a low-intensity variant of conventional war.[12] Population-centric methods, on the other hand, target the local population, including through "hearts and minds" campaigns, on the logic that insurgents can be defeated by winning over the favor (hearts) and opinions (minds) of local civilians.[13] Following Mao Zedong's timeworn maxim that insurgents are fish swimming in the civilian sea, population-centric approaches assume that addressing the population, as the insurgents' essential political environment, is the best approach to isolating and defeating an insurgency.[14] Terrain-centric methods emphasize the need to control and pacify restive areas in order to deny the enemy physical safe haven.[15] Still other approaches stress the importance of manipulating local elites,[16] promoting better leadership to effectively respond to dynamic battlefield conditions,[17] or advocating for a context-specific approach that combines elements of different approaches.[18]

Irrespective of these disagreements over strategy, to varying degrees all postcolonial counterinsurgency approaches recognize the underlying importance of an effective local government to provide stability and solidify gains.[19] Enemy-centric tactics that aim to eliminate insurgents, for example, will have short-lived success unless local institutions are in place to provide stability following military operations. This dynamic was painfully demonstrated after the first and second battles in Fallujah, Iraq. Despite incurring significant casualties in intense fighting, Fallujah nevertheless continually reverted back to being an insurgent safe haven.[20] As one frustrated US Marine in Fallujah expressed,

we're going to kill a lot of people, probably three quarters of which will be civilians. We'll get some insurgent bad guys, but what we will do is have pissed off everybody and their f**king mother, especially like the real people that matter, these f**king sheikh guys and then the clerics. So short term we may have struck with the hammer, but long term we're going to get hit with a million different nails.[21]

Without an effective local government in place to capitalize on gains, the utility of high-risk strategies to forcefully attack insurgents – at a cost of military and civilian lives – is questionable. As Senator John McCain commented, "I worry ... we're playing a game of whack-a-mole here."[22]

Furthermore, population-centric approaches that depend on isolating civilians from insurgents are similarly fated to fail without an effective local government to provide citizens with long-term political incentives to oppose insurgents. This dynamic was similarly evident in Iraq, where American officials watched Prime Minister Maliki reverse course on several key policies that the USA had implemented to win over Iraqi Sunnis. Maliki's anti-Sunni sectarian approach left an isolated Sunni population susceptible to insurgent propaganda, and ISIS, a malignant resurgent offshoot of the al-Qaeda insurgency, soon took root in Iraq and Syria.[23] The critical requirement of a coherent government to translate military gains and political openings into lasting victories against insurgents has been long recognized as a critical component of counterinsurgency theory. As Robert Thompson reasoned in 1966 regarding Vietnam and Malaya, "without a reasonably efficient government machine, no programs or projects, in the context of counter-insurgency, will produce the desired results."[24] Local regimes are a decisive participant in the complex contest between insurgent and counterinsurgent. As a result, understanding the factors that motivate local allies to comply with (or defy) policies proposed by intervening forces is a critical component of the complex counterinsurgency puzzle.

The Mismatched Priorities of Local and Intervening Allies

In the postcolonial context, even though local allies exercise only partial sovereignty, with their territorial control under attack by insurgents and their state propped up by foreign forces, they nonetheless typically have sovereign legal authority over state institutions. This legal authority offers certain benefits but also poses challenges to foreign patrons aiming to employ local clients to implement security policies. The theory of counterinsurgency intervention assumes that the political benefits of backing an independent, sovereign local regime to legitimize the intervention and serve as a proxy for the foreign ally will outweigh the disadvantage to the latter of relinquishing some control over the design and implementation of COIN policy by having to coordinate, as opposed to dictate policy.

When postcolonial counterinsurgency missions rely on successful, legally sovereign local allies as a key component legitimizing their military mission, intervening militaries are effectively rendering these local proxies politically pivotal to operational success. Reflecting on President George W. Bush's approach toward the Iraqi Prime Minister, for example, Bush's National Security Adviser Stephen Hadley recalled that Bush said, "'I've got to be his best friend. I've got to be his counselor and aid in helping him succeed as prime minister of Iraq, because if he doesn't succeed, U.S. policy isn't going to succeed.' And in some sense [according to Hadley], Bush's view was, 'If he doesn't succeed, I don't succeed.'"[25] Although intervening forces typically take the lead in military operations, winning in the long term is inextricably tied to the success of the local regime.

As such, local and intervening allies are bound together by an imperative security objective to defeat insurgents, but their preferences on how best to achieve this outcome, and how much risk local officials should accept in the process, will inevitably differ. Local allies seek foreign assistance that will help them secure their hold on power and bolster their wealth and capabilities.[26] Local counterinsurgents will rely on and exploit external COIN partners to provide expertise, money, firepower, and manpower to serve these ends. As Walter Ladwig observed, the local ally "will try to manipulate the dynamics of the relationship with its patron to maximize the amount of political, economic, or military assistance it receives while simultaneously seeking to avoid surrendering its autonomy."[27] In 1968, US Secretary of Defense Clark Clifford expressed this sentiment in colorful terms. According to notes by his assistant, Clifford commented that South Vietnamese President Nguyen Van Thieu "wants the war to go on forever. His gov't is getting richer & richer. [Clifford is] mad & wants us to get out ... We should get on with the negotiations & start pulling out. Thieu will just 'pee away' our substantial military victory."[28] According to Clifford, Saigon was focused on accumulating resources and maintaining power and sought to avoid being coerced by its American partners – so much so that Clifford believed Thieu was willing to make decisions that were harmful to the long-term collective war effort.

Importantly, while both intervening and local counterinsurgency partners seek to expand local fighting capabilities and weaken enemies, there are short-term incentives for local counterinsurgents to remain vulnerable, or at least to appear weak, in order to shift the burden to intervening partners and make a credible case for increased aid from them. Speaking to a US Congressional delegation in 2006, Afghan Defense Minister Abdul Wardak, for example, referred to "the large amounts being spent in Iraq [and] said that if 'we were a failure, we would have received much more aid'."[29] Similarly, Afghan President Hamid Karzai candidly teased a group of US senators, saying, "Never before has a country welcomed a foreign presence so much ... reconstruction is something we complain about, but inside we are very happy with it. We complain because we want more money from you."[30] This dynamic was also noted by Robert Keohane, who observed that in alliance politics "weakness does not entail only liabilities; for the small power, it also creates certain bargaining assets."[31] The worse shape the local regime is in, the more assistance it can squeeze out of donors and partners.[32] This provides local allies with additional incentives to refuse reforms promoted by intervening partners, such as anti-corruption reform that would entail short-term personal risks for long-term institutional stability, because appearing vulnerable and in need of foreign assistance can be quite lucrative. Reform and compliance with policy requests from intervening patrons is likely to eventually be rewarded with greater autonomy, but also greater risks, more responsibility, and less funding, a potentially unattractive proposition.

For their part, intervening forces seek to employ local allies as proxies to implement the larger partner's foreign policy agenda. Foreign powers in COIN interventions are typically concerned with fortifying their positions in regional or geopolitical contexts, while minimizing costs and casualties. While the leaders of local regimes are focused on their personal physical and political survival, intervening forces are less concerned about which particular local figure is in power, so long as the institutions of the regime endure and carry on the counterinsurgency effort.

These discrepancies in priorities are a persistent source of friction between local and intervening counterinsurgency allies. Richard Holbrooke, US Special Envoy to Afghanistan and Pakistan, admitted that there is "constant tension in relationships like this one between what the Americans want and what the local officials want."[33] For example, intervening allies may pressure local partners to take risks by instituting anti-corruption reforms or entering power-sharing arrangements that would potentially weaken their personal hold on power. This distresses local officials, who may fear that rivals, insurgents, or even foreign patrons will take advantage of their vulnerable state to replace them.[34] Fearing that they are in a politically tenuous position and disposable to their foreign partners, local allies are risk-averse and disinclined to implement painful long-term reforms that may create short-term vulnerabilities.

Speaking to this dynamic, when John Kenneth Galbraith, then US Ambassador to India, was invited by President Kennedy to comment on South Vietnamese President Ngo Dinh Diem, Galbraith declared, "Diem will not reform either administratively or politically in any effective way. That is because he cannot. It is politically naïve to expect it. He senses that he cannot let power go because he would be thrown out."[35] More recently, Afghan power brokers and international donors selected Afghan President Hamid Karzai during the Bonn Conference in December 2001 in the hope that he could unite disparate factions in Afghanistan due to his ability to compromise and provide concessions in a system largely constructed on patronage.[36] Had Karzai adhered to subsequent US prescriptions on reform and rule of law, thereby dismantling the patronage networks that brought him to power, he would have potentially undermined the foundations of his own political power and the reason he was attractive in the first place. Given these structural disparities in priorities between intervening and local counterinsurgents, it is not surprising that there is significant, persistent tension in these partnerships, with local allies seeking to leverage the resources and capabilities of patrons to stay in power, while patrons aim to leverage local partners to serve as conduits for their foreign policy agendas. Under these difficult conditions, how can intervening forces motivate local compliance with reforms they deem vital for success? What are the critical factors affecting the likelihood of local allied compliance in large-scale counterinsurgency interventions?

Theory: What Motivates Local Compliance?

Here I offer four hypotheses drawn from principal–agent models and alliance politics to test their usefulness in explaining local compliance in counterinsurgency interventions. By doing so I aim to explore critical political dimensions of counterinsurgency partnerships and contribute to our collective understanding of the behavior of local partners. Specifically, I identify four primary independent variables to explain the likelihood that local allies will comply with requests from intervening partners in counterinsurgency interventions: (1) the *capacity* of the local ally to fulfill the request, (2) whether or not the *interests* of the two allies converge or diverge over the policy, (3) the extent to which the intervening power *depends* on the local ally to implement the policy, and (4) a significant *enemy threat*. The remainder of this chapter introduces four hypotheses regarding the influence of each of these four variables on the likelihood of compliance. These hypotheses also propose conditions under which intervening patrons will likely be able to persuade or coerce local partners toward compliance, as well as, alternatively, other circumstances under which intervening forces are likely to have little leverage to compel reform.

The concept of "theory" is here defined as a "reasoned and precise speculation" about a particular set of observed and puzzling events.[37] I do not claim that these four variables can comprehensively explain every relevant facet of asymmetric alliances or counterinsurgency partnerships. Rather, when thoughtfully considered, these variables can be used to formulate a series of associated hypotheses to explain some variation in local COIN compliance across a percentage of interventions. Exceptions and additional factors that affect compliance are detailed in specific case study chapters, which consider how particular contexts have affected compliance. Details regarding how each variable was measured are offered in Chapter 3.

Capacity

Capacity is a threshold issue for small, local allies responding to demands from larger intervening partners. The local regimes in question invariably have limited capacity. If they were stronger, they would not be fighting an insurgency, or at a minimum, they would be able to fight the insurgency on their own without a large-scale foreign military intervention. If a local ally lacks the capacity to implement the requested policy, then noncompliance is likely. However, the converse is not always true: even if the small ally is able to implement the request, this does not necessarily mean it is likely to comply. Additional factors also influence compliance, as reflected in the hypotheses set forth in this chapter.

Embattled regimes can have capacity failings for a variety of reasons, with limited institutional and bureaucratic ability to carry out tasks being the most obvious and common source of inadequacy. But it is also important to recognize that low-intensity violence, such as the typical harassment operations adopted by insurgents, can further strain the capacity of local allies to implement policies. War-related complications vary; when they intensify, this may make compliance especially challenging. Such moments of friction are an unavoidable problem in warfare. Clausewitz famously described friction as "the force that makes the apparently easy so difficult ... Countless minor incidents – the kind you can never really foresee – combine to lower the general level of performance, so that one always falls short of the intended goal."[38] Furthermore, insurgents often explicitly intend to disrupt normal local government functions and disturb counterinsurgency protocols. As David Galula noted, "Promoting disorder is a legitimate objective for the insurgent. It helps to disrupt the economy, hence to produce discontent; it serves to undermine the strength and the authority of the counterinsurgent. Moreover, disorder – the normal state of nature – is cheap to create, and very costly to prevent."[39]

Interests

If the local capacity condition is satisfied and local proxies are able to fulfill the request, a second set of structural factors becomes relevant in determining the likelihood of compliance. These factors start with the interest of local partners in the request being proposed. This variable builds on work in principal–agent theory, which explains how one actor (the principal or patron) can provide incentives to another (the agent or client) to undertake tasks on the former's behalf.[40] The approach has been employed to model numerous issues in politics and economics, including insurance markets, bureaucratic politics, congressional oversight, and asymmetric security alliances. In the context of this study, the principal–agent approach is helpful for appreciating the asymmetric nature of partnerships between local and intervening COIN allies and the dependency of local allies on their external patrons for support. It also helps account for the informational asymmetry that favors local allies, who have a distinct advantage in understanding and predicting their own preferences and the intricacies of the local political landscape.

A rich literature has emerged using patron–client models to explain how an external counterinsurgent (as the principal or patron) can motivate a local ally (the agent or client) to act on its behalf. Walter Ladwig, for example, analyzes US interventions in the Philippines (1947–53), Vietnam (1957–63), and El Salvador (1979–92). He argues that local compliance was more likely if the USA required these allies to meet certain policy conditions prior to receiving assistance, concluding that conditionality in proposing requests is key to local compliance.[41] This finding builds on earlier work by Douglas J. Macdonald, who analyzes US interventions in the Philippines (1949–53), Vietnam (1960–3), and China (1945–8). Macdonald makes a distinction between demanding a "quid pro quo" relationship that trades aid for policy implementation, as opposed to "relatively unconditional use of commitments" as a factor shaping US leverage in bargaining with client governments.[42] Adding to this body of work, contributors to an edited volume by Eli Berman and David Lake examine nine cases in which foreign principals attempted to implement security policy through local proxies in South Korea (1950–3), Denmark (1940–5), Colombia (1990–2010), Lebanon and Gaza (1975–2017), El Salvador (1979–92), Pakistan (2001–11), Palestine (1993–2017), Yemen (2001–11), and Iraq (2003–11). Berman and Lake offer a principal–agent model demonstrating that when powerful principals "use rewards and punishments tailored to the agents' domestic political context, proxies typically comply."[43] Furthermore, the more the interests of allies diverge, the more "higher-power incentives – larger rewards and punishments" must be applied to pressure compliance.[44]

Building on the logic that local compliance is less likely when the interests of local and intervening counterinsurgency partners diverge, I argue that incentives or threats are required to coerce compliance in this circumstance. The diplomatic record demonstrates that without incentives, local allies will ignore policies requested by intervening forces if these policies violate the local ally's divergent priorities. For example, in Sri Lanka, Colombo refused to comply with a request from Indian forces that Sri Lanka incorporate elements of a Tamil militia formed by Indian forces, the Civil Volunteer Force (CVF), into military reserves and state police forces.[45] Indian intelligence agencies had used CVF forces to extend Indian influence in Tamil territory and to undermine the Liberation Tigers of Tamil Eelam (LTTE) insurgency by providing an alternative to it.[46] Threatened by the idea of an additional armed Tamil group that could potentially infiltrate the ranks of their own security forces, Colombo refused to comply, and eventually permitted LTTE insurgents to eliminate CVF forces once India withdrew.[47]

As detailed in the case study chapters, the interests of counterinsurgency allies typically diverge over issues such as how to pay for programs, whether to enact reforms that empower political opposition groups, or whether to crack down on corruption networks that profit the local ally. The partners' respective interests tend to converge over development programs, military action, and economic restructuring programs, but there are exceptions based on the particular context of the case.[48]

Principal–agent models largely assume that if the interests of principals and agents converge, local agents are likely to comply.[49] Noncompliance, according to this logic, is a product of divergent interests and a lack of coercive incentives that raise the costs for noncompliance. However, this approach overlooks a substantial literature on asymmetric alliances and free riding, including seminal work on unequal burden sharing in the North Atlantic Treaty Organization (NATO) by Mancur Olson and Richard Zeckhauser. These authors reason that when the interests of small and large security allies align over security policy, smaller partners can be expected to evade and shift burdens to wealthy partners.[50] There

is pervasive evidence of free-riding behaviors in counterinsurgency interventions, supporting the idea that the logic presented by Olson and
Zeckhauser regarding burden sharing in asymmetric partnerships should
also be considered in a counterinsurgency context.[51]

In 1967, for example, US officials asked their partners in Saigon to
expand benefits to South Vietnamese veterans in order to boost morale in
Saigon's military and to combat the practice of keeping wounded soldiers
listed on active duty in order to provide veterans with compensation – a
practice that severely distorted active-duty unit estimates.[52] Officials in
Saigon agreed, commenting on the need to improve their efforts to
provide services to veterans and the benefits of such policies.[53] Yet
Saigon failed to comply, prompting US Military Assistance Command
Vietnam (MACV) to step in and provide certain supplemental benefits
for local veterans.[54] Both the MACV commander and the US ambassador decided not to pressure Saigon to comply with this as well as an
array of manpower issues out of concern it would undermine the tenuous
"political stability" of Saigon.[55] Where the interests of local and intervening allies converge, there are often incentives for local forces to free
ride, allowing proxies to reap the benefits of a particular policy without
incurring proportional costs. This is unsurprising since smaller allies are
notorious for free riding on the efforts of larger partners. Therefore it is
important to examine interests in the context of additional factors that
may promote local compliance or defiance.

Furthermore, "interests" is a multifaceted variable and can be measured in a number of different ways that go beyond convergence or
divergence over policies. Also important, for example, is the intensity
of concern on either side. Several scholars examining alliances have
framed negotiations as a battle of wills, predicting that the ally with the
most intense interest in the issue being negotiated has an advantage and
will prevail.[56] According to this logic, the partner with the most at stake
will push harder in the bargaining process and will likely succeed as a
result.

However, applying a framework that associates high interest with
leverage is not terribly helpful in the context of counterinsurgency partnerships. Because local allies are often in a battle for their survival, these
smaller partners would be expected to have higher interest in most of the
issues being negotiated. Intervening forces do not have as much at risk,
despite substantial investments and geopolitical and reputational concerns. These types of interventions are what Jonathan D. Caverley refers
to as wars "consistent with the strong state's grand strategy but not
essential to it."[57] However, although local allies can be expected to have
greater intensity of interest, there is nevertheless significant variation in

compliance. As detailed later in this chapter, Saigon, for example, vociferously argued against particular concessions in peace negotiations, but the USA nevertheless was able to coerce its South Vietnamese partners into agreeing.

Thus, in order to measure a multifaceted factor such as "interests," I make several assumptions about the actors involved. First, I assume that local allies want to stay in power. Second, I assume that each party to the alliance prefers to have the other do as much of the heavy lifting as possible. Third, I assume that local allies have local political interests to consider in addition to pressures from intervening allies. The fourth assumption is that the intervening force would like the local ally to thrive, but also seeks to minimize the costs of the intervention.

Interests and the Disinclination to Make Foreign Aid Conditional on Local Compliance

Because principal–agent models identify the importance of interest convergence between principals and agents as a factor influencing policy compliance, several such models advocate that aid from patrons should be conditional on local compliance in order to raise the costs of noncompliance.[58] Ladwig, for example, explains that aid conditionality "tries to shape the client's behavior by making delivery of assistance contingent on a client's prior implementation of a patron's preferred policies."[59] However, as advocates for such approaches often acknowledge, though it may be a promising means to compel compliance, withholding assistance from local partners in retaliation for defiant local behavior is a strategy that patrons have often decided is ill advised, especially in COIN interventions where intervening forces have publicized a significant commitment to local partners.[60] In large-scale counterinsurgency interventions, policymakers representing intervening forces have frequently recognized that their ability to issue threats to uncooperative local partners is hampered by the structural reality that their missions are hanging on the longevity and success of the local allied regime. In a review of the US pacification effort in Vietnam, Chester L. Cooper and his colleagues at the Institute for Defense Analyses reflected:

Every US ambassador to Saigon since 1954 has grappled with the problem of extracting commitments for improved military, political, and economic performance from South Vietnam's leaders. And having gotten such commitments, American officials have struggled to assure meaningful implementation. With the passage of time and the increase in the American commitment there was a concomitant increase in Washington's stake in effective GVN performance. The ability to influence the Vietnamese consequently became a matter of increasing urgency, but in the last analysis, Americans had to rely on the carrot rather than the stick. Threats to hold back or cancel aid became increasingly ineffectual with the growing GVN awareness that Washington had almost as much to lose as Saigon. There was probably no greater source of frustration for American officials serving in Vietnam.[61]

Threats to withhold assistance to local allies to pressure compliance with policy requests can be risky if intervening forces have made a significant commitment to the survival of a local partner and have politically tethered operational success to the vitality of their local client. Consequently, threats issued to uncooperative local allies by intervening forces can lack credibility, since harming local allies would also harm the long-term objectives of the intervention and the self-interests of intervening forces. W. Patrick Lang, former head of the Middle East section of the US Defense Intelligence Agency, argued in 2008 that US allies in Baghdad would doubt the credibility of US threats to withhold aid should they fail to follow through on promised benchmark reforms, because "realistically [the Iraqis] can figure out that the chances we would pull the plug and leave is just about zero."[62] Similarly, when asked why Washington couldn't threaten US-allied Afghan President Hamid Karzai in light of uncooperative behavior and antagonistic anti-US statements, US Ambassador to Afghanistan Ronald Neumann quipped,

You know that scene in the movie "Blazing Saddles," when Cleavon Little holds the gun to his own head and threatens to shoot himself? The argument that we could pull out of Afghanistan if Karzai doesn't do what we say is stupid. We couldn't get the Pakistanis to fight if we leave Afghanistan; we couldn't accomplish what we've set out to do. And Karzai knows that.[63]

There are structural reasons why threats to significantly undermine local partners in retaliation for defiance may be a risky means to motivate compliance in counterinsurgency.

While intervening forces have on occasion resorted to hardline approaches with local allies, at times they voice concern that these tactics can put future negotiations with local partners at risk. Officials fear that pressing allies too hard over one particular policy can either shut down negotiations or undermine the local ally's receptivity to other proposals. As detailed in the Pentagon Papers in 1964:

Ambassador Taylor had returned to South Vietnam on December 7 and immediately set about getting the GVN to undertake the reforms we desired, making clear to both the civilian and military leaders that the implementation of phase II was contingent on their efforts to revive the flagging war effort and morale in the South. For his efforts, he was rewarded with a military purge of the civilian government in late December and rumored threats that he would be declared persona non grata. The political crisis boiled on into January with no apparent solution in sight in spite of our heavy pressure on the military to return to a civilian regime. And, while Taylor struggled with the South Vietnamese generals, the war effort continued to decline.[64]

Taylor was not the first US ambassador to Vietnam rebuffed by Saigon in response to high-pressure tactics to reform. Robert Komer described how "Ambassador Durbrow (1957–1961) pressed [South Vietnamese President] Diem so hard on corruption, reform and other issues that he was almost declared *persona non grata*. By late 1960, when his repeated efforts proved mostly unavailing, Durbrow began urging pressure on Diem and warning that alterative leadership might be needed."[65] In Afghanistan Hamid Karzai was also notorious for stonewalling American policymakers who issued threats in retaliation for his obstinacy, reportedly snapping, "If you and the international community pressure me more, I swear that I am going to join the Taliban."[66]

At the heart of these threats and the difficulties coercing partners is the question of local legal sovereignty that separates colonial from postcolonial COIN alliances. As Berman et al. note, "Despite their nominal sovereignty, fragile and failed states do not – almost by definition – control all of their territory."[67] Yet the diplomatic record demonstrates that local proxies do not need to control all their territory to leverage their legal sovereignty and authority over local state structures in negotiations with intervening forces. Though their sovereignty may be contested and incomplete, so long as intervening patrons reliably approach local counterinsurgency clients as materially dependent but legally sovereign governments, local proxies will leverage those opportunities. As Robert Komer commented, a significant "constraint on use of leverage [with Saigon] was that, no matter how deeply it became committed, the U.S. almost always saw itself as in an advisory and supporting role vis-à-vis a sovereign GVN."[68]

Of course, due to their overwhelming superiority in resources, intervening forces can deny local partners legal sovereignty if they so choose, dominating their proxies as the Soviets did in Afghanistan (detailed in Chapter 8). They can also opt to overthrow defiant proxies, as did the USA in the 1963 coup against South Vietnamese President Ngo Dinh Diem. However, the Soviet annexation of the Afghan state and the Diem coup demonstrate the limits, rather than the promise, of disregarding local legal sovereignty in an intervention. The institutional dependency of Kabul on Moscow was a liability, and over time it delegitimized the Afghan communists, which aided the insurgency.[69] And while Diem was notoriously obstinate, isolated, and difficult, the instability left in the wake of his assassination came at a high political cost. Indeed, it deepened American involvement, as US officials felt obligated to take increasing action to stabilize South Vietnam in the aftermath of the coup against Diem.[70] According to the Pentagon Papers, "As the nine-year rule of Diem came to a bloody end, our complicity in his overthrow heightened our responsibilities and our commitment in an essentially leaderless Vietnam."[71] Following Diem's ouster there were six successive changes of government, creating such political volatility that in 1964 that the primary American political goal became stabilization of the regime in Saigon at practically any cost.[72] Reflecting on the lessons learned from the Diem coup, Robert Komer noted that "the destabilizing consequences when we did acquiesce in Diem's ouster made us doubly cautious, while a stable political environment became doubly important as our troop commitment grew."[73] Therefore, in addition to interests, the political dependency of intervening forces on local forces is also a vital component. The legal authority of local regimes makes them an indispensable tool to actualize governance policies – a dynamic that limits the power of intervening forces and affects patterns in local compliance with reforms.

Dependency

An intervening ally depends on its local ally to carry out certain tasks, given the local regime's legal state authority and the outside power's wish to validate and reinforce this local sovereignty. This dependency complicates politics between local and intervening forces. But, like interests, dependency varies from case to case according to the particular policy being proposed and whether or not it falls under the jurisdiction of the local regime. For some policies, such as firing corrupt officials from local office or passing legislation, due to legal sovereignty, intervening forces must work through local proxies, while for other policies, such as undertaking military operations or providing funding, intervening forces have the ability to carry out the request unilaterally if they so choose. Existing models fail to account for this variation; yet it stands to reason that local allies will be better positioned to refuse to comply with requests when they are indispensable for implementing the policy, while finding it harder to defy patrons when foreign forces have alternative means of implementation.

In Afghanistan, for example, US officials pressured President Hamid Karzai to dismiss Ismail Khan, a powerful regional leader, from his position as Minister of Water and Energy in 2009 due to corruption and incompetence.[74] According to one US State Department assessment, Khan was considered "the worst" of Karzai's personnel choices, an ignoble distinction considering the complaints US advisors voiced about Karzai's other ministers.[75] While pressing Karzai to remove Khan, US Ambassador Karl Eikenberry raised the possibility of withholding aid if Karzai failed, stating that "all members of the U.S. Congress expressed great concern over the long-term costs of Afghanistan, especially during the current financial crisis. If incompetent and corrupt ministers were appointed, it would provide a good reason for them to limit funding," a warning that prompted Karzai to agree to "further consider this choice."[76] But this pressure proved ineffectual, as Karzai ultimately ignored the US request and Khan maintained his position for another four years.[77] US requests to remove Paktya Governor Juma Khan Hamdard and Governor Abdul Basir Salangi in Parwan for similar reasons also went unfulfilled.

Interestingly, though, Karzai complied with a concomitant US request to remove Kapisa Governor Ghulam Qawis Abu Bakr after General David Petraeus reportedly presented evidence that Abu Bakr was not only corrupt but associated with a 2009 suicide attack that killed American personnel. This attack would justify unilateral US action against Abu Bakr had the Karzai administration failed to act.[78] However, while Karzai complied with the US demand that Abu Bakr be removed from office, Afghan officials never charged him with any crime, suggesting that the Afghan administration did not consider Abu Bakr a significant security threat requiring arrest. By preempting American action and removing Bakr from office, the Karzai administration potentially protected Bakr from more significant punishment by averting US detention. Existing models that focus on principals providing incentives to motivate proxies to comply can be strengthened by considering variables that make threats of punishment more or less credible, or that make allies more or less likely to respond to threats, such as the unilateral ability to implement the policy without them. The unilateral ability of the USA to pursue Abu Bakr for violence against US personnel, for example, provided leverage in negotiating with the Karzai administration that American officials did not have regarding the other three figures. Those officials remained in power, as Kabul was the only entity with legal authority to remove them from office for corruption or incompetence.

The dependency of patrons on local clients, then, varies depending on the nature of the policy requested and on the ability of intervening forces to threaten unilateral action or not. This sheds light on one way for patrons to craft specific policy approaches incentivizing local compliance without threatening to withdraw aid or providing additional funding-based inducements. Threatening to unilaterally implement a particular policy without local input is likely to be more palatable for diplomats representing intervening forces because such approaches are limited to a particular issue and thus avoid doing far-reaching harm by undermining the regime more broadly. Such threats to leave local partners behind should they fail to participate in a proposed policy provides leverage that can be used to motivate compliance, as local allies would typically rather not be isolated or undermined when their foreign partner implements a policy unilaterally. Additionally, threats of unilateral policy enforcement are more likely to be credible than threats to withhold funding, since it can be difficult for local allies to believe that intervening forces will readily undermine their own mission by weakening local partners through a withdrawal of support.

Indeed, the diplomatic record demonstrates numerous occasions on which this policy-specific, "either you do it or we will" logic was effective in motivating local compliance. The process of negotiating peace terms to end the US involvement in Vietnam provides a powerful example of how small allies can be coerced when the larger partner has the capacity to carry out a proposal unilaterally over strong local opposition. During the negotiating process, Saigon vociferously insisted that North Vietnam withdraw its troops at the same time US forces withdrew.[79] Stridently rejecting US requests to drop this demand, in October 1968 South Vietnamese President Nguyen Van Thieu retorted, "You are powerful, you can say to small nations what you want. All Vietnamese know our life depends on US support, but you cannot force us to do anything against our interest. This negotiation is not a life or death matter for the US but it is for Vietnam."[80] Yet Thieu nevertheless eventually acquiesced, and in 1973 he reluctantly agreed to this concession at the insistence of his American partners.[81]

Thieu's initial refusal to agree to make concessions to communist forces motivated the Americans to negotiate with North Vietnam without Saigon officials at the table. According to historian Larry Berman, there was a "pattern of exclusion for the next four years – Kissinger negotiating an American troop disengagement with the North Vietnamese while informing [South Vietnamese President] Thieu only after the fact."[82] In 1973 National Security Advisor Henry Kissinger wrote to President Nixon, "With respect to Thieu's reaction, it is clear to me that he will not yield short of his fully realizing that he is being given absolutely no alternative. In this respect I believe the certainty of our initialling the agreement without him if necessary is the only way to accomplish this."[83]

Therefore, it is imperative to consider the dependencies and whether or not intervening forces have the ability to pressure local allies by threatening unilateral action. Like interests, dependencies manifest in multiple ways in the politics of counterinsurgency partnerships. The local ally is dependent on intervening forces for its immediate survival, while the intervening partner, due to the norm of local legal sovereignty and legitimacy, often depends on the local partner to implement particular policies. This concept of dependency acknowledges that the duties and functions of allies are not perfectly substitutable. In-country legislative reform, for example, fundamentally requires the participation of the host regime as the legally sovereign authority over legislation. This dependency provides local partners with leverage when negotiating these particular requests. For example, in November 2007, US Ambassador Ryan Crocker and General David Petraeus asked Iraqi Prime Minister Maliki to extend legal immunity to US contractors in Iraq, a privilege foreign contractors had enjoyed in Iraq since 2003.[84] Popular Iraqi outcry over the violent activities of several foreign contractors, however, had hardened Iraqi opinion against contractor immunity. Washington was dependent on Baghdad to fulfill the request, since by 2007 the United States had no authority over Iraqi law. Furthermore, there was nothing the United States could do to compel Iraq to offer continued protection of contract personnel. The Iraqis did not comply. The 2008 Status of Forces Agreement specified: "Iraq shall have the primary right to exercise jurisdiction over United States contractors and United States contractor employees."[85] In February 2011, a British contractor was sentenced to twenty years in prison in Iraq for the murder of two foreign workers.[86]

Intervening forces that depend on small allies to implement particular policies will observe that this dependency provides local clients with substantial influence over the outcome of these policies. However, policy dependency on local regimes does not automatically produce noncompliance. Policies that in-country partners anticipate will serve their interests by providing a particular benefit are likely to be seen favorably and result in compliance. Therefore, dependency on the local ally to implement counterinsurgency policy influences the local ally's decision to comply with or defy policies, but the effect is indirect. Dependency as a variable explaining local compliance is best understood in combination with an analysis of the interests of both allies. A simplistic model that ignores the interaction between dependency and interests will not accurately describe the pressures influencing the behavior of local clients in COIN wars.

Interaction between Interests and Dependency

The political dynamics of counterinsurgency alliances are complex. Variables such as interests and dependencies promote or discourage cooperation *when combined*. This indirect pathway is logical but not necessarily obvious. Overburdened policymakers coping with these difficult partnerships have not had the opportunity to observe such large-scale trends in compliance. Scholars, for their part, have acknowledged

Table 2.1 *Interaction between dependency and interests: effect on local compliance*

		Is unilateral action possible for intervening forces? (dependency)	
		Yes	No
Interests of allies	Converge	Noncompliance (Free riding)	Compliance (Harmony)
	Diverge	Compliance (Leverage)	Noncompliance (Inaction)

the importance of interests but have not specified how they might interact with dependencies. Table 2.1 models how these variables combine to affect the likelihood that local allies will comply with the demands of intervening forces.

If the interests of allies diverge over a given policy proposal, and the intervening state is unable to implement the requested policy unilaterally, compliance is unlikely. Under these circumstances, there is little to motivate the local ally to implement a request that it would rather not see carried out, and the policy will probably not be implemented (*Inaction*). Because intervening forces do not have alternatives in such cases, they depend on local partners for policy implementation; this in turn means that the local ally can decide whether or not the policy will be executed. There are multiple examples of these circumstances in counterinsurgency wars, including Indian requests that Sri Lanka cut off independent negotiations with Tamil militants, or US demands that Iraq renew legislation granting legal immunity to foreign contractors. These kinds of requests typically have gone unfulfilled.

A reverse relationship is hypothesized when the participation of the local ally is not required for the policy proposed. Under such circumstances, local partners would rather contribute in order to have a say in the implementation of the policy and to avoid being isolated or undermined when the larger ally carries out the policy unilaterally (*Leverage*). As discussed in Chapter 4 on Iraq, one example is Baghdad's decision to begrudgingly comply, at least in part, with US demands to match funding to the Sons of Iraq (SOI) program and incorporate a percentage of SOI members into the government.[87] The Shi'a-dominated regime in Baghdad was wary of the SOI program, which had armed local Sunnis pledging to resist al-Qaeda insurgents. Even though promised SOI salary payments from the government were frequently delayed, and there were widespread reports that Baghdad was discriminating against SOI elements, eventually the Iraqi government went along with aspects of the US request during the period when the United States still occupied Iraq. This enabled the Iraqis to control, and later marginalize, the program.[88]

If, on the other hand, the allies' interests converge, and the local ally must participate in order for the request to be fulfilled, compliance is likely (*Harmony*). Cooperation is required for the desired policy to materialize. One example is the May 1987 request by India that Sri Lankan forces in the northeast retreat and stay in their barracks.[89] In order to give Indian forces the opportunity to approach the Tamils, New Delhi encouraged Sri Lankan forces to withdraw peacefully in order to set the stage for Indian-led negotiations. An organized retreat from entrenched positions by the Sri Lankan military was not a task Indian officials could arrange without Colombo's participation. At this point in time, Colombo hoped that the Indians could broker a solution to the Tamil insurgency, and they were willing to let Indian forces take the lead, at least for the moment.[90]

Lastly, if the interests of allies converge and the intervening force can implement the policy on its own, the host ally is unlikely to comply (*Free riding*). Local partners can instead free ride on the efforts of intervening forces, benefiting from the activity without paying the costs. For example, in Afghanistan in April 2009, the US government requested that Kabul create "longer and better coordinated opening hours at Af-Pak border crossing points."[91] The government of Afghanistan could expect to profit from the increased customs revenue from additional cross-border traffic made possible by longer hours at crossing points. US forces would also benefit from greater border efficiency. NATO troops in land-locked Afghanistan rely on resupply routes through either Pakistan or the Northern Distribution Network, a series of circuitous routes that run through various Central Asian states including Russia, Kazakhstan, Uzbekistan, Georgia, Azerbaijan, Kyrgyzstan, and Tajikistan.[92] Frustrated with border inefficiencies and a lack of Afghan initiative, in late 2009 the USA opened two Border Coordination Centers (BCCs). While, as Patrick Kelley and Scott Sweetser argued, "a U.S.-heavy solution [to border coordination is] not necessarily the ideal for a facility with aspirations to multi-national cooperation," but waiting for a "process-oriented" solution incorporating Afghan and Pakistani partners "would almost guarantee failure."[93] The BCCs were commanded by a US colonel or lieutenant colonel, while "the Afghan National Army has provided representatives more-or-less as a show of goodwill; and the

Afghan Uniformed Police and Afghan Border Police are represented by officers who essentially drop-in as a part of their daily beat in the local neighbourhood."[94] The Afghans benefited, but due to the shared interest of both allies in the request and US ability to implement the policy without Afghan participation, Kabul did not have the motivation to comply. They were able to free ride on US efforts.

Significant Enemy Threats

Studies in alliance politics have detailed the importance of enemy threats as a factor that reliably affects interalliance bargaining dynamics. This includes work by Patricia Weitsman, who observed that the wartime dependency of large states on smaller partners "grants significant leverage to smaller states during wartime operations."[95] Importantly, scholarship on alliance relations tends to argue that acute enemy threats will lead to more cohesive, cooperative security partnerships because enemy threats diminish the importance of shifting burdens to partners or maintaining autonomy vis-à-vis a powerful patron.[96] Of course all insurgent action threatens local counterinsurgents in some capacity. However, an organized, acute enemy offensive, like the Tet Offensive in Vietnam, is likely to pose such a substantial threat that it will increase the likelihood of local ally compliance with foreign partner demands. First, behaviors such as free riding become less attractive to local allies when faced with a security crisis. Interalliance cohesion is likely to increase in the face of an acute enemy threat that shifts the focus of local allies toward defeating insurgents and away from other priorities such as maintaining as much autonomy as possible vis-à-vis intervening forces and limiting political risk from internal rivals. This is a long-standing observation in work on alliances. As Evan Resnick has noted, allies are more likely to work closely together "if the level of external threat to the alliance exceeds the level of internal or intra-alliance threat rather than vice-versa."[97]

I theorize that a significant threat is likely to draw allies together, inspiring local allies to participate in efforts to bolster security rather than leaving security in the hands of larger allies. To borrow a concept from economics: a spike in demand will increase the value of a commodity, providing suppliers with an advantage. In wartime, an acute crisis such as an enemy offensive increases the value of security, and the intervening forces supplying that good are able to charge a higher price from weaker allies for protection. That "price," demanded in exchange for security, might be policy reforms that local actors had previously resisted. Additionally, host regimes are not unified, rational actors. As in all political organizations, there are factions, infighting, and bureaucratic constraints. A security crisis can unify disparate local groups, creating opportunities to fulfill requests put forth by intervening forces that might otherwise be stalled by domestic infighting. In essence, this variable is an external shock that orients the local ally toward the joint counterinsurgency security effort.

Hypotheses

We may summarize the proposed hypotheses as follows:

1. Local allies with insufficient resources, capabilities, or institutional capacity to fulfill a given request from an intervening ally are less likely to comply with that request.
2A. If the interests of allies *diverge* over a given policy proposal, the local ally is *less* likely to comply if the policy requires its participation.
2B. If the interests of allies *diverge* over a given policy proposal, the local ally is *more* likely to comply if the policy does not require its participation.
3A. If the interests of allies *converge* over a given policy proposal, the local ally is *more* likely to comply if the policy requires its participation.

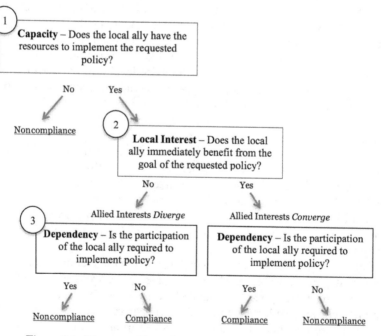

Figure 2.1 Diagram of hypotheses 1–3

3B. If the interests of allies *converge* over a given policy proposal, the local ally is *less* likely to comply if the policy does not require its participation.

4. A significant, acute enemy offensive will make local allies more likely to comply with requests from intervening allies.

Figure 2.1 diagrams the hypotheses outlined above, showing interactions between the three variables of capacity, interests, and dependency.

Summary

Local allies in counterinsurgency wars are often ridiculed as the puppets of intervening militaries. In other instances they are derided as corrupt tyrants, wielding disproportionate influence over larger allies and damaging counterinsurgency objectives. Both narratives are too simple. Coercing local allies in counterinsurgency wars is a complex undertaking and requires a nuanced understanding of the factors influencing the cooperative or defiant behavior of local allies. At times wealthy patrons will be able to sway local partners, while at other times they will be unpersuasive.

This book explores why local allies behave in certain ways in response to policy requests from intervening allies. It identifies four variables, namely capacity, interests, dependency, and significant enemy offensives, to explain variation on the dependent variable, the compliance of local counterinsurgency allies. In particular, I hypothesize that there is an interaction effect between interests and dependency that creates incentives for local allies to comply with, or defy policies proposed by foreign patrons. These variables and hypotheses are considered within the context of nine large-scale counterinsurgency wars. Table 2.2 summarizes the proposed theory.

While the theory explains much of the observed variation, it cannot account for all factors that affect compliance outcomes. Priorities and pressures may shift depending on an array of variables: geographic region, the government officials involved, and bureaucratic preferences, among others. For example, although opium eradication in Afghanistan was at one time a US priority, it was later set aside by US officials in key areas for fear of alienating rural populations and was no longer promoted by American advisers.[98] This study provides a broad overview of trends across COIN partnerships, but many localized factors that have been omitted also influence the likelihood of compliance. Future work can address these factors, as well as further document, specify, and test the effectiveness of particular coercive diplomatic strategies.

Table 2.2 *Summary of arguments*

	Conditions making compliance by local allies **more** likely	Conditions making compliance by local allies **less** likely
Capacity (Capabilities)	High local ally capacity	Low local ally capacity
Interests and Dependency	Interests of allies diverge, and the request does not require the participation of the local ally Interests of allies converge, and the request requires the participation of the local ally	Interests of allies diverge, and the request requires the participation of the local ally Interests of allies converge, and the request does not require the participation of the local ally
Acute Enemy Threat	Substantial, acute enemy threat exists	No substantial, acute enemy threat exists

Notes

1 Sarah Sewall. "Introduction to the University of Chicago Press Edition," in *The U.S. Army/Marine Corps Counterinsurgency Field Manual*, US Army and Marine Corps (University of Chicago Press, 2007), xxi. Regarding a definition for counterinsurgency (COIN), I follow a basic notion that COIN is a military-political effort to suppress an insurgency. This concept is fairly well captured in *FM 3-24*, which defines COIN as "those military, paramilitary, political, economic, psychological, and civic actions taken by a government to defeat insurgency." US Army and Marine Corps, *Counterinsurgency*, *FM-3-24*, chapter 1, 2.

2 George C. Herring, *America's Longest War: The United States and Vietnam 1950–1975*, 3rd ed. (New York: McGraw-Hill Companies, 1995), 159, citing Robert Shaplen, *The Lost Revolution: The U.S. in Vietnam, 1946–1966* (New York: Harper & Row, 1966), 186.

3 See Table 3.1 in Chapter 3 for outcome in counterinsurgency interventions from 1945 to 2018 with 1,000+ foreign military deaths. Only Angola and Sri Lanka are clear successes, and in the Sri Lankan case counterinsurgency victory was not due to the intervention of Indian troops; see Chapter 7 for more information. Also see Watts et al., *Countering Others' Insurgencies*, Table 3.2, 53.

4 Ivan M. Arreguin-Toft, *How the Weak Win Wars: A Theory of Asymmetric Conflict* (Cambridge University Press, 2006); T. V. Paul, *Asymmetric Conflicts: War Initiation by Weaker Powers* (Cambridge University Press, 1994). Advocates for population-centric counterinsurgency argue that wealthy militaries have excessively relied on military solutions to political problems, neglecting the strategic importance of the population; see US Army and Marine Corps, *Field Manual FM 3-24 MCWP 3-33.5 Insurgencies and Countering Insurgencies*; Ved P. Nanda, "The 'Good Governance' Concept Revisited," *The Annals of the American Academy of Political and Social Science* 603 (January 2006):

269–83; Julian Paget, *Counter-Insurgency Operations: Techniques of Guerrilla Warfare* (New York: Walker, 1967); Robert Thompson, *Defeating Communist Insurgency: The Lessons of Malaya and Vietnam* (New York: F. A. Praeger, 1966); Peter W. Chiarelli and Patrick R. Michaelis, "Winning the Peace: The Requirement for Full-Spectrum Operations," *Military Review* 85, no. 4 (August 2005): 4–17; Andrew J. Enterline, Emily Stull, and Joseph Magagnoli, "Reversal of Fortune? Strategy Change and Counterinsurgency Success by Foreign Powers in the Twentieth Century," *International Studies Perspectives* 14, no. 2 (May 1, 2013): 176–98. Critics of population-centric counterinsurgency argue such approaches are amorphous and rely on ill-defined goals. See Gian Gentile, *Wrong Turn; America's Deadly Embrace of Counterinsurgency* (New York: The New Press, 2013); Michael A. Cohen, "The Myth of a Kinder, Gentler War," *World Policy Journal* 27, no. 1 (May 3, 2010): 75–86; Douglas Porch, *Counterinsurgency: Exposing the Myths of the New Way of War* (Cambridge University Press, 2013); Jacqueline L. Hazelton, "The 'Hearts and Minds' Fallacy: Violence, Coercion, and Success in Counterinsurgency Warfare," *International Security* 42, no. 1 (July 1, 2017): 80–113.

5 Jonathan K. Graff, "United States Counterinsurgency Doctrine and Implementation in Iraq," ARMY Command and General Staff College – Fort Leavenworth, June 18, 2004, www.dtic.mil/docs/citations/ADA428901; Robert Egnell, "Lessons from Helmand, Afghanistan: What Now for British Counterinsurgency?" *International Affairs* 87, no. 2 (March 1, 2011): 297–315; Steven Metz, "Learning from Iraq: Counterinsurgency in American Strategy," US Army War College, Strategic Studies Institute, January 2007, www.dtic.mil/docs/citations/ADA459931; James Pritchard and M. L. R. Smith, "Thompson in Helmand: Comparing Theory to Practice in British Counter-Insurgency Operations in Afghanistan," *Civil Wars* 12, no. 1–2 (January 1, 2010): 65–9; Warren Chin, "Examining the Application of British Counterinsurgency Doctrine by the American Army in Iraq," *Small Wars & Insurgencies* 18, no. 1 (March 1, 2007): 1–26; Peter Viggo Jakobsen, "Right Strategy, Wrong Place – Why NATO's Comprehensive Approach Will Fail in Afghanistan," UNISCI Discussion Papers, no. 22 (2010), www .redalyc.org/resumen.oa?id=76712438006; Michael O'Hanlon, "America's History of Counterinsurgency," Washington, DC: Brookings Institution, June 6, 2016, www.brookings.edu/wp-content/uploads/2016/06/06_counter insurgency_ohanlon.pdf.

6 Jason Lyall and Isaiah Wilson III, "Rage against the Machines: Explaining Outcomes in Counterinsurgency Wars," *International Organization* 63, no. 1 (January 1, 2009): 67–106.

7 M. L. R. Smith and David Martin Jones, *The Political Impossibility of Modern Counterinsurgency: Strategic Problems, Puzzles, and Paradoxes* (New York: Columbia University Press, 2015).

8 Ioannis Koskinas, "President Karzai Is the One to Blame," *Foreign Policy*, June 20, 2014, https://foreignpolicy.com/2014/06/20/president-karzai-is-the-one-to-blame/; Marc Lynch, "How Can the U.S. Help Maliki When Maliki's the Problem?," *Washington Post*, June 12, 2014, sec. Monkey Cage, www.washingtonpost .com/news/monkey-cage/wp/2014/06/12/iraq-trapped-between-isis-and-maliki/.

9 There are important and notable exceptions. See Byman, "Friends Like These"; Ladwig, *The Forgotten Front*; Berman and Lake, *Proxy Wars*.

10 For studies that address the compliance of clients or proxies with demands from patrons in foreign policy, see Christopher P. Carney, "International Patron–Client Relationships: A Conceptual Framework," *Studies in Comparative International Development* 24, no. 2 (Summer 1989): 42–55; Ladwig, "Influencing Clients in Counterinsurgency"; Ladwig, *The Forgotten Front*; Berman and Lake, *Proxy Wars*.

11 Octavian Manea, Interview with Dr. David Kilcullen, *Small Wars Journal*, November 7, 2010, http://smallwarsjournal.com/jrnl/art/interview-with-dr-david-kilcullen.

12 For examples see Craig A. Collier, "Now That We're Leaving Iraq, What Did We Learn?," *Military Review* 88 (October 2010): 91–2; Ian F. W. Beckett, *Roots of Counterinsurgency: Armies and Guerrilla Warfare, 1900–1945* (London: Blandford Press, 1988).

13 David Galula, *Counterinsurgency Warfare: Theory and Practice* (Westport, CT: Praeger, 2006); US Army and US Marine Corps, *Counterinsurgency, FM 3-24*.

14 Mao Tse-Tung, *On Guerrilla Warfare*, trans. Samuel B. Griffith (Santiago: BN Publishing, 2007), 93.

15 David Kilcullen, "Counterinsurgency Seminar 07," Small Wars Center of Excellence, September 26, 2007, 10, http://smallwarsjournal.com/blog/coin-seminar-dr-david-kilcullen; Ann Marlowe, "The Picture Awaits: The Birth of Modern Counterinsurgency," *World Affairs* 172, no. 1 (2009): 64–73.

16 Hazelton, "The 'Hearts and Minds' Fallacy."

17 Mark Moyar, *A Question of Command: Counterinsurgency from the Civil War to Iraq* (New Haven, CT: Yale University Press, 2009).

18 Kilcullen, *Counterinsurgency* , 7.

19 Ladwig, *The Forgotten Front*, 21.

20 US Department of State, US Embassy Baghdad, "Fallujans Mobilized for Election amid Increased Tension in City," 05BAGHDAD4971, December 13, 2005; Camp, *Operation Phantom Fury*; Bing West, *No True Glory: A Frontline Account of the Battle for Fallujah* (New York: Random House Publishing Group, 2011); Jeremy Shapiro, "The Latest Battle in Fallujah Is a Symbol of the Futility of US Efforts in Iraq," *Vox*, May 25, 2016, www.vox.com/2016/5/25/11750054/battle-fallujah-iraq; "Fallujah, Again," *The Economist*, May 28, 2016, www.economist.com/news/middle-east-and-africa/21699461-why-retaking-jihadist-stronghold-has-become-priority-fallujah-again.

21 James R. Arnold, *Americans at War: Eyewitness Accounts from the American Revolution to the 21st Century* (Santa Barbara, CA: ABC-CLIO, 2018), 1019.

22 United States Senate Armed Services Committee, *Iraq, Afghanistan, and the Global War on Terrorism: Hearings before the Committee on Armed Services*, 109th Congress, 2nd Session, § Armed Services Committee, 2006, 18.

23 Zaid Al-Ali, "How Maliki Ruined Iraq," *Foreign Policy* (blog), June 19, 2014, http://foreignpolicy.com/2014/06/19/how-maliki-ruined-iraq/; Boghani, "David Petraeus"; Patrick Cockburn, *The Rise of Islamic State: ISIS and the New Sunni Revolution* (London: Verso Books, 2015).

24 Thompson, *Defeating Communist Insurgency*, 51; also cited in Ladwig, *The Forgotten Front*, 21.

25 Stephen Hadley, "Stephen Hadley: How Bush Started – and Ended – the Iraq War," interview by Sarah Childress, *PBS Frontline*, July 29, 2014, www.pbs .org/wgbh/frontline/article/stephen-hadley-how-bush-started-and-ended-the-iraq-war/.

26 Daniel Byman, *Going to War with the Allies You Have: Allies, Counterinsurgency, and the War on Terrorism* (Carlisle Barracks, PA: US Army War College Strategic Studies Institute, 2005), 27; Steven Metz, "Unruly Clients: The Trouble with Allies," *World Affairs* 172, no. 4 (2010): 49–59; Vanda Felbab-Brown, "Counterinsurgency, Counternarcotics, and Illicit Economies in Afghanistan: Lessons for State-Building," in *Convergence: Illicit Networks and National Security in the Age of Globalization*, ed. Michael Miklaucic and Jacqueline Brewer (Washington, DC: National Defense University Press, 2013), 189–209.

27 Ladwig, *The Forgotten Front*, 32.

28 George M. Elsey, "Notes of Meeting" (US Department of State, Office of the Historian, November 5, 1968, Document 195, *Foreign Relations of the United States, 1964–8, Volume VII, Vietnam, September 1968–January 1969*, Johnson Presidential Library, George M. Elsey Papers, Van De Mark Transcripts [1 of 2]), https://history.state.gov/historicaldocuments/frus1964-68v07/d195.

29 US Department of State, US Embassy Kabul, "CODEL Hayes Meets Karzai, Wardak," 06KABUL2723, June 15, 2006.

30 US Department of State, US Embassy Kabul, "Karzai Urges CODEL McCain to Support Zardari and Welcome Increase in U.S. Forces," 08KABUL3237, December 21, 2008.

31 Keohane, "The Big Influence of Small Allies," 162.

32 As Glenn Snyder observed, theories on alliances need to acknowledge that "the relative influence of allies turns not simply on their relative military strength and potential, as is often assumed, but on their comparative dependence on, or need for, each other's aid, which is a more complex notion." Snyder, *Alliance Politics*, 31. See also Walt, *The Origins of Alliances*, 43–4.

33 George Packer, "Statecraft as Psychiatry," *The New Yorker*, May 11, 2010, www.newyorker.com/news/george-packer/statecraft-as-psychiatry.

34 Troubled that the USA did not favor him for reelection, Afghan President Hamid Karzai accused the USA of conspiring against him, including colluding with the Iranians to support alternative candidates, leading US Ambassador Karl Eikenberry to describe him as "paranoid and weak." US Department of State, US Embassy Kabul, "Karzai on the State of U.S.-Afghan Relations," 09KABUL1767, July 7, 2009.

35 John Kenneth Galbraith, *The Selected Letters of John Kenneth Galbraith* (Cambridge University Press, 2017), 205. Diem is an interesting example, since the Kennedy administration took the unusual step of condoning a coup that led to Diem's overthrow and death. In Diem's case, there is a brutal irony in that his resistance to adopting the reforms recommended by the USA was motivated by his fear of losing power, yet his inflexibility ultimately hastened his demise because the USA had become so frustrated by his obstinacy that it approved the anti-Diem coup.

36 Joshua Partlow, *A Kingdom of Their Own: The Family Karzai and the Afghan Disaster* (New York: Knopf Doubleday Publishing Group, 2016), 51–2; James Dobbins, *After the Taliban: Nation-Building in Afghanistan* (Washington, DC: Potomac Books, 2008), 89.

37 Gary King, Robert O. Keohane, and Sidney Verba, *Designing Social Inquiry* (Princeton University Press, 1994), 19.

38 Carl von Clausewitz, *On War*, trans. Michael Eliot Howard and Peter Paret, reprint (Princeton University Press, 1989), 121, 119.

39 Galula, *Counterinsurgency Warfare*, 6.

40 Laffont and Martimort, *The Theory of Incentives*, 2–4, 146–7; Miller, "The Political Evolution of Principal-Agent Models," 203–4; Ladwig, "Influencing Clients in Counterinsurgency," 5.

41 Ladwig, *The Forgotten Front*, chapter 3. See pp. 76–84.

42 Macdonald, *Adventures in Chaos*, 6.

43 Berman and Lake, *Proxy Wars*, 4.

44 Ibid., 5.

45 "Decisions of the Security Co-ordination Group Regarding Security of the Tamils, Colombo, October 8, 19; November 1, 3, 1989 – Minutes of the Special Committee Appointed by the Security Co-ordination Group to Discuss the Details and the Numbers Required for the Citizens Volunteer Force, Provincial Police and the Armed Services," in *India–Sri Lanka: Relations and Sri Lanka's Ethnic Conflict Documents – 1947–2000*, ed. Avtar Singh Bhasin, Vol. IV, Document 934 (New Delhi: Indian Research Press, 2001), 2408–10.

46 Gunaratna, *Indian Intervention in Sri Lanka*, 355–6.

47 Ibid.; Robert Oberst, "A War without Winners in Sri Lanka," *Current History* 91, no. 563 (March 1992): 129. See also "Extract from the Statement of the Sri Lankan Minister of Foreign Affairs and Minister of State for Defence Ranjan Wijeratne in Parliament during the Debate on Foreign Affairs," March 31, 1989, in *India–Sri Lanka: Relations and Sri Lanka's Ethnic Conflict Documents – 1947–2000*, ed. Avtar Singh Bhasin, Vol. IV, Document 886 (New Delhi: Indian Research Press, 2001), 2356–8.

48 In the Iraq war Washington and Baghdad had divergent interests regarding economic reforms. The USA was seeking market-based solutions, while Iraq favored more familiar statist economic solutions.

49 Ladwig, *The Forgotten Front*, 13; Berman and Lake, *Proxy Wars*, 4.

50 Olson and Zeckhauser, "An Economic Theory of Alliances"; Mancur Olson, *The Logic of Collective Action: Public Goods and the Theory of Groups, Second Printing with New Preface and Appendix*, revised ed. (Cambridge, MA: Harvard University Press, 1971); Bruce M. Russett and John D. Sullivan, "Collective Goods and International Organization," *International Organization* 25, no. 4 (October 1, 1971): 845–65; John R. Oneal, "The Theory of Collective Action and Burden Sharing in NATO," *International Organization* 44, no. 3 (July 1, 1990): 379–402; Oneal and Diehl, "The Theory of Collective Action and NATO Defense Burdens," 373–96; Wallace J. Thies, *Friendly Rivals: Bargaining and Burden-Shifting in NATO* (New York: Routledge, 2003); Ellen Hallams and Benjamin Schreer, "Towards a 'Post-American' Alliance? NATO Burden-Sharing after Libya," *International Affairs* 88, no. 2 (March 1, 2012): 313–27.

51 Barbara Elias, "The Likelihood of Local Allies Free-Riding: Testing Economic Theories of Alliances in US Counterinsurgency Interventions," *Cooperation and Conflict* 52, no. 3 (September 1, 2017): 309–31.

52 US Department of State, "Memorandum for Walt Rostow, from William Leonhart, 'Blueprint for Vietnam'," September 11, 1967, NLJ 87-50, Lyndon B. Johnson Library; Jeffrey J. Clarke, *Advice and Support: The Final Years, 1965–1973* (Washington, DC: US Government Printing Office, 1988), 226.

53 US Department of State, "Viet-Nam Political Situation Report," November 18, 1967, NLJ 94-480, LBJ Presidential Library.

54 US Department of State, "For the President from Bunker, Saigon 16850," January 24, 1968, NLJ 96-207, LBJ Presidential Library, 14–15.

55 Clarke, *Advice and Support*, 151. Furthermore, according to US Senator Edward Kennedy, the outcome of a US-designed program to support scholarships for South Vietnamese veterans to study at US universities failed because of Saigon's decision to undermine the program, noting,

> [W]e asked the government of South Vietnam to select some qualified men for this opportunity. The list they gave us consisted mainly of relatives of government officials. When we discovered this, we asked them to find other men, unrelated to them, but after the second list came in, it was discovered that all of the new applicants had been made to promise a percentage of their scholarship payments to the officials who chose them.

See US Department of State, "Sen. Edward Kennedy's Report on His Recent Trip to Vietnam," January 27, 1968, NLJ 94-117, LBJ Presidential Library.

56 Snyder, *Alliance Politics*, 199.

57 Jonathan D. Caverley, "The Myth of Military Myopia: Democracy, Small Wars, and Vietnam," *International Security* 34, no. 3 (January 1, 2010): 121.

58 Ladwig, "Influencing Clients in Counterinsurgency"; Ladwig, *The Forgotten Front*; Berman and Lake, *Proxy Wars*.

59 Ladwig, "Influencing Clients in Counterinsurgency," 100. Ladwig also specifies that the strategy of conditionality is likely most effective in situations of limited intervention, a commitment to a client that falls short of large-scale intervention. Ladwig, *The Forgotten Front*, 78.

60 Ladwig, *The Forgotten Front*, 81–3; Berman and Lake, *Proxy Wars*, 295.

61 Cooper et al., "The American Experience with Pacification in Vietnam," 23.

62 Lionel Beehner and Greg Bruno, "What Are Iraq's Benchmarks?," *Council on Foreign Relations*, March 11, 2008, www.cfr.org/iraq/iraqs-benchmarks/p13333.

63 Helene Cooper, "In Leaning on Karzai, U.S. Has Limited Leverage," *The New York Times*, November 11, 2009, sec. International / Asia Pacific, www.nytimes.com/2009/11/12/world/asia/12karzai.html.

64 The Pentagon Papers, "[Part IV. C. 2. c.] Evolution of the War. Military Pressures against NVN. November– December 1964," 1/1969 1967, vii, Series: Report of the Office of the Secretary of Defense Vietnam Task Force, 6/1967 – 1/1969, Record Group 330: Records of the Office of the Secretary of Defense, 1921–2008, US National Archives, https://catalog.archives.gov/id/5890501; see also Komer, "Bureaucracy Does Its Thing," 26; Office of the

Secretary of Defense Vietnam Task Force, "United States–Vietnam Relations, 1945–1967," January 1971, vii.

65 Komer, "Bureaucracy Does Its Thing," 22.

66 Alissa J. Rubin, "Karzai's Words Leave Few Choices for the West," *The New York Times*, April 4, 2010, sec. Asia Pacific, www.nytimes.com/2010/04/05/world/asia/05karzai.html.

67 Berman et al., "Introduction: Principals, Agents, and Indirect Foreign Policies," in Berman and Lake, *Proxy Wars*, 8.

68 Komer, "Bureaucracy Does Its Thing," 33.

69 See Chapter 8 and Lake, *The Statebuilder's Dilemma*.

70 "In making the choice to do nothing to prevent the coup and to tacitly support it, the U.S. inadvertently deepened its involvement. The inadvertence is the key factor." The Pentagon Papers, "[Part IV. B. 5.] Evolution of the War. Counterinsurgency: The Overthrow of Ngo Dinh Diem, May–Nov. 1963," 1/1969 1967, Series: Report of the Office of the Secretary of Defense Vietnam Task Force, Record Group 330: Records of the Office of the Secretary of Defense, 1921–2008, US National Archives, www.archives.gov/research/pentagon-papers.

71 Ibid., viii.

72 Komer, "Bureaucracy Does Its Thing," 24.

73 Ibid., 32.

74 US Department of State, US Embassy Kabul, "Karzai on Elections and the Future: September 1 Meeting at the Palace," 09KABUL2681, September 3, 2009.

75 US Department of State, US Embassy Kabul, "The New Cabinet: Better but Not Best," 09KABUL4070, December 19, 2009. See also US Department of State, US Embassy Kabul, "Karzai on Elections and the Future: September 1 Meeting at the Palace."

76 US Department of State, US Embassy Kabul, "Karzai on ANSF, Cabinet, and 2010 Elections," 09KABUL4027, December 16, 2009.

77 Kenneth Katzman, "Afghanistan: Politics, Elections, and Government Performance" (Congressional Research Service, RS21922, February 19, 2010), 7–8; Christian Neef, "Return of the Lion: Former Warlord Preps for Western Withdrawal," *Spiegel Online*, September 23, 2013, sec. International, www.spiegel.de/international/world/afghan-warlords-like-ismail-khan-prepare-for-western-withdrawal-a-924019.html.

78 Office of the Special Inspector General for Afghanistan Reconstruction (SIGAR), "Corruption in Conflict: Lessons from the U.S. Experience in Afghanistan," September 2016, www.dtic.mil/dtic/tr/fulltext/u2/1016886 .pdf, 57; Maria Abi-Habib, "U.S. Blames Senior Afghan in Deaths," *Wall Street Journal*, April 1, 2012, sec. World News, www.wsj.com/articles/SB10001424052702303404704577311522824172282; US Department of State, US Embassy Kabul, "Corrected Copy: IDLG Director Popal Meets with Ambassador: Moving Ahead on the District Delivery Program," 10KABUL570_a, February 15, 2010.

79 John Prados, "The Shape of the Table," in *The Search for Peace in Vietnam, 1964–1968*, ed. Lloyd C. Gardner and Ted Gittinger (College Station: Texas A&M University Press, 2004), 363.

80 Randall Woods, *LBJ: Architect of American Ambition* (New York: Simon & Schuster, 2006), 874; Richard H. Immerman, "'A Time in the Tide of Men's Affairs,' Lyndon Johnson and Vietnam," in *Lyndon Johnson Confronts the World: American Foreign Policy 1963–1968*, ed. Warren I. Cohen and Nancy Bernkopf Tucker (Cambridge University Press, 1994), 91.

81 Prados, "The Shape of the Table," 366–8; Lien-Hang T. Nguyen, "Cold War Contradictions: Toward an International History of the Second Indochina War, 1969–1973," in *Making Sense of the Vietnam Wars: Local, National, and Transnational Perspectives*, ed. Mark Philip Bradley and Marilyn B. Young (Oxford University Press, 2008), 234; Pierre Asselin, *A Bitter Peace: Washington, Hanoi, and the Making of the Paris Agreement* (Chapel Hill: University of North Carolina Press, 2002), 170–80.

82 Larry Berman, *No Peace, No Honor: Nixon, Kissinger, and Betrayal in Vietnam* (New York: Touchstone, 2002), 54.

83 Henry Kissinger, "Message from the President's Assistant for National Security Affairs (Kissinger) to President Nixon," US State Department, Office of the Historian, January 11, 1973, Document 266, *Foreign Relations of the United States, 1969–76, Volume IX, Vietnam, October 1972–January 1973*, National Archives, Nixon Presidential Materials, NSC Files, Kissinger Office Files, Box 28, HAK Trip Files, HAK Paris Trip Hakto 1–48, January 7–14, 1973. Top Secret; Flash; Sensitive; Exclusively Eyes Only. Sent via Kennedy, https://history.state.gov/historicaldocuments/frus1969–76v09/d266. See also the account of negotiations in Asselin, *A Bitter Peace*, 162. Kissinger informs Nixon, "The only way to bring Thieu around will be to tell him flatly that you will proceed, with or without him."

84 US Department of State, US Embassy Baghdad, "Implementation of Recommendations on Personal Protective Services: Status Report Update #1," 07BAGHDAD4001, December 10, 2007.

85 "Agreement between the United States of America and the Republic of Iraq on the Withdrawal of United States Forces from Iraq and the Organization of Their Activities during Their Temporary Presence in Iraq" (US Department of State, November 27, 2008), 10, www.state.gov/documents/organization/122074.pdf.

86 Michael S. Schmidt, "Immunity Gone, Contractor in Iraq Sentenced to Prison," *The New York Times*, February 28, 2011, sec. World / Middle East, www.nytimes.com/2011/03/01/world/middleeast/01iraq.html.

87 US Department of State, US Embassy Baghdad, "CG and CDA Discuss Foreign Fighters and Syria, Turkey and the PKK, and UNSCR with PM," 07BAGHDAD3911, December 2, 2007.

88 Office of the SIGIR, "Quarterly Report to Congress," October 30, 2008, 49. See also Office of the SIGIR, "Sons of Iraq Program: Results Are Uncertain and Financial Controls Were Weak," SIGIR 11-010, January 28, 2011.

89 "Press Release of the Indian High Commissioner in Colombo," July 6, 1987, in *India–Sri Lanka: Relations and Sri Lanka's Ethnic Conflict Documents – 1947–2000*, ed. Avtar Singh Bhasin, Vol. III, Document 714 (New Delhi: Indian Research Press, 2001), 1933. See also "Question in the Sri Lankan Parliament Regarding the Implementation of the Indo-Sri-Lanka Peace Accord of July 1987," June 20, 1989, in *India–Sri Lanka: Relations and*

Sri Lanka's Ethnic Conflict Documents – 1947–2000, ed. Avtar Singh Bhasin, Vol. IV, Document 914 (New Delhi: Indian Research Press, 2001), 2385–8.

90 "Question in the Sri Lankan Parliament Regarding the Implementation of the Indo-Sri-Lanka Peace Accord of July 1987."

91 US Department of State, US Embassy Kabul, "Economic Agenda Items for Af-Pak Trilateral Commission," 09KABUL943, April 15, 2009.

92 Thomas P. Kelly, "The Northern Distribution Network and the Baltic Nexus," remarks at the Commonwealth Club, Washington, DC, January 20, 2012, US Department of State, https://2009-2017.state.gov/t/pm/rls/rm/182317.htm.

93 Patrick Kelley and Scott Sweetser, "The Spaces in between: Operating on the Afghan Border (or Not)," *Small Wars Journal*, April 1, 2010, 4, smallwarsjournal.com/blog/journal/docs-temp/404-kelley.pdf.

94 Ibid.; Office of the SIGAR, "Quarterly Report to Congress," April 30, 2010, 53, www.sigar.mil/pdf/quarterlyreports/2010-04-30qr.pdf.

95 Patricia A. Weitsman, *Waging War: Alliances, Coalitions, and Institutions of Interstate Violence* (Stanford University Press, 2013), 35.

96 P. Terry Hopmann, "International Conflict and Cohesion in the Communist System," *International Studies Quarterly* 11, no. 3 (1967): 212–36; William R. Thompson and David P. Rapkin, "Collaboration, Consensus, and Détente: The External Threat-Bloc Cohesion Hypothesis," *The Journal of Conflict Resolution* 25, no. 4 (1981): 615–37; Patricia A. Weitsman, *Dangerous Alliances: Proponents of Peace, Weapons of War* (Stanford University Press, 2004), 24–7; Evan N. Resnick, "Hang Together or Hang Separately? Evaluating Rival Theories of Wartime Alliance Cohesion," *Security Studies* 22, no. 4 (October 1, 2013): 672–706.

97 Resnick, "Hang Together or Hang Separately?," 673.

98 Rod Nordland, "U.S. Turns a Blind Eye to Opium in Afghan Town," *The New York Times*, March 20, 2010, www.nytimes.com/2010/03/21/world/asia/21marja.html.

3   Methodology
Wars, Documents, and Data

> The future historian may shed light on why the Saigon Government, assisted unstintingly by the most powerful nation in the world, was not able to dispose of the ragtag Viet Cong and its major ally North Vietnam, a country of 16 million people. But to us the confrontation has been bewildering and frustrating – especially frustrating.
> —Chester L. Cooper, *The Lost Crusade: America in Vietnam*

Counterinsurgency alliances are dynamic institutions that affect and are affected by local and international politics. In order to balance these contextualized and systemic influences that create alliance dynamics, I use detailed data from specific interalliance interactions to compare and analyze alliances in nine counterinsurgency wars with large-scale third-party interventions from 1962 to 2010. For five wars, including the USA in Afghanistan, Iraq, and Vietnam, as well as India in Sri Lanka and the USSR in Afghanistan, thousands of US, Indian, and Soviet government documents describing the day-to-day workings of alliance relations were dissected using qualitative and quantitative methods to reveal larger systemic alliance dynamics. For these 5 wars, I identified 460 specific policy requests from intervening forces to local political partners and built a database to track local compliance. The coding rules for each variable are detailed in this chapter. For the remaining four wars, namely Vietnam in Cambodia, Egypt in Yemen, Cuba in Angola, and Syria in Lebanon, the lack of primary source documents released by these regimes prevents quantitative analysis. These four wars are explored qualitatively by relying on secondary historical accounts or primary source observations from outside actors such as US intelligence agencies.

These nine wars represent all cases of post–WWII counterinsurgency wars with 1,000+ foreign military casualties identified by the Rand Corporation's 2008 study on counterinsurgencies, "War by Other Means,"[1] which updated James Fearon and David Laitin's 1999 dataset "Ethnicity, Insurgency and Civil War."[2] From these sources, I found twenty-eight, or approximately a third, of post–WWII insurgency wars

Table 3.1 *Post–WWII counterinsurgency wars with 1,000+ foreign military casualties*

Conflict	Years	Foreign COIN intervener	Outcome
Yemen	1962–70	Egypt	Mixed
South Vietnam	1964–73	USA	Insurgency victory
Angola	1975–91	Cuba	COIN victory
Lebanon	1975–90	Syria	Mixed
Cambodia	1979–91	Vietnam	Mixed
Afghanistan	1979–89	USSR	Insurgency victory
Sri Lanka	1987–2009	India	COIN victory[a]
Afghanistan	2001–ongoing	USA	Ongoing
Iraq	2003–11	USA	Mixed[b]

Notes: [a] In 2009, the Sri Lankan military defeated the insurgent Tamil movement, long after the March 1990 withdrawal of Indian forces from the conflict. The Indian intervention is considered a strategic failure. See Chapter 7.
[b] Upon US withdrawal in 2011, Iraq appeared to be a counterinsurgency victory, but is coded as a mixed outcome since the Islamic State, an outgrowth of the al-Qaeda insurgency, successfully conquered significant Iraqi territory in 2014.

had foreign forces intervene on the side of the counterinsurgent; 9 incurred over 1,000 foreign deaths.[3]

Focusing on large-scale foreign military intervention isolates counter-insurgency alliances with a high level of alliance commitment. Other types of interventions, such as colonial wars, indirect military interventions, or covert operations, may have unique influences on the behavior of allies. The 1,000-military-death threshold is drawn from the Correlates of War (COW) dataset on civil wars, since insurgencies are frequently considered a subset of civil wars.[4] Furthermore, the data separates into 2 categories, with a notable gap between 83 foreign military deaths (Somalia) and the next highest casualty count, 1,150 (Sri Lanka). Taking this 1,000+ battle-death metric, I find nine conflicts of the initial twenty-eight with foreign military intervention that meet the criteria (see Table 3.1).[5]

For the five wars that rely on primary source documentation (Vietnam, both Afghanistan cases, Iraq, and Sri Lanka), I systematically analyzed thousands of high-level government materials from intervening forces in order to identify requests made to the primary leadership of the local allied regime. Individual policy requests from intervening allies are the primary unit of analysis, while the wars are the universe of cases from which these observations are drawn. Government documents from local regimes are not used in this initial stage of identifying policy requests from intervening forces, due to the lack of availability of primary source records from allies such as Saigon, Baghdad, or Kabul.

To identify requests from intervening forces to local allies, over 12,500 documents (25,000+ pages) were analyzed, yielding several hundred documents related to policy requests made to allies. I identify 460 requests in Vietnam, Afghanistan–USA, Afghanistan–USSR, Iraq, and Sri Lanka by searching through all cables for the specific titles most often used to describe the local allied regime, such as "GOI," "Government of Iraq," "GIRoA," "Government of the Islamic Republic of Afghanistan," "GVN," "South Vietnam," "Colombo," the names of high-ranking officials likely included in discussions, as well as verbs commonly found in diplomatic documents that describe bargaining: "request," "press," "push," "demand," "work," "reform," "ask," "urge," "insist," "assist." Not all 4,000+ cables identified by searching for these terms were relevant in identifying requests from intervening allies, but casting a wide net better ensured capturing relevant observations.

The sources for the documents used to identify the requests of intervening allies to local partners are diplomatic agencies, such as the US Department of State, instead of military bureaucracies. Therefore, this book does not provide an exhaustive account of all policies demanded from local partners by intervening forces. While there are some potential costs to relying on documents from the civilian diplomatic corps, as the military serves as the preeminent institution in military intervention, this intentional focus serves a variety of purposes. First, there are significantly more materials from civilian bureaucracies available to the public, allowing greater depth of analysis. Second, because high-ranking military personnel, such as General David Petraeus in Iraq, were often present in meetings and discussions documented in diplomatic cables, I would not expect military sources to contradict the findings offered that rely on diplomatic documents and therefore would not anticipate the omission of military records to sway the findings presented here. Third, relying on data from traditional diplomatic sources, as opposed to intelligence and defense agencies operating in exceptionally insecure environments, is more likely to enable the model offered here to extend more broadly to other kinds of alliances, including traditional peacetime security partnerships. Fourth, this focus on diplomatic institutions enables an exploration of a neglected component of counterinsurgency wars focused on political engagement, not only military coordination. Also, embassies tend to be the center of high-level diplomatic engagement with local allies, containing high-level instructions regarding coordination and reform as well as reports on bargaining processes and outcomes.

As a precaution against bias from relying on diplomatic cables to identify requests, whenever possible the data was correlated with public-source materials, such as the 2007 "benchmarks" for Iraq.[6] However, for the sake of consistency and validity, only primary source diplomatic documents were used to compile the list of requests to allies from intervening forces. Other sources, including media reports, statements from local allies, or existing scholarship, were used to determine and triangulate compliance outcomes with the requests identified in primary source diplomatic correspondence.

While not exhaustive compilations of all policy demands from intervening forces in these wars, the requests identified capture a significant portion of requests discussed in diplomatic channels. Of course, issues arise in particular wars regarding the reliability of document sources. These issues are mentioned in case study chapters, as well as in Appendix B. These qualifications aside, the documents and methods utilized provide unique insights into the complex alliances. There are significant benefits and limitations to archival materials. As outlined by George and Bennett, the researcher "should not assume that going to primary sources and declassified government documents alone will be sufficient to find the answers to his or her research questions."[7] For this reason, primary source documents are used to identify initial requests from intervening forces, and are used as one source of data for understanding what took place once a particular request was presented to local allies. Whenever there are limitations to the available sources of documents, any potential bias introduced by those source limitations is noted.[8]

One interesting component of identifying requests from intervening powers to local allies in documents was differentiating between general statements about desired outcomes, such as "stability" or "security," as opposed to a specific policy request, such as holding elections that aimed to achieve an outcome like "stability." A hopeful statement about an improved situation is not specific enough for an ally to implement. For example, consider the Soviet statement to Afghan President Karmal that he should "do everything to revive and develop the national economy, to raise the standard of living of the population and, first of all, of all workers and peasants."[9] What is "everything?" How exactly is Karmal to "develop" the national economy and raise the standard of living for workers? A higher living standard and economic expansion are possible consequences of robust economic policies, but are not policies themselves. Furthermore, such categories of issues are so broad they would be difficult to analyze as meaningful data. Requests are limited to specific policy recommendations that can be identified and coded.[10] Consider, for example, that in a different meeting, the Soviets were more specific with their Afghan partners, proposing they adopt new policies emphasizing foreign trade to expand the Afghan economy.[11] Determining whether or not the Afghans implemented new foreign trade policies provides a more reliable metric about compliance with Soviet demands than general trends in economic growth that may not indicate much about the alliance or Afghan decision-making.

Additionally, there is no guarantee that the policies advocated by the intervening ally will accomplish what was intended. Whether the Afghans implement new policies on foreign trade is a different issue from whether Soviet policies designed to expand the Afghan economy were effective. I measure compliance, not strategic effectiveness, which makes differentiating between "requests" and "goals" all the more important. Local allies can do exactly as requested, yet still fail to achieve the expected outcome. Such cases are instances of compliance, even though the end result did not achieve the advantage that was imagined. The failure of certain policies to produce results is a problem with strategy – a topic set aside here, in part because many other counterinsurgency scholars address it in detail.

Once specific demands were identified, I analyzed compliance outcomes. This information was gathered again through primary source materials from diverse agencies and institutions, where available, as well as media reports, research publications, and historical accounts. Sometimes the outcome of the demand was self-evident. For example, if the USA asked its Iraqi partners to construct a school, the existence of that school on a map with photographs and media announcements would provide evidence. The demand was then coded, tracking relevant conditions of the request, such as the interests of allies, as well as the compliance outcome. Quantitative statistical analysis enabled investigation of patterns across requests, wars, and alliances. Adopting this dual approach, which utilizes both qualitative and quantitative tools, offers opportunities for understanding aspects of a complex political relationship across wars without losing the complex context of interalliance relations detailed in primary source documents.

An example from data collected may be helpful in illustrating the process. Below is an example from the US war in Vietnam. The Vietnam chapter systematically uses primary source documents from the US war to construct a unique compliance database tracking (1) requests made by the USA on the Government of Vietnam in Saigon (GVN), (2) the outcome of these requests, and (3) structural issues and characteristics of requests that may impact compliance outcomes. Tracking these elements provides a better understanding of their potential importance and impact on coercion, compliance, and interalliance bargaining dynamics.

A Step-by-Step Example, from the US War in Vietnam

Step 1 – *Analyze statements from the intervening government out-lining its policy goals for the local regime (policy documents)*

The USA was directly involved in Vietnam from 1954 to 1975. But the American commitment to a noncommunist South Vietnam went through varying stages, a phenomenon that led some to refer to the US intervention in Southeast Asia as "The Vietnam Wars."[12] Nevertheless, the war most Americans know as "Vietnam" lasted from 1965 to 1973. In this study I began collecting data in January 1964, a year before the first sizable introduction of US troops. There are several methodological reasons for choosing to start the analysis in 1964. The Tonkin Gulf Resolution was passed in August 1964 signaling the start of the escalation that led to 184,300 US troops being deployed in 1965, a force that suffered 1,369 combat deaths that year alone.[13] Starting the analysis a year ahead of substantial troop arrivals helps ensure thorough analysis that captures as much data as possible, including interactions that prepare for the 1965 offensive, without including requests made before the USA was fully committed to Vietnam, a form of intervention outside the scope of this study. Ultimately, beginning a year ahead of the troops helps gather a more complete picture of USA–GVN relations and interalliance bargaining space.[14]

The first step in analysis entails identifying specific demands from the USA to the GVN required relying on the twelve volumes of the Foreign Relations of the United States (FRUS) on the US war on Vietnam 1964–73. FRUS is produced by the US Department of State Office of the Historian, and contains

declassified records from all the foreign affairs agencies. FRUS volumes contain documents from presidential libraries, the departments of state and defense, the National Security Council, intelligence agencies, the Agency for International Development, and other foreign affairs agencies as well as the private papers of individuals involved in formulating U.S. foreign policy. In general, the editors choose documentation that illuminates policy formulation and major aspects and repercussions of its execution.[15]

Each volume contains the text of several hundred of the most important declassified policy documents for a given time period in order to provide a definitive documentary resource on an important history event for scholars and government officials to use for reference. FRUS is considered one of the foremost resources for US national security policy materials.

Due to the size and importance of the war in Vietnam, there are twelve volumes of materials on the issue, organized chronologically, collectively containing roughly 11,000 pages of material. Using a combination of the bound volumes published by the US Government Printing Office, online copies of these volumes, as well as copies of original documents when material was only referenced (but not reproduced) by the FRUS collection, it was possible to identify 105 unique demands from the USA to the GVN over the course of the war.

Step 2 – *Determine the compliance outcome of the given demand, and the relevant conditions for the demand*

Once specific policy requests from the USA to Saigon have been identified it is possible to trace what happens to each of those policy requests. This requires carefully monitoring compliance as well as the critical factors that seem to influence compliance. Most of the time, evidence regarding compliance was found in primary source documents. For example, in November 1964 the US asked Saigon to "Initiate a program for the improvement of the port of Saigon and the Saizon River channel."[16] The process of improvement and compliance (or lack thereof) is

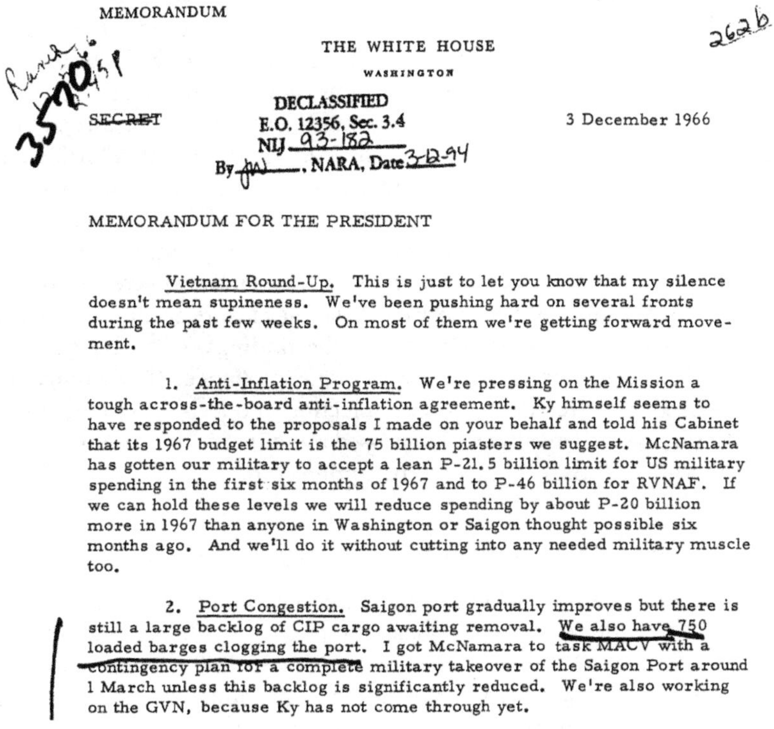

MEMORANDUM

THE WHITE HOUSE

WASHINGTON

SECRET

DECLASSIFIED
E.O. 12356, Sec. 3.4
NIJ 93-182
By ____, NARA, Date 3-12-94

3 December 1966

262b

MEMORANDUM FOR THE PRESIDENT

     Vietnam Round-Up. This is just to let you know that my silence doesn't mean supineness. We've been pushing hard on several fronts during the past few weeks. On most of them we're getting forward movement.

     1. Anti-Inflation Program. We're pressing on the Mission a tough across-the-board anti-inflation agreement. Ky himself seems to have responded to the proposals I made on your behalf and told his Cabinet that its 1967 budget limit is the 75 billion piasters we suggest. McNamara has gotten our military to accept a lean P-21.5 billion limit for US military spending in the first six months of 1967 and to P-46 billion for RVNAF. If we can hold these levels we will reduce spending by about P-20 billion more in 1967 than anyone in Washington or Saigon thought possible six months ago. And we'll do it without cutting into any needed military muscle too.

     2. Port Congestion. Saigon port gradually improves but there is still a large backlog of CIP cargo awaiting removal. We also have 750 loaded barges clogging the port. I got McNamara to task MACV with a contingency plan for a complete military takeover of the Saigon Port around 1 March unless this backlog is significantly reduced. We're also working on the GVN, because Ky has not come through yet.

Figure 3.1 1966 Memorandum for the President – evidence for monitoring local compliance with US requests in Vietnam discussed in Figure 3.1, a memo from December 1966, declassified by the Lyndon Johnson Presidential Library in 1994.

In additional to qualitative analysis, this information was then coded in a database following the rules outlined in this chapter.

Requests of Intervening Allies

When debating what to request from local allies, intervening forces often reflect on the strengths, weaknesses, and limited capacity of these partners, which creates a selection bias influencing the population of potential requests to local allies as intervening forces tailor their requests to actions considered within the grasp of their local partners. This selection effect likely contributes to the consistency in the rates of compliance across conflicts as well as the high percentage of requests pertaining to political reforms as opposed to other topics such as military operations.

However, note that this selection effect is not problematic for the findings offered. First, intervening states frequently have no choice but to ask for difficult reforms if those actions are critical for the war effort, or if the intervening ally is dependent on local allies to adopt the given policy. Second, in retrospect, several intervening powers including the Soviets in Afghanistan and the Americans in Vietnam were naïve (especially early on in their interventions), underestimating the difficulty of certain tasks and overestimating the ally's ability to reform. Third, a bias in the sample toward selection of requests the intervening ally believed their ally was capable of implementing would not be unique to any particular intervention. This dynamic of tailoring requests to the capabilities of local allies when choosing policies would be expected to persist across interventions, and to endure into similar future scenarios. Finally, such a bias does not adversely impact the findings, since I argue that lacking capacity tends to lead toward noncompliance. If anything, this bias created by intervening forces selecting requests estimated to be within the grasp of local partners causes the effect between noncompliance and limited local capacity to be underestimated.

Local Ally Compliance – The Dependent Variable

Specific policy demands from the intervening state to the local ally are identified in order to measure local compliance or defiance with these specific requests. Consider, for example, New Delhi's 1987 request that Colombo "hold elections to [the] provincial council of northern and eastern [province] before 31.12.87." The outcome being explained is Colombo's compliance with the request, with compliance being defined as the act or process of conforming, submitting, or adapting to a desire, demand, proposal, or regimen, or to coercion. To undertake quantitative analysis that provides a perspective on trends in compliance to consider alongside the qualitative document-based work, substantive, material compliance is divided into three basic categories, full compliance, partial compliance, and noncompliance, coded on a continuous scale as: 0 = noncompliance, 1 = partial compliance, 2 = full compliance. Of course these simple categories do not capture the full range of potential compliance outcomes, which can fall anywhere along a wide spectrum and reflect a variety of complex processes. As such, the data quantified should be considered within the context of the qualitative analysis provided in the case study chapters that can better account for complexity.

As such, the following guidelines were used for quantitatively coding the outcome of each of the 460 policy requests identified: "full compliance" was assigned to a request when the local allied regime successfully fulfilled *all* major components of the request that were specified by the intervening ally. Minor aspects of the request may have gone uncompleted or been altered by the local ally, but because all major components were fulfilled, intervening forces are largely satisfied with the actions of their partner regarding the given request, as indicated by documentary evidence. "Partial compliance" captures moments where local allies fulfilled *some* of the major components of the request, while leaving other parts unfulfilled. The intervening ally is satisfied with aspects of the response of its ally, while unsatisfied about others. Lastly, "noncompliance" was assigned to requests when the local ally failed to fulfill the major components of the request. Minor aspects may be satisfied, but the primary components of the request have gone unfulfilled and the larger ally is unsatisfied with the actions of its in-country partners.

These are rudimentary categories covering a broad range of outcomes. "Partial compliance," after all, could mean a small or significant amount of compliance, lying somewhere in between noncompliance and fulfillment.[17] Nevertheless, relying on these simple, rudimentary categories is the best way to ensure the validity of the data. The subjects measured (compliance, interests, dependencies, etc.) are not naturally prone to quantification. There is no physical or standardized unit to measure. As one set of scholars observed, "There is no one objective standpoint from which social scientists can study a social phenomenon like compliance."[18] The categories assigned in this study are best conceptualized as "most," "some," and "less" compliance. For the sake of simplicity, however, these categories are referred to as "full," "partial," or "non" compliance, since it would be confusing to refer to an ally providing "most compliance." It is more clear to say "full compliance" to indicate a high degree of cooperation, despite the fact that there is no objective marker for what constitutes "full," "partial," or "non" compliance. Such an "objective" standard does not exist. These categories are relational, conceptualized relative to one another. There may be disagreement, for example, about whether a given outcome should be considered "full" or "partial" compliance, but that debate would have little bearing on the findings offered because despite variation in what is considered "full," "partial," or "non" compliance, there would be agreement regarding which outcomes contain more or less compliance. Labeling one category "more" or "less" for any given outcome will correspond with the general scale given for "full," "partial" or "non" compliance utilized here. The findings offered are not based on what is partial compliance, but on what conditions make compliance more or less likely.

Therefore, given the nature of this information, instead of attempting to assign a complex scheme of quantitative values to the spectrum of potential compliance outcomes, it is more appropriate to rely on broad categories, investigating compliance in terms of the relation of outcomes to one another, instead of a particular value. Thomas Thayer discussed a similar issue regarding quantitative analysis of data in the Vietnam War:

Sir Josiah Stamp (1880–1941) had a few pertinent words on the subject: "The government are very keen on amassing statistics. They collect them, raise them to the nth power, take the cube root and prepare wonderful diagrams. But you must never forget that every one of these figures comes in the first instance from the village watchman, who just puts down what he damn pleases." Perhaps. But the village watchman often pleases to tell the truth, and in any case he probably reports about the same way most of the time. So one must learn to look for a constant bias in reporting. The individual numbers may not be completely accurate, but the trends and changes in relationships among them may reveal quite a bit about what is going on in the village and how that village compares with other villages.[19]

Therefore, in addition to qualitative methods accounting for complexity, quantitative methods help provide an expansive perspective on compliance by local allies with requests made by a foreign intervening ally by noting for each request if there was full (more) compliance, partial (less) compliance, or noncompliance, based on the available evidence. Consider the following three requests made by the USA early in the US war in Afghanistan as examples. In January 2002, Washington asked its partners in Kabul to make sure the selection committee for the Loya Jirga conference tasked with writing constitutional protocol included minorities and women in order to better protect their interests in the future post-Taliban Afghan state.[20] President Karzai followed through, and three women and multiple ethnic factions were represented, which was not difficult considering the Northern Alliance that dominated political life was mostly comprised of ethnic minorities.[21] This request was coded as "full compliance."

Contrast that example with the outcome of a request from the Americans just a few days earlier that weapons and current military factions come under the control of the Interim Administration in Kabul.[22] Efforts were made by Kabul to incorporate factions into the Ministry of Defense, while programs such as Disbandment of Illegal Armed Groups (DAIG) were established to complement US and UN-run programs on Disarmament, Demobilization, and Reintegration (DDR). Nevertheless. localized armed groups remained exceedingly powerful in Afghanistan, and various militias continued to control territory, making it difficult for Kabul to implement disarmament and reintegration.[23] This request was coded as "partial compliance."

Lastly, consider the November 2002 request that all Afghan police trainees complete an anti-narcotics self-certification, similar to the requirements for employment with the United States Agency for International Development (USAID). The Government of the Islamic Republic of Afghanistan (GIRoA) conducted drug testing, and recruits were vetted through the Afghan Criminal Investigation Department at the Ministry of Interior, but a self-certification was not required. The utility of such certifications was questionable anyway, considering that over 70 percent of recruits were illiterate and oversight of the training programs was transferred to the US military in 2005.[24] This request was not fulfilled by the Afghans and was therefore coded as "noncompliance."

What Determines Compliance – The Independent Variables

Multiple factors affect the likelihood that local allies will comply with requests from intervening partners. The independent variables considered include the capacity of local allies, the dependency of intervening forces on local partners to implement the policy, the convergence or divergence of the interests of allies regarding the proposed policy, as well as other salient factors such as the issues addressed in the request.

Capacity – Are There Institutional Weakness, Difficulties from the Wartime Environment or Local Internal Politics Limiting the Capacity of Local Allies?

By definition, states that have partnered with foreign militaries to control their own territories are lacking governance capacity. If they had greater capabilities, which produced greater capacity, they would likely avoid large-scale foreign military interventions to secure their regimes. This variable, *capacity*, accounts for whether the compliance process has been affected by local allies lacking the means to undertake the requested activity. If there is a clear indication from documents or reliable historical accounts that compliance with a request has been affected by a lack of capacity, usually political, economic, or bureaucratic, the request was coded as capacity limitations = yes (1). If there was no indication or discussion of a lack of capacity affecting compliance processes, the variable was coded as capacity limitations = no (0). Consider, for example, the March 1979 Soviet request to the Kabul that Afghanistan seal its border with Pakistan.[25] Once the Soviets took over military activity in Afghanistan, they realized the impossibility of this task due to terrain and cross-border demographics. Fifty thousand Soviet soldiers assigned to the task could not prevent arms and militants from infiltrating into Afghanistan.[26] It was a task far beyond the reach of Afghan communist forces.[27]

However, a discussion of potential capacity limitations complicating a request does not necessarily indicate that the request will go unfulfilled. Requests with capacity concerns can be fulfilled, in full or in part, even if there is a signal that the compliance process may be affected by capacity limitations. Consider, for example, the US request to Kabul that poppy elimination programs focus on Helmand Province.[28] As the request was made, violence in Helmand was on the rise, along with protests against counternarcotics programs. Helmand was titled "the most difficult province for ISAF [the International Security Assistance Force]," and US policymakers noted the situation was straining the capacity of poppy elimination teams.[29] Nevertheless, the Afghan government pushed forward and enforced some eradication programs focusing on Helmand including preplanting outreach programs, leading to partial compliance with the request.[30]

Local allies can fail to find the capacity to implement a policy requested by intervening forces for a variety of reasons. The most common capacity failure related to institutional shortcomings. However, there are other ways that capacity becomes an issue, such as the war exhausting resources that would otherwise been sufficient to implement the policy. Whereas one is an internal failing, the other is an external stressor. Both create capacity failures. Therefore, to better measure the variety of ways local capacity failings can emerge, I produced different variables to reflect these pathways, all under the general category of capacity limitations, yet easily separated in order to allow testing for any unique effects of each pathway on the dependent variable.

The first, *wartime environment*, tracks occasions when complications from the war impacted the capacity of local allies to comply. To borrow the concept of friction from Clausewitz, "everything in war is simple, but the simplest thing is difficult. The difficulties accumulate and end by producing a kind of friction."[31] Simple requests can be complicated by conditions of the war. This variable controls for these complications that may include violence, combat operations, refugee influxes, and offenses by the enemy. Despite the fact that war is a constant, complications arising from war are not uniform. Friction is unpredictable; sometimes it paralyzes organizations, sometimes it leaves operations unscathed.

An example of the war influencing conditions relating to capacity and compliance can be observed in the following example. In 2009, the USA asked Kabul to work on establishing longer, and better-coordinated hours at several border crossings with Pakistan to facilitate trade and ease the burden of transporting goods into Afghanistan.[32] But in November 2011, US aircraft attacked two Pakistani border posts, killing twenty-four Pakistani soldiers. In response, Pakistan shut down entry points into Afghanistan, including the routes specified in the demand to have longer hours, only to reopen them months later in July 2012.[33] Compliance with the American request that Kabul coordinated schedules with the Pakistanis was complicated by a US military action in a way that could not have been foreseen. The request, and indeed the border itself, were affected by wartime complications. For this request and ones with similar dynamics, this variable was coded as wartime environment (complications) = yes (1). For other requests where there was no evidence of a direct impact of military action on the compliance environment, the variable was coded as wartime environment (complications) = no (0).

Another pathway for capacity failures besides institutional failure and complications from the wartime environment are *local internal politics*, a variable accounting for internal political conflict in local allied regimes. The local ally is not a unitary actor. As with any political organization, internal disputes can affect policy outcomes. In such cases, the necessary infrastructure may be in place to execute the action requested, but the collective political will is absent. When there was evidence of a specific local political issue influencing the compliance environment, the variable was coded as local internal politics = yes (1), and when such evidence was not apparent, the request was coded as local internal politics = no (0).

Of course internal divisions undoubtedly exist on the side of the intervening ally as well. However, because requests from intervening forces are drawn only from very high-level decision documents that have been passed along to international allies, the internal debate among organizations within the intervening power has been filtered as it has risen to the top of the bureaucracy. For the most part, the final product, the result of those policy debates, is where coding for this project begins, which excludes the debate within the bureaucracies of the intervening force. There are a few exceptions where internal politics in an intervening ally has influenced compliance outcomes, and those moments are noted with examples in the subsequent chapters on specific wars.

Internal political turmoil for local allies can influence compliance outcomes. Legislatures can block executive efforts to comply with requests from large allies, while political rivals can employ nationalistic rhetoric to threaten decision makers to refuse a request from intervening forces, and bureaucratic organizations can work at cross-purposes in order to diminish the effectiveness of a rival bureaucracy. Consider, for example, the request discussed in Chapter 7. In 1988, New Delhi demanded that Colombo prevent individuals settled in northeast Sri Lanka after 1983 from voting.[34] However, there was notable popular Sinhalese pressure not to acquiesce, as not only would it weaken Sinhalese representation, but would also exclude individuals who had turned eighteen since 1983, therefore violating long-standing electoral laws.[35] Colombo caved to the pressure and did not comply.

The variables of local internal politics and wartime environment are monitoring separate phenomena that can affect local capacity; however, because they were observed much less frequently than the more common form of institutional capacity failings, and because both variables monitor occasions when outside factors complicate the capacity of local allies to cope with policy demands, an additional composite variable was created to combine them, *wartime environment or local internal politics*. A variety of statistical models were employed occasionally using the composite variable and, at other times, using the unique variables on internal local politics and wartime environment. Models always specify which data was used and why.

Interests – Do the Interests of Allies Align or Diverge over Policies?

The second primary independent variable, *interests*, considers whether a proposed policy could potentially create *short-term political and/or economic benefits* for local allies. For example, in 2008, the USA requested that Kabul pursue aggressive policies to register *hawalas*, the informal bankers that conduct the majority of financial transactions in Afghanistan. Kabul could potentially gain a short-term benefit from complying with this request because requiring *hawalas* to be licensed would open up new sources of revenue and information for the government, and potentially generate new opportunities for kickbacks. Requests that have potential short-term benefits were coded as short-term benefits = yes (1).

Not all requests provide potential short-term benefit. Consider US requests for increased reporting on human rights abuses in Afghanistan. Human rights reporting might provide long-term positive benefits for Kabul, but under the given conditions of instability, the request offered few short-run benefits. Even reports of human rights abuses perpetuated by enemy forces highlight the lack of governance in Afghanistan, which reflects poorly on Kabul. Questions of potential long-term benefits were not as helpful in explaining variation in compliance outcomes; almost all requests had some potential long-term benefit, but variation in short-term payoffs was a more immediate influence on the decision-making process of local partners. If no short-term benefit for local allies was evident, the request was coded as short-term benefits = no (0).

In addition to potential short-term benefits, *potential short-term political or economic costs* were monitored. Requests that have potential short-term costs were coded as short-term costs = yes (1). One example is the November 2002 request from Washington that Kabul pay for its ministries of Justice and Interior, instead of relying on the supplementary funding from international donors that had been covering the operating costs of these ministries.[36] This request imposed high short-term financial costs on US allies in Kabul.

Requests can, of course, provide both potential costs and benefits. The 2006 US request that the Afghans ratify the UN Convention Against Corruption is an example. Kabul could have potentially benefited politically in the short term by adopting measures countering the unpopular endemic corruption plaguing Afghanistan. But Kabul would also endure short-term costs from anti-corruption measures holding Afghan policymakers accountable for any disreputable practices.

The potential short-term costs and benefits of a request were coded in simple terms as yes (1) or no (0) despite the fact that not all costs and benefits are equal. Some requests may pose very high costs or very low benefits. But quantifying varying levels of benefits would be exceedingly difficult and likely unreliable for several reasons. First, although economic costs can be quantified in terms of money, they are not easily measured here because these are *potential* costs (or benefits). How much policies may actually cost or benefit the local ally is at best a rough estimate and cannot be expressed as a numerical figure. Second, political costs or benefits are not easily quantified. Third, costs and benefits are dynamic. They shift based on changing circumstances, which is a characteristic exacerbated by shifts in political and economic environments during a war. Nonetheless, this complexity in interests is considered in the qualitative work discussed in the case study chapters.

In order to gather additional information on costs and benefits and to analyze their influence on the likelihood of compliance, a composite variable was created that takes both potential short-term costs and short-term benefits into consideration. Both costs and benefits after all are balanced in the decision-making process of local policymakers, and therefore should also be considered in balance with one another in estimating interests. This composite potential costs and benefits variable was termed *convergence/divergence of interests* because it approximates whether the interests of the local and intervening allies match or deviate over a given request. The variable assumes the intervening ally has an interest in the request since it is the one making the request. The interests of local allies are then estimated by considering what the evidence indicates about costs and benefits. If the record indicates that, for the most part, local allies expressed that short-term benefits outweighed costs, the variable was coded as interests = converge (1). If the evidence regarding the request indicated local policymakers felt that potential costs outweighed benefits, the request was coded as interests = diverge (0). This is a highly useful variable because it estimates how costs and benefits were both considered by the local ally. Of course, many calculations of costs and benefits are conducted outside the official record, and are difficult to fully document. Additionally, political calculations can be very complex and can shift depending on changes in circumstances. However, this variable relies on the existing evidence, including media reports, public statements by local actors, and political analysis of the evolving situation by intervening forces. It focuses on critical local policymakers in the decision-making process. The variable is a general estimate of interests, and is designed to paint a broad picture through quantitative analysis. The variable is not supposed to provide a full accounting for political interests in a complex wartime environment, which would be unreasonable to try to quantify.

To describe both the purpose and limits of the variable measuring convergence/divergence of interests, consider the 2009 US request that Kabul extend service hours at certain Afghanistan–Pakistan border crossings to facilitate trade and ease the burden of transporting goods into Afghanistan.[37] This request posed both potential costs and benefits to the regime in Kabul. On the one hand, extended hours would change current bureaucratic practices as well as potentially disrupt profitable black-market practices since Afghan officials were reportedly able to take advantage of closed border crossings to smuggle goods. However, extended hours at border checkpoints would also create opportunities for increased taxes on imports from trade and transport volume as well as new opportunities for kickbacks.[38] Based on the evidence, the potential short-term costs and benefits to powerbrokers in Kabul, the request was coded as allies' interests converging because longer hours at official border crossings would immediately increase import tax revenue without necessarily creating obstacles that black markets could not easily shift to accommodate. This was observed at the Torkham Gate border crossing between Kabul and Peshawar, for example, where province Governor Gul Agha Sherazi was accused of levying a second, unofficial tax at the official border crossing that profited the governor and his network.[39] Taken in total, the regime in Kabul had more to gain than lose in the short term by the request.

Dependency – Can Intervening Allies Unilaterally Implement Policies?

Dependency measures whether or not local allies are required to participate in order to fulfill a request, or whether intervening forces, can, if they so choose, accomplish the task independently. In counterinsurgency wars, intervening allies frequently make requests to local allies that, if necessary, intervening forces can implement independently, because the participation of local allies: (1) provides local legitimacy for the policy and (2) promotes the development of local institutions, which are both potentially strategically useful. The variable tracking this unilateral capability by intervening forces is important because it measures dependencies within the alliance. As discussed in Chapters 1 and 2, because intervening forces are often promoting local sovereignty in order to legitimize the intervention, they are required to negotiate (not dictate) certain policies as local allies have legal control of state institutions. This means powerful foreign patrons are at times inextricably dependent on local partners to implement certain reforms, which can provide local allies leverage. In order to simplify this variable, dependency was coded as either intervening unilateral capability = 1, indicating that the foreign force has independent capability to complete the task at hand in the request and therefore is not dependent on its ally regarding this particular issue, or intervening unilateral capability = 0, indicating the intervening ally is indeed dependent on local allies for implementing the request.

One example is the September 1968 US request that Saigon begin to pay for the "Lien Minh" program, an initiative designed to unify anti-communist political groups in Vietnam.[40] Since the USA had been funding the program and had the financial capacity to continue to do so, Washington could implement the policy (funding the program) without the participation of the Saigon government. The Central Intelligence Agency (CIA) recommended the Vietnamese to take over funding because "without a direct GVN input (and, hence, vested interest) there will always be the risk of the program's being considered, even in Thieu's eyes, an American scheme the Vietnamese are indulging."[41] This request was coded as unilateral capability = yes (1) because the Americans had the capability to implement their request for program funding without the participation of Saigon. It was a request made out of preference for Vietnamese participation, not because their action was necessary.

This is in contrast to Washington's 2007 request to Baghdad that the regime "cease sectarian appointments and politically motivated prosecutions,"[42] or the 2003 request that Kabul pass anti-money-laundering legislation.[43] By the nature of these requests, the participation of the allied regimes in Baghdad and Kabul is required due to the authority they command as legally sovereign governments. Washington is dependent on Baghdad to participate in stopping sectarian appointments and on Kabul for passing legislation. These requests were therefore coded as unilateral capability = no (0).

However, coding whether the foreign force has unilateral capability to implement a request was not always clear-cut. Consider a more ambiguous example to illustrate some of the potential complexities in coding unilateral capability. While the USA was asking the Government of Iraq (GOI) to cease "sectarian appointments and politically motivated prosecutions," as discussed in the example provided above, Washington was also asking Baghdad to seek the "vigorous prosecution of government and security officials who break the law."[44] For the request that Baghdad cease sectarian appointments, the USA had to rely on Baghdad, because in 2007 the USA could not unilaterally make personnel decisions for Iraqi policymakers. The USA could provide incentives with the hopes of encouraging less sectarian processes, but Washington was were nevertheless dependent on action by its Iraqi partners to mitigate sectarian appointments.

The request regarding prosecuting Iraqi security officials that break the law touches on important questions about American jurisdiction and Iraqi legal sovereignty. At the time of the request, the USA had independent law enforcement capabilities in Iraq. Americans were routinely detaining and prosecuting Iraqi nationals suspected of terrorist acts against Iraqi or US forces. Although the USA was unable to prosecute Iraqi government and security officials under Iraqi law, US institutions were in place in 2007 to detain Iraqi government officials suspected of crimes, in particular those against US forces. In theory, therefore, the institutions existed for the USA, if so motivated, to take over responsibility for "government and security officials who break the law." This request was therefore coded as unilateral capability = yes (1). If the request had specified prosecution under Iraqi law, or if the Americans had less legal infrastructure at work in Iraq, the USA would not have the ability to do so without Iraqi participation, and the request would have been coded unilateral capability = no (0).

Therefore, two factors are considered. First, does the intervening ally have the relevant resources, personnel, means, or institutions in place that can take over the roles being requested in the demand? If the foreign power has independent institutions required to fulfill the demand, it is less dependent on local allies. In 2007, even though the USA could theoretically prosecute Iraqi government officials, it could not pass Iraqi law or make Iraqi political decisions about sectarian appointments.

Second, the concept of the status quo is a critical component to coding the dependency of intervening forces on local partners for implementing a requested policy. What is the situation at the moment the request is being made that would impact whether or not intervening allies can fulfill the request without the participation of local forces? There are times when intervening forces may be managing particular domestic issues for local allies, as legal sovereignty can be incomplete at times. Under such conditions intervening forces may have unilateral capability to implement a request, whereas at other moments in the war, local actors are more independent and not readily influenced by foreign officials. Which ally is in control of the relevant organizations and how does that affect which actors can have direct influence over the state of affairs? In 2004, under the Coalition Provisional Authority, the USA controlled Iraqi personnel decisions. However, by 2007, it had lost that ability and therefore was dependent on Iraqi decision-making, sectarian or otherwise.

Significant Enemy Threats – Is There a Substantial, Acute Danger?

Lastly, a variable measuring *significant enemy threats* was generated, but interestingly was only observed in the US intervention in Vietnam. As detailed in Chapter 6, the Tet Offensive inspired a surge in local compliance with US requests. While Tet was the only example of this dynamic across the nine wars examined, it signals that a significant, acute enemy threat such as Tet might in future interventions emerge as an additional factor motivating local cooperation. A significant offensive is a distinct phenomenon from routine friction of operating in a warzone as described in the variable related to capacity, *wartime environment*. Because there is a difference between routine operations and emergency conditions, a separate variable was produced to monitor *significant enemy threats* (enemy threat = yes [1] or no [0]). The Tet Offensive in Vietnam had a markedly different effect on compliance when contrasted with daily operations in a counterinsurgency war. Explanations and examples of how this variable affected particular US-promoted policies are detailed in Chapter 6.

Issues Areas – What Topics Do Requested Policies Address?

In addition to capacity, interests, dependency, and significant enemy threats, each request was coded to track the *general issue areas*. This enables testing if local allies had distinct interests in different topics, or if certain issues were prone to compliance or noncompliance. There are six categories of issue areas:

1. *Development*, defined as projects or activities intended to support economic growth and provide social services, including land reform, school construction, reconstruction of damaged urban areas, micro-finance strategies, and assistance to refugees and veterans.
2. *Economic reform*, defined as actions intended to change economic policies. These requests include changes to exchange rates, currency manipulation, banking sector reforms, and tax policy.
3. *Military strategy*, defined as actions intended to guide military forces in the execution of the war effort, including policies regarding the distribution and buying back of weapons and strategic decisions about troop placements.
4. *Military reform*, defined as actions intended to change military policies and institutions, including expanding/shrinking security forces, increasing authority for military commanders, and reforming command and control protocols.
5. *Political reform*, defined as actions intended to change government policies and institutions. These requests could include electoral issues, law enforcement, governance, protections for minorities, bureaucratic protocols, constitutional reforms, and reconciliation programs.
6. *Political-military counterinsurgency strategy*, defined as actions intended to implement counterinsurgency strategy. These include COIN projects, negotiated ceasefire agreements, and pacification activities.

The data also contains essential information to contextualize the described variables, including coding the year the request was made, based on the available evidence. Allies can ask for requests repeatedly if they are not complied with in a timely manner and if they are deemed important to the war effort. The earliest communication of that request to local allies was coded as the initial year of the request. Additionally, the year of compliance was also coded, and again recorded as the earliest date based on the available evidence. Lastly, each chapter contains a short section discussing particular methodological questions related to that war and the partnerships examined.

Notes

1 David Gompert et al., *War by Other Means – Building Complete and Balanced Capabilities for Counterinsurgency Study – Final Report* (Santa Monica, CA: RAND Corporation, 2008), www.rand.org/pubs/monographs/MG595z2.

2 James D. Fearon and David D. Laitin, "Ethnicity, Insurgency, and Civil War," *American Political Science Review* 97, no. 1 (2003): 75–90.

3 Colonial counterinsurgency (COIN) wars such as Algeria (1954–62) and Mozambique (1962–74) have been coded as wars *without* foreign COIN military intervention even though they might be commonly considered as interventions. There are several reasons for excluding these wars. First, a colonial project is fundamentally different from a foreign policy mission that has an established exit goal – especially in terms of the dynamics of the alliance between intervening or "foreign" versus "local" COIN partners. In colonial wars "local" state institutions are officially part of the intervening state apparatus. Furthermore, by definition in colonial wars outside powers are fighting to preserve control of these contested areas as their territorial holdings – a dynamic that changes definitions of "local" and "intervening." In colonial wars powerful states are fighting not as a foreign power, but as the internationally and locally recognized state authority of the territory. Second, I intend to offer insight into future counterinsurgency interventions, which is best served by omitting colonial-era political and military dynamics.

4 www.correlatesofwar.org. Note that these are battlefield deaths, not casualties specified to either the insurgency or counterinsurgency. This study borrows this 1,000+ death threshold as a metric for casualties incurred by intervening forces.

5 There are five COIN conflicts where there is insufficient public data available on foreign military deaths but where foreign forces were engaged. Mozambique (1976–92) fought with the aid of Zimbabwean, Zambian, and Tanzanian troops. Tajikistan (1992–7) fought with the aid of Russian and Uzbek troops. Ethiopia in the Ogaden War (1977–9) fought with the aid of Russian, Cuban, and South Yemeni troops. Laos (1959–73) fought with the aid of Thai troops. And lastly, in the complex war in the Democratic Republic of the Congo (1998–2003) Namibian, Angolan, Zimbabwean, and Chadian troops supported the counterinsurgency.

6 Beehner and Bruno, "What Are Iraq's Benchmarks?."

7 Alexander George and Andrew Bennett, *Case Studies and Theory Development in the Social Sciences* (Cambridge, MA: MIT Press, 2005), 97.

8 See Appendix B.

9 "Soviet Briefing on the Talks between Brezhnev and B. Karmal in Moscow," October 29, 1980, Top Secret, Central Committee Foreign Department Bulletin, Budapest, Hungary, Cold War International History Project, The Woodrow Wilson International Center for Scholars, http://digitalarchive .wilsoncenter.org/document/112500. Original source, National Archives of Hungary (MOL), M-KS 288 f. 11/4391.o.e.

10 Certain policy requests have multiple components. Consider, for example, US requests that Kabul fund *and* expand the Community Defense Initiative (CDI). Since the diplomatic correspondence treats these two actions (funding and expanding the CDI) as a unified CDI-related issue, it was coded as a single request. Kabul funding, but not expanding, or expanding, but not funding the CDI would have been coded as partial compliance with a single request. If the intervening force considers the proposed program as a cohesive unit in terms of creating a budget and allocating personnel, it is coded as a single request.

11 "CPSU CC Politburo Decision," January 28, 1980, with Report by Gromyko-Andropov-Ustinov-Ponomarev, January 27, 1980, Top Secret, No. P181/34, To Comrades Brezhnev, Andropov, Gromyko, Suslov, Ustinov, Ponomarev, Rusakov. Cold War International History Project, The Woodrow Wilson International Center for Scholars, "Documents on the Soviet Invasion of Afghanistan," CPSU Politburo session of January 28, 1980, E-Dossier No. 4, November 2001, 58–60.

12 Marilyn Blatt Young, *The Vietnam Wars, 1945–1990* (New York: HarperCollins, 1991).

13 "Vietnam Conflict – U.S. Military Forces in Vietnam and Casualties Incurred: 1961 to 1971," No. 402, US Bureau of the Census, Statistical Abstract of the United States 1971 (92nd Annual Edition), US Department of Commerce, Library of Congress, Washington, DC, No. 4-18089, 253. Also US Department of Defense, *Military Personnel Historical Reports*, 2010, http://siadapp.dmdc.osd.mil/personnel/MILITARY/history/309hist.htm.

14 Furthermore, many of the requests made by US officials in 1964 are requested again in later years. This means these data points would have been analyzed in the study in an identical fashion under a different start date.

15 For more information see http://history.state.gov/historicaldocuments/about-frus.

16 US Department of State, US Embassy Saigon, "Telegram from the Embassy in Vietnam to the Department of State," Saigon, November 9, 1964 – 7 p.m. Central Files, POL 1-1 VIET S. Top Secret; Priority; LIMDIS. Saigon Embassy Files: Lot 68 F 8, *Foreign Relations of the United States, Volume I, Vietnam, 1968*, Document 408. November 9, 1964, http://history.state.gov/historicaldocuments/frus1964–68v01/d408.

17 Carney, "International Patron–Client Relationships," 45.

18 Christine Parker and Vibeke Nielsen, "The Challenge of Empirical Research on Business Compliance in Regulatory Capitalism," *Annual Review of Law and Social Science* 5 (October 23, 2009): 64.

19 Thomas Thayer, "How to Analyze a War without Fronts: Vietnam, 1965–72," *Journal of Defense Research* 7B, no. 3 (1975): 768–9.

20 US Department of State, "Next Steps in the Political Process, President Bush to President Karzai" (Washington, DC, The National Security Archive), January 18, 2002.

21 Drude Dahlerup and Anja Taarup Nordlund, "Gender Quotas: A Key to Equality? A Case Study of Iraq and Afghanistan," *European Political Science* 3, no. 3 (2004): 91–8; Cary Gladstone, *Afghanistan Revisited* (New York: Nova Publishers, 2001), 9; Congressional Record, "Afghanistan Stabilization and Reconstruction: A Status Report," Hearing before the Committee on Foreign Relations, United States Senate, 108th Congress, 2nd Session, January 27, 2004 (Washington, DC: US Government Printing Office, 2004), 10.

22 US Department of State, "The Formation of Afghan Military and Police – U.S. President Bush to Hamid Karzai," January 19, 2002.

23 Angelo Rasanayagam, *Afghanistan: A Modern History* (London: I. B. Tauris, 2005), 272; Kenneth Katzman, "Afghanistan: Politics, Elections and Government Performance," (Congressional Research Service, RS21922, January 20, 2012), 37.

24 US Department of State, "Letter of Agreement on Police and Justice Projects," November 29, 2002, 02STATE244042 (Washington, DC: The National Security Archive); Combined Security Transition Command – Afghanistan, "Afghan National Police (ANP) Vetting and Recruiting Presentation," 2005, http://edocs.nps.edu/AR/topic/misc/09Dec_Haskell_appendix_II.pdf; US Department of State, Department of Defense Inspectors General, "Interagency Assessment of Afghan Police Training and Readiness, Department of State Report No. ISP-IQO-07-07, Department of Defense Report No. IE-2007-001," November 2006, http://oig.state.gov/documents/organization/76103.pdf.

25 "Special File Record of Conversation of L. I. Brezhnev with N. M. Taraki," March 20, 1979, History and Public Policy Program Digital Archive, http://digitalarchive.wilsoncenter.org/document/111282.

26 Steve Coll, *Ghost Wars: The Secret History of the CIA, Afghanistan, and Bin Laden, from the Soviet Invasion to September 10, 2001* (New York: Penguin, 2005), 158.

27 David Isby, *Russia's War in Afghanistan* (London: Osprey Publishing, 1986), 8; Lester W. Grau, *The Bear Went over the Mountain: Soviet Combat Tactics in Afghanistan* (New York: Psychology Press, 1996), 75; Anthony James Joes, *Victorious Insurgencies: Four Rebellions That Shaped Our World* (Lexington: University Press of Kentucky, 2010), 196, 249.

28 US Department of State, US Embassy Kabul, "Scenesetter: U.S.-Afghan Strategic Partnership Talks In Kabul – March 13," 07KABUL804, March 8, 2007.

29 US Department of State, US Embassy Kabul, "Brokering Eradication Consensus in Helmand," 07KABUL1045, March 27, 2009.

30 US Department of State, US Embassy Kabul, "Helmand Province – Poppy Eradication Force's Pre-planting Campaign Has Good Start," 07KABUL3135, September 18, 2007.

31 Clausewitz, *On War*, 119.

32 US Department of State, US Embassy Kabul, "Economic Agenda Items for Af-Pak Trilateral Commission."
33 Rob Crilly, "Pakistan Permanently Closes Borders to NATO after Air Strike," *The Telegraph*, November 28, 2011, sec. World News, www.telegraph.co.uk/news/worldnews/asia/pakistan/8919960/Pakistan-permanently-closes-borders-to-Nato-after-air-strike.html; CNN News, "Pakistan Reopens NATO Supply Routes to Afghanistan," July 3, 2012, www.cnn.com/2012/07/03/world/asia/us-pakistan-border-routes/index.html; Jon Boone, "Pakistan Border Closure Will Have Little Effect on NATO's Afghanistan Campaign," *The Guardian*, November 27, 2011, sec. World News, www.guardian.co.uk/world/2011/nov/27/paki stan-border-nato-afghanistan-supplies.
34 "Points of Verbal Message of the Indian Prime Minister Rajiv Gandhi and the Reaction of the Sri Lankan President J. R. Jayewardene Conveyed through the Indian High Commissioner J. N. Dixit," Colombo, January 13, 1988, in *India – Sri Lanka: Relations and Sri Lanka's Ethnic Conflict Documents – 1947–2000*, ed. Avtar Singh Bhasin, Vol. IV, Document 793 (New Delhi: Indian Research Press, 2001), 2184–5.
35 Gunaratna, *Indian Intervention in Sri Lanka*, 349. See also "TOP SECRET note of the RAW Agent to the Chief of the *Sri Lankan* Intelligence and Security RE: LTTE's Surrender of Weapons," Colombo, June 20, 1988, in *India – Sri Lanka: Relations and Sri Lanka's Ethnic Conflict Documents – 1947–2000*, ed. Avtar Singh Bhasin, Vol. IV, Document 835 (New Delhi: Indian Research Press, 2001), 2263. A complementary example from the US intervention in Iraq is also illustrative. See US Department of State, US Embassy Baghdad, "Legal Ambiguity in Baghdad Governance Structures and Political Violence," 07BAGHDAD2040, June 20, 2007. The USA asks the GOI to legally describe Baghdad's status as the capital of Iraq. This is a more complicated request than it initially seems. Anna Lamberson, "A Capital Law for Baghdad: A Governance Framework for Iraq's Ancient Capital," *State and Local Government Review* 43 (August 2011): 151–8.
36 US Department of State, "Letter of Agreement on Police and Justice Projects."
37 US Department of State, US Embassy Kabul, "Economic Agenda Items for Af-Pak Trilateral Commission."
38 US Department of State, US Embassy Kabul, "Powerbroker and Governance Issues in Spin Boldak," 10KABUL467, February 7, 2010; US Department of State, US Embassy Kabul, "EXBS Afghanistan Advisor Monthly Border Management Initiative Reporting Cable – April 2007," 07KABUL1731, May 24, 2007; Amie Ferris-Rotman, "NATO Races to Secure Violent, Porous Afghanistan-Pakistan Border," *Reuters*, September 2, 2011, www.reuters.com/article/2011/09/02/us-afghanistan-pakistan-border; Nathan Hodge, "Afghans Probe Corruption at Borders," *Wall Street Journal*, December 10, 2012, sec. World News, http://online.wsj.com/article/SB100014241278873240240045788171410335390372.html.
39 Hodge, "Afghans Probe Corruption at Borders"; Justin Mankin, "Rotten to the Core," *Foreign Policy*, May 10, 2011, www.foreignpolicy.com/articles/2011/05/10/rotten_to_the_core; Julius Cavendish, "In Afghanistan War, Government Corruption Bigger Threat than Taliban," *Christian Science*

Monitor, April 12, 2010, www.csmonitor.com/World/2010/0412/In-Afghani stan-war-government-corruption-bigger-threat-than-Taliban.

40 "Memorandum for Director of Central Intelligence Helms," September 12, 1968, Washington DC. *Foreign Relations of the United States, 1964–1968, Volume VII, Vietnam, September 1968–January 1969*, Document 11, https:// history.state.gov/historicaldocuments/frus1964-68v07/d11.

41 Ibid.

42 US Department of State, US Embassy Baghdad, "The New Joint Campaign Plan for Iraq," 07BAGHDAD2464, July 25, 2007.

43 US Department of State, "Objectives for Certification to the Government of Afghanistan," June 6, 2003.

44 US Department of State, US Embassy Baghdad, "The New Joint Campaign Plan for Iraq."

4   The USA in Iraq

How can we get to self-sufficiency if we do not have control?
—Saad Qindeel, Member of The Iraq Transitional National Assembly

Most scholarship on the US war in Iraq addresses either the decision to invade, the rise of the insurgency, or US strategic successes and failures. Surprisingly few studies have examined the dynamics of the US-Iraqi alliance, despite the fact that this partnership was a critical political component of the costly US mission in Iraq.

Summary Findings

There are several notable aspects of the US-Iraqi partnership during the American intervention. First, as hypothesized in Chapter 2, and similar to other wars examined, Iraqi compliance was affected by the convergence or divergence of US and Iraqi interests, interacting with US dependency on Iraq to implement particular reforms. Second, Baghdad had an exceptionally low level of compliance with US requests regarding economic reforms. This resistance to US financial advice was partially due to a combination of deficient Iraqi institutional capacity, a lack of US institutional influence over economic policies, and divergent economic philosophies between US and Iraqi officials. Iraq's proclivity for statist solutions clashed with the American promotion of free market reforms. In other interventions, including the USA in Afghanistan and Vietnam, economic-related requests tended toward high rates of compliance when compared to other issues such as political reform or military strategy.[1]

Third, US control of the Iraqi state from March 2003 to June 2004 under the Coalition Provisional Authority (CPA) was in certain aspects similar to the Soviet takeover of the Afghan state from 1980 to 1989. This initial era in US-Iraqi politics stands in contrast to US relations in Iraq following the handover of legal sovereignty to Iraqis in June 2004, which fundamentally altered the alliace. Furthermore, in Iraq, Washington negotiated policies with two local political authorities: the Shi'a-dominated Government of Iraq (GOI) in Baghdad, and the

Table 4.1 *Timeline of the US–Iraqi alliance during the US intervention*

March–April 2003	The USA launches operations against Iraq. The successful US ground invasion captures Baghdad on April 9.
May 2003	On May 1, US President George W. Bush declares the cessation of major hostilities in Iraq despite widespread disorder on the ground. On May 23, L. Paul Bremer, head of the transitional governing body in Iraq known as the Coalition Provisional Authority (CPA), orders the removal of Baathists from the government and disbands the Iraqi Army.
June–August 2003	Escalating violence in Iraq steadily expands, marking the beginning of the insurgency.
December 2003	Coalition forces capture Saddam Hussein, ten miles south of Tikrit.
2004	Toward the end of March, the Sunni insurgency in the center and west escalates while a radical Shiite uprising erupts in Baghdad and the south. On June 28, sovereignty is turned over from the CPA to the new Iraqi government headed by interim Prime Minister Iyad Allawi.
2005	In April 2006, United Iraqi Alliance names Nouri al-Maliki as Prime Minister, who forms a unity government with Iraqi Kurds and Sunnis in the following month. However, sectarian violence continues. Saddam Hussein is executed in December.
2007	US forces collaborate with Sunni leaders to combat militants working with al-Qaeda in Iraq. Also known as the Sunni Awakening, this strategy leads to a significant decline in insurgent violence.
2009	On February 2009, US President Obama announces plans to withdraw US combat brigades from Iraq by August 2010. Toward the end of June, US forces begin to withdraw from Baghdad.
2010	Iraq holds parliamentary elections in March and the Parliament approves a coalition cabinet with Nouri al-Maliki as Prime Minister.
2011	In December, US soldiers withdraw from Iraq.

Kurdish Regional Government (KRG) in Erbil. This is unique. During the US occupation the KRG was effectively independent, requiring a separate US diplomatic envoy.[2] In the case of Iraq there were at least two entities the USA treated as the "local regime" – one Shi'a-dominated, one Kurdish.

Fourth, local Iraqi partners allied with the USA had similar rates of compliance with Washington's requests when compared with other counterinsurgency (COIN) partnerships, but interestingly, Iraq had the lowest rate of full compliance of all interventions analyzed in this study: 35/106 (33.0 percent) of requests were fully complied with, 32/106 (30.2 percent) partially complied with, and 38/106 (35.8 percent) went unfulfilled.[3] Furthermore, American policymakers expressed frustration with

Baghdad's pattern of passing legislation, yet failing to implement the laws promulgated, a process of essentially publically agreeing to reform without following through. The Iraqi Council of Representatives (COR) working with the Iraqi Prime Minister would pass new legislation, yet little would materialize to implement the law. For example, in March 2008, Baghdad signed the UN Convention Against Corruption (UN-CAC), yet by December 2008 it had not implemented the majority of required provisions. The US Embassy reported that GOI Deputy Prime Minister "Salih's chief of staff told us in November he believed that by signing the agreement Iraq had fully met its provisions. There are 166 provisions in the UN-CAC that require Iraqi action in order to be fully compliant with the convention, of which ACCO [the US Embassy's Anti-Corruption Coordinator's Office] estimates Iraq has completed about a third."[4]

Due to significant shifts in US institutional involvement in Iraq over the course of the US intervention this chapter starts with outlining the evolution of Iraqi legal sovereignty. Iraqi autonomy shifted the US-Iraqi alliance and limited the tools Washington could use to compel reform. The chapter then offers a summary of patterns in Iraqi compliance and an analysis of the proposed theory to explain these observations.

The US–Iraqi Counterinsurgency Partnership (2003–2011)

American involvement in Iraq evolved from maintaining de facto legal authority over Iraqi political institutions under the CPA during the initial fourteen months of the war, to increasingly promoting Baghdad's legal sovereignty from 2004 to 2011, leading the USA to withdraw its forces in late 2011, expecting the regime in Baghdad to defend Iraq's territory, a task it failed at as the Islamic State (ISIS) insurgency conquered and controlled several major Iraqi cities from 2013 to 2018. Virtually all aspects of the US intervention aside from initial military operations did not go as the USA planned, including the American-Iraqi partnerships.

The Coalition Provisional Authority

On January 20, 2003, roughly two months before the US invasion of Iraq, the US Department of Defense created the Office of Reconstruction and Humanitarian Assistance (ORHA) for Iraq. ORHA was intended to serve as a catchall organization addressing a wide variety of potential postinvasion developments. The Pentagon "assumed no more than a 2 or 3 month period" would be required to transfer sovereignty to friendly Iraqis.[5] ORHA would handle this transition, and generally "deal with the post-Saddam conditions in Iraq."[6] During the run-up to the invasion, policymakers in Washington assumed Iraq would remain stable following the initial US offensive. Iraq was not only potentially well-resourced through oil revenues, but also had relatively modern and well-developed bureaucracies. Many key officials in the Pentagon and White House officials expected to remove high-level Saddam loyalists without having to engage in "nation building."[7] These expectations soon proved to be thoroughly unreasonable.

A much more costly and complicated political situation emerged after US troops captured Baghdad. Years of sanctions and resource mismanagement had gutted Iraq's once reasonably well-functioning bureaucracies. The assumption that Iraq's ministries could function without high-ranking Ba'athists also failed to appreciate the nature of government administration under Saddam Hussein. As Andrew Rathmell observed, "The nature of the bureaucratic authority structures in the Saddamite state meant that the removal of the ministers did not simply allow their subordinates to take over and carry on. Power and authority in the Saddamite system had been too centralized to allow competent subordinates to emerge."[8] The removal of high-ranking Ba'athists created gutted, ineffective bureaucracies and contributed to the expanding anti-American insurgency.

The Bush administration was looking for a quick transition to Iraqi control after the US invasion. On April 1, Secretary of Defense Donald Rumsfeld wrote to the president, "The new regime is going to be a free Iraqi government, not a U.S. military government."[9] The plan was untenable. As military personnel reviewing US performance in Iraq summarized, "chaos on the ground threw the plan for a rapid political transfer to an interim Iraqi authority into confusion."[10] ORHA was scrapped for a more substantial US-run state administration for Iraq, the Coalition Provisional Authority (CPA), led by Ambassador L. Paul Bremer.[11] As veteran diplomat James Dobbins summarized, under the CPA, Bremer "exercised supreme executive, legislative and judicial powers" over Iraq.[12]

For fourteen months starting in May 12, 2003 the CPA was the government of Iraq. As the CPA Deputy Chief of Local Governance Support noted, "the operating decisions were all being made by the [US] military and the CPA."[13] Exactly how the CPA fit into the US government was unclear. Ambassador Bremer was a presidential envoy communicating directly with the White House, while also part of the Department of Defense, supposedly under Defense Secretary Rumsfeld.[14] In Bremer's opinion it was "not entirely clear that the CPA was a U.S. government entity" at all.[15] The US Office of Management and

Budget wanted to monitor CPA spending, like that of any US executive branch agency. The CPA, however, was operating Iraqi public funds. As such, US Army Legal Services Agency defended the CPA's status as a legal entity outside the US administration.[16] Clayton McManaway, a close aide to Bremer, claimed that the "CPA was the Iraqi government; it was not an American entity. [Even though] many American policy-makers, including the Pentagon and Office of Management and Budget, didn't see it that way."[17] Bremer attempted to promulgate and enforce laws and build a new Iraqi state with an American staff, acting under American authority. He was eager to create an Iraqi advisory body to weigh in on decisions and help legitimize the CPA. As a result, the Iraqi Governing Council (IGC), whose members were appointed by the CPA, was formed on July 13, 2003. Although the IGC had no official legislative or executive authority, it became increasingly influential.[18]

The CPA's notorious first order banned Ba'athist party members from government.[19] When Bremer turned administration of the de-Ba'athification order over to the IGC, headed at the time by Ahmed Chalabi, the Council announced "new rules" that would have "more depth and [affect] more people."[20] Bremer feared this was a sign that influential Shi'a Iraqi exiles like Chalabi might be inclined to use their newfound positions of power in post-Saddam Iraqi politics for anti-Sunni policies. For Bremer and President Bush, this retaliatory action confirmed the necessity of America's "extended occupation." According to Bremer, the CPA had to "walk the cat back in the spring of 2004,"[21] attempting to reinstate thousands of professionals who had been fired under the extended de-Ba'athification order, including hundreds of teachers, but significant political damage had already been done.[22]

Similar to the Soviet approach in Afghanistan, the Americans in Iraq positioned CPA officials at the top of Iraqi bureaucracies to run Iraqi the state. These advisers acted as chief operating administrators issuing agency policies.[23] The CPA had offices of varying specialties designed to establish governance, including security, interior affairs, oil, civil affairs, economic development, regional operations, and communications. And as the Council on Foreign Relations specified, CPA officials "were charged with actually running their respective [Iraqi] ministries."[24] By August 2003 the IGC had appointed twenty-five interim ministers, lowering the profile of CPA managers, but CPA officials maintained control of ministerial budgets and retained veto authorization over agency decisions.

However, unlike the Soviets in Afghanistan, the Americans in Iraq under the CPA did not stay, nor did they systematically purge lower levels of the Iraqi bureaucracy. In fact, once the CPA announced that it

nal Americans in advisory roles within the GOI, focusing on
ance, electricity, oil, and justice.[47] However, American
f this "surge" indicate that US advisers were no longer
ay-to-day GOI activities. In fact, by 2008 many US officials
Iraqi ministries were having a difficult time regularly meeting
GOI partners due to security constraints and the resistance of
ninisters toward American agendas.[48] In September
icans were reporting, for example, "The Joint Planning Com-
'C) and the Joint Reconstruction Operations Center (JROC)
l are fast becoming Iraqi-driven institutions … The JPC and
go from being largely USG-led to completely Iraqi-led by
s."[49] Unsurprisingly, at this time the USA was reporting
difficulty negotiating certain key issues with GOI partners.
the Iraqi constitutional review, General David Petraeus com-
it has proven difficult to urge movement from the GOI on
oncerns, let alone politically charged issues like new legisla-
2008 Baghdad had consolidated into a political entity capable
ng its legal sovereignty to oppose American agendas.
Iraqi ministers were granted authority over the activities of their
after the June 2004 transfer of sovereignty, but embedded
advisers remained influential for several years after this transfer.
nce in the GOI varied widely, depending on the ministry, but
decreased across Iraqi bureaucracies over the course of the US
on. The ministries of defense and oil were the most significantly
by US advising. But by late 2007 when the Iraqi military was
d to Iraqi control, US advisers were considered "visitors," rather
orities. As the US Embassy summarized in 2008, "we have less
than in 2004–2006, and will have even less influence … Our
desired outcomes influence and shape Iraqi debates, but we can
dictate the exact shape of the outcomes."[51] The post-CPA US
role in Iraqi civilian ministries was more invasive than what the
pursued in advising Vietnam or Afghanistan, but certainly less
e or permanent than the Soviet approach in Afghanistan.
his background on the extent of US involvement in Iraqi
making institutions it is interesting to explore patterns in Iraqi
ce or defiance of US requests across the intervention. I analyzed
)0 US primary source documents (2,500+ pages) to identify
que demands from the US government to Iraqi allies in the
003–11. These documents were collected from three sources.
examined declassified documents located on US government
websites, including Secretary Rumsfeld's documents in the
Collections Library.[52] A second source of material came from

was handing over government administration to Iraqi officials, enforcement of CPA orders became a problem for Washington. "Iraqi lawyers
and judges began to procrastinate in implementing or interpreting [CPA]
law, preferring the established Iraqi version, even if it contradicted the
Bremer Orders."[25] In a style aptly described by James Scott in *Weapons of
the Weak*, Iraqis engaged in repeated small acts of defiance against US
advisers, incrementally imposing boundaries on American power brokers
and drawing lines defining for Washington what the USA could and
could not accomplish in Iraq.[26]

The Iraqi Interim Government

In 2004, as violence in Iraq increased, the USA pushed to transfer
sovereignty to Iraqis. Washington hoped transferring power to local allies
would weaken the insurgency. As early as September 2003, Secretary of
Defense Rumsfeld communicated "enthusiasm for the concept of
granting sovereignty as soon as possible to the Council or some other
group of Iraqis."[27] Rumsfeld's statement also demonstrates the uncertainty, even six months after US operations commenced, about who
might assume authority in Iraq, and what that power transition might
look like.

On June 28, 2004, two days ahead of schedule, the CPA, which had
promulgated over 100 orders over the course of fourteen tumultuous
months, formally transferred authority to the Iraqi Interim Government
(IIG). In a rushed and rather unceremonious public announcement, at
10.26 a.m. Bremer ended the era of direct US administrative governance
of the Iraqi state. Two hours later Bremer was on a flight and John
Negroponte presented his credentials as US Ambassador to Iraq to the
IIG.[28] The twenty-six Iraqi ministries operating at the time had "been
shifting to full Iraqi control" since June 1, 2004.[29] This meant the Iraqi
ministers appointed by the IIG were officially independent of American
legal authority, but the same American advisers that had been running
Iraqi ministries under the CPA largely stayed put. "Most of the CPA's
former senior advisors (now known as senior consultants) … would
continue to provide technical and operational reconstruction assistance
to the Iraqi ministries."[30] In addition, Bremer had vetted the Iraqi
ministers. Reflecting on the June 28 ceremony, former US National
Security Advisor Zbigniew Brzezinski said, "This is not a transfer of
power, a handover to a sovereign government. We are transferring
limited authority to a satellite government, a satellite government that
is still to establish its legitimacy."[31]

was handing over government administration to Iraqi officials, enforcement of CPA orders became a problem for Washington. "Iraqi lawyers and judges began to procrastinate in implementing or interpreting [CPA] law, preferring the established Iraqi version, even if it contradicted the Bremer Orders."[25] In a style aptly described by James Scott in *Weapons of the Weak*, Iraqis engaged in repeated small acts of defiance against US advisers, incrementally imposing boundaries on American power brokers and drawing lines defining for Washington what the USA could and could not accomplish in Iraq.[26]

The Iraqi Interim Government

In 2004, as violence in Iraq increased, the USA pushed to transfer sovereignty to Iraqis. Washington hoped transferring power to local allies would weaken the insurgency. As early as September 2003, Secretary of Defense Rumsfeld communicated "enthusiasm for the concept of granting sovereignty as soon as possible to the Council or some other group of Iraqis."[27] Rumsfeld's statement also demonstrates the uncertainty, even six months after US operations commenced, about who might assume authority in Iraq, and what that power transition might look like.

On June 28, 2004, two days ahead of schedule, the CPA, which had promulgated over 100 orders over the course of fourteen tumultuous months, formally transferred authority to the Iraqi Interim Government (IIG). In a rushed and rather unceremonious public announcement, at 10.26 a.m. Bremer ended the era of direct US administrative governance of the Iraqi state. Two hours later Bremer was on a flight and John Negroponte presented his credentials as US Ambassador to Iraq to the IIG.[28] The twenty-six Iraqi ministries operating at the time had "been shifting to full Iraqi control" since June 1, 2004.[29] This meant the Iraqi ministers appointed by the IIG were officially independent of American legal authority, but the same American advisers that had been running Iraqi ministries under the CPA largely stayed put. "Most of the CPA's former senior advisors (now known as senior consultants) … would continue to provide technical and operational reconstruction assistance to the Iraqi ministries."[30] In addition, Bremer had vetted the Iraqi ministers. Reflecting on the June 28 ceremony, former US National Security Advisor Zbigniew Brzezinski said, "This is not a transfer of power, a handover to a sovereign government. We are transferring limited authority to a satellite government, a satellite government that is still to establish its legitimacy."[31]

The Iraqi Transitional Government and Government of Iraq

How much influence American officials had over Iraqi government institutions after the transfer of legal sovereignty in June 2004 varied depending on the US and Iraqi bureaucracies involved. Until 2010, for example, the USA maintained authority over Iraqi detainee facilities, detention centers, and inmates, primarily due to the lack of capability of Iraqi security forces to cope with the problem and US concerns about the threat that released prisoners could pose to American forces.[32] American policymakers were wary of surrendering the intelligence services to Iraqi authorities for fear that information and intelligence capabilities would be used for sectarian vendettas.[33] The USA also maintained direct military command and control over Iraqi military forces for more than four years. US Secretary of Defense Donald Rumsfeld found this troublesome because it undermined notions of Iraqi sovereignty. Rumsfeld promised the president, "the GOI will assume command and control of the Iraqi Army not later than June 2007."[34] Yet, due to the poor security situation, the last Iraqi division was not transferred to the Iraqi-led government of Iraq until November 3, 2007.[35]

While the Department of Defense handed over control of Iraqi troops in an effort to promote Iraqi sovereignty and self-sufficiency, mounting security problems prompted US personnel to simultaneously promote US involvement in the regime in Baghdad to augment immediate GOI performance. In November 2006 the Secretary felt the Pentagon needed to "aggressively beef up the Iraqi MOD [Ministry of Defense] and MOI [Ministry of Interior], and other Iraqi ministries critical to the success of the ISF [Iraqi Security Forces] – the Iraqi Ministries of Finance, Planning, Health, Criminal Justice, Prisons, etc. – by reaching out to U.S. military retirees and Reserve/National Guard volunteers."[36] The Pentagon sought to strengthen Iraqi ministries by embedding increasing numbers of US personnel while also recognizing the importance of transferring power to Iraqi authority. These measures were often contradictory and worked at cross-purposes.

The number of US officials assigned to Iraq ministries varied. The US Department of State was assigned to advise ten civilian ministries: Oil, Electricity, Planning, Water, Health, Finance, Justice, Municipalities and Public Works, Agriculture, and Education. In 2007 US State Department advisory teams varied in size from twenty officials in the Ministry of Oil and eighteen in Finance to three in Agriculture.[37] These teams "typically interact[ed] with the minister, deputy minister, or department director levels."[38] US advisers would present "options" to the Iraqi ministry, indicating the decision was left to the Iraqis, but under significant American guidance.[39]

The US Department of Defense (DOD) took a different approach. In charge of advising the Iraqi ministries of defense and interior, which controlled the Army and Police respectively, the DOD, under the Multi-national Security Transition Command – Iraq (MNSTC–I), was more intrusive than the State Department and the United States Agency for International Development (USAID). In 2007, for example, it had placed 215 American officials in the 2 Iraqi security ministries (Defense and Interior). According to a Government Accountability Office (GAO) report on US efforts to augment Iraqi ministerial develop-ment, these officials "advise[d] Iraqi staff about establishing plans and policies, budgeting, and managing personnel and logistics, among other things. According to MNSTC–I advisors, they work with their Iraqi counterparts on a daily basis to develop policies, plans, and procedures."[40]

US advisers also had varying influence across Iraqi ministries depending on the relationships that developed. According to *The New York Times*, the twenty or so Americans advising the Ministry of Oil reportedly "played an integral part in drawing up contracts between the Iraqi government and five major Western oil companies to develop some of the largest fields in Iraq."[41] State Department officials denied they chose which companies received Iraqi contracts, yet stated that Ameri-cans regularly "provided template contracts and detailed suggestions on drafting the contracts."[42] Released State Department documents collab-orate that description of US engagement.[43] Nevertheless, multiple oil companies have questioned the "sovereign" Iraqi governance at the Ministry of Oil, noting that the biggest development contracts were awarded to US and other Western companies.[44]

US control over key Iraqi ministerial budgets such as intelligence spending in the early years of the war, had lasting effects on Iraqi deci-sion-making.[45] Until 2007 the USA had final authority over the majority of Iraqi state security apparatuses, including Multi-National Force – Iraq (MNF–I), select Iraqi Army units, and private security details. With rates of violence skyrocketing from 2004 to 2007, American control of security organizations created a bargaining chip for Washington. Just weeks after the transfer of authority, for example, US officials wrote, "the IIG knows that it has to consult with us on constructing a state of emergency law since its enforcement could involve coalition forces."[46]

From 2004 to 2008 the influence of US advisers in Iraq gradually decreased. However, in early 2008, the White House ordered a "surge" of US development-oriented advisers, aiming to place at least seventy-five additional Americans in advisory roles within the GOI, focusing on health, finance, electricity, oil, and justice.[47] However, American accounts of this "surge" indicate that US advisers were no longer directing day-to-day GOI activities. In fact, by 2008 many US officials assigned to Iraqi ministries were having a difficult time regularly meeting with their GOI partners due to security constraints and the resistance of various ministers toward American agendas.[48] In September 2007 Americans were reporting, for example, "The Joint Planning Commission (JPC) and the Joint Reconstruction Operations Center (JROC) in Baghdad are fast becoming Iraqi-driven institutions ... The JPC and JROC will go from being largely USG-led to completely Iraqi-led by early 2008."[49] Unsurprisingly, at this time the USA was reporting increasing difficulty negotiating certain key issues with GOI partners. Discussing the Iraqi constitutional review, General David Petraeus commented, "it has proven difficult to urge movement from the GOI on practical concerns, let alone politically charged issues like new legislation."[50] By 2008 Baghdad had consolidated into a political entity capable of leveraging its legal sovereignty to oppose American agendas.

In sum, Iraqi ministers were granted authority over the activities of their ministries after the June 2004 transfer of sovereignty, but embedded American advisers remained influential for several years after this transfer. US influence in the GOI varied widely, depending on the ministry, but gradually decreased across Iraqi bureaucracies over the course of the US intervention. The ministries of defense and oil were the most significantly impacted by US advising. But by late 2007 when the Iraqi military was transferred to Iraqi control, US advisers were considered "visitors," rather than authorities. As the US Embassy summarized in 2008, "we have less influence than in 2004–2006, and will have even less influence ... Our views on desired outcomes influence and shape Iraqi debates, but we can no longer dictate the exact shape of the outcomes."[51] The post-CPA US advisory role in Iraqi civilian ministries was more invasive than what the USA had pursued in advising Vietnam or Afghanistan, but certainly less aggressive or permanent than the Soviet approach in Afghanistan.

With this background on the extent of US involvement in Iraqi decision-making institutions it is interesting to explore patterns in Iraqi compliance or defiance of US requests across the intervention. I analyzed over 1,000 US primary source documents (2,500+ pages) to identify 106 unique demands from the US government to Iraqi allies in the period 2003–11. These documents were collected from three sources. First, I examined declassified documents located on US government agency websites, including Secretary Rumsfeld's documents in the Special Collections Library.[52] A second source of material came from declassified documents released to the National Security Archive Iraq project. Last, the largest source of data for this chapter was US State Department documents published in the *Cablegate* database released by *Wikileaks*.[53] In November 2010 251,287 US Department of State cables were publicly posted despite the majority remaining classified. For more on the usefulness and limitations of this data, please see Appendix B. Note that most of the released cables were dated between 2002 and 2010, including 6,677 from the US Embassy in Baghdad. These documents provide extraordinary access to the interactions between the USA and the Iraqis partnered with the USA.

In addition, the Special Inspector General for Iraq Reconstruction (SIGIR), an independent US agency established by Congress in October 2004 to monitor the use of US reconstruction funds in Iraq, was helpful in clarifying compliance outcomes.[54] Since the USA would regularly provide funding for programs it asked Iraqi allies to adopt, SIGIR reports provided reliable accounts of Iraqi performance. Another helpful resource for finding information on compliance outcomes was a set of reports the US Department of Defense was required to produce according to House Conference Report 109-72, entitled "Measuring Stability and Security in Iraq."[55] Produced quarterly from July 2005 through July 2010, these lengthy reports monitored progress, growth, and developments in Iraqi governance, frequently addressing Iraqi progress on US policy requests.[56]

Combining these sources, documents analyzed date from March 1, 2003 to December 31, 2011.[57] The 106 US requests to Iraqi partners date from July 14, 2004 to February 23, 2010, providing relatively comprehensive coverage of demands made throughout the war.[58] Note that since the USA governed Iraq from March 2003 to June 28, 2004 under the CPA, there were no demands from the USA to the "Iraqi government," since there was no appreciable independent Iraqi government.

The "local regime" with which the USA negotiated throughout the Iraq War was different than local allies discussed in other counterinsurgency interventions in one notable aspect. There were *two* autonomous political regimes in Iraq. One was the Government of Iraq (GOI), based in Baghdad. The Shi'a-dominated GOI was the primary US partner in Iraq. Yet, there was a second entity, the Kurdish Regional Government (KRG), the governing body of the predominantly Kurdish northern region of Iraq. Officially the KRG is a regional government operating under Baghdad's authority. In practice, however, de facto regional autonomy after the Persian Gulf War (1990–1) meant that the Kurds exercised independence in the northern region. In addition, the history of ethnic antagonism – most notably genocidal offensives against the

Kurds during the Iran–Iraq War also meant that the KRG had little interest in closer relations with Baghdad during the US intervention. The emergence of the Islamic State in 2014 only solidified the Kurdish position on self-defense, autonomy, and Kurdish territorial sovereignty. During the US intervention, if Washington sought policies in northern Iraq, it had to deal directly with the KRG, as higher-ups in the GOI did not govern the Kurdish region. I treat US requests made to the heads of state in the KRG and those made to the GOI separately because this is the approach Washington adopted in its diplomatic relations in Iraq. The USA sent high-level delegations to negotiate with two Iraqi governmental entities, one Shi'a-dominated, one Kurdish.

The vast majority of US requests, 103/106, were directed at the GOI. Only 5/106 of US requests to Iraq located in the documents were for the KRG. Interestingly, only two US policy requests were presented to both the GOI and KRG. On July 25, 2007, US Ambassador Ryan C. Crocker and MNF–I Commanding General David Petraeus agreed in the "New Joint Campaign for Iraq" that it would be best to ask Baghdad to agree to delay a constitutional referendum on the status of Kirkuk, the oil-rich province at the heart of the governance dispute between Baghdad and Erbil regarding Kurdish autonomy and oil revenues.[59] However, in a cable from the US Embassy dated that same day, US Ambassador Ryan C. Crocker discussed a meeting with KRG President Massoud Barzani on July 19, in which the Ambassador asked the KRG President to work toward "transparent implementation" of a referendum.[60] These are two different messages, exemplifying the depths of sectarian divisions in Iraq that motivated the USA to craft multiple, even contradictory messages to different groups. Kurdish officials were frustrated with continual delays in establishing the status of Kirkuk, and in response, the US Ambassador emphasized communications between the KRG and GOI, sidestepping the delicate issue of delaying the Kirkuk referendum with the KRG. However, with Baghdad the US Ambassador was more direct, asking the Shi'a-dominated regime to postpone the referendum in order to defer a political showdown between Baghdad and Erbil.

What Kinds of Policies Did the USA Ask for in Iraq?

The USA proposed a wide variety of requests to Baghdad and Erbil, including addressing sectarian policies, personnel decisions, banking policies, command and control protocols, pension reform, constitutional referendums, hydrocarbon legislation, and reforming the Iraqi security forces. Table 4.2 provides a summary of US requests to Iraqi partners identified in US Department of State records.

Table 4.2 *Summary of US requests to Iraqi allies*

	Year	US request to Iraqi allies
1.	2004	Keep al-Jazeera's office in Baghdad open
2.	2005	Reconstruction of Fallujah
3.	2005	Initiate civil service reforms
4.	2005	Agricultural reform – increased privatization
5.	2005	State banking reform program
6.	2005	WTO accession
7.	2005	Nonimplementation of the Arab League Boycott
8.	2005	TNA approval of US/Iraq Joint Commission on Reconstruction and Economic Development (JCRED) bilateral agreements
9.	2005	Urge prominent Iraqi Sunni Arabs to engage in the draft constitution
10.	2005	Increased reconstruction funds for Fallujah
11.	2006	All militias need to be under the control of the state
12.	2006	MOI hold officials responsible for human rights abuses
13.	2006	Fund militia reintegration programs through GOI budget
14.	2006	Seize assets from former regime elements
15.	2006	Reverse decision to prohibit imported poultry products
16.	2006	Constitutional Review Committee
17.	2006	De-Ba'athification legislation
18.	2006	Legislation on procedures to form semiautonomous regions
19.	2006	Legislation on high electoral commission, provincial elections law
20.	2006	Amnesty legislation
21.	2006	Militia disarmament program
22.	2006	Support committees for the Baghdad Security Plan
23.	2006	Three brigades for Baghdad
24.	2006	Increased authority for Iraqi military commanders
25.	2006	Take measures to promote "even-handed" law enforcement
26.	2006	Baghdad Security Plan – do not protect sectarian actors
27.	2006	Measures to reduce sectarian violence and militia control
28.	2006	Establish planned joint security stations
29.	2006	Increase Iraqi security forces
30.	2006	Protections for minority political parties
31.	2006	Allocate and spend $10 billion in Iraqi revenues for reconstruction and services
32.	2006	Do not take actions undermining members of the Iraqi Security Forces
33.	2006	Issue letters of credit for contracts for the Public Distribution System
34.	2006	Accept MOI/MOD budget of $8 billion
35.	2006	Pass Hydrocarbon Law[a]
36.	2007	Compromise over funding for the Kurdish Peshmerga[a]
37.	2007	Budget conferences
38.	2007	National Reconciliation Plan and legislation
39.	2007	Train more prison guards and detainee officers
40.	2007	Remove yellowcake uranium from Tuwaitha
41.	2007	Budget execution
42.	2007	Fill cabinet posts
43.	2007	Clarify Baghdad's status and bureaucratic hierarchy
44.	2007	Provincial Powers Legislation

Table 4.2 (*cont.*)

	Year	US request to Iraqi allies
45.	2007	Letter requesting assistance removing yellowcake uranium from Tuwaitha
46.	2007	Establish "three plus one" group
47.	2007	Pass Revenue Management Law
48.	2007	Provision of nonsecurity, postsurge basic services
49.	2007	Clarify command and control in Samarra
50.	2007	Ensure rule of law complex continues operating
51.	2007	Pay Kuwait for fuel tankers
52.	2007	Transparent implementation of Constitutional Article 140[b]
53.	2007	Delay Article 140 Referendum
54.	2007	More authority delegated to the Council of Representatives and Presidency Council
55.	2007	Enact elections law (and provincial elections law)
56.	2007	Speedier processing of Coalition and Iraqi detainees and selective releases of detainees
57.	2007	Constitutional referendum establishing organizational hierarchy with national, regional, and local authorities
58.	2007	Prosecution of law-breaking officials
59.	2007	Increased negotiated ceasefires with armed Sunni and Shia groups
60.	2007	Put fuel meters on all depots
61.	2007	Fund agribusiness
62.	2007	Support vocational training
63.	2007	Expand microfinance and SME lending
64.	2007	Continue fuel subsidy reductions, normalize fuel prices in accordance with IMF recommendation
65.	2007	Construction of storage facilities to support private-sector fuel imports
66.	2007	Implement new social safety net
67.	2007	Gradually increase electricity prices
68.	2007	Reformed pension plan
69.	2007	Implement an Iraqi-led system to make cash transactions safer
70.	2007	Comply with financial disclosure regulations
71.	2007	Cease sectarian appointments and politically motivated prosecutions
72.	2007	Establish an Anti-Corruption Institute
73.	2007	Implement reforms called for by the standby arrangement with the IMF
74.	2007	Provide Spokesman Sheekly with protection
75.	2007	KRG provide temporary refuge to 60 Palestinian families[b]
76.	2007	Send letter to UN requesting new resolution
77.	2007	Match US funding for Concerned Local Citizens (CLCs)
78.	2007	Continue supporting CLCs, incorporating a percentage into Iraqi Security Forces (ISF)
79.	2007	Do not pass law revoking immunities for contractors
80.	2007	Implement the 2006 National Investment Law
81.	2008	Legal protections for investors
82.	2008	Establish the National and Provincial Investment Commissions (NIC and PIC)
83.	2008	National energy strategy
84.	2008	Reform the Public Distribution System (PDS)

Table 4.2 (*cont.*)

	Year	US request to Iraqi allies
85.	2008	Ministry of Finance to have dedicated funding, including Ministry of Electricity funds
86.	2008	Counterterrorism Law
87.	2008	Iraqi Army to protect central bank
88.	2008	Legislation to enforce UN sanctions on those threatening stability in Iraq
89.	2008	Provide list of sanction designees
90.	2008	Lobby for Russian support of UN Resolution
91.	2008	Actively defend Iraqi Christian community
92.	2008	Humane treatment for Mujahedin e-Khalq residents of Camp Ashraf
93.	2008	Authorize new Status of Forces Agreement
94.	2008	Minister of Electricity to sign the Parsons Brinckerhoff contract
95.	2008	Accept transfer of Iraqi Naval Station Guantánamo Bay (GTMO) detainees
96.	2009	Negotiate with Kuwait at April 28–29 UN Compensation Commission Meeting (UNCC)
97.	2009	Promotion of Weapons of Mass Destruction (WMD) nonproliferation
98.	2009	United Nations Convention Against Corruption (UNCAC)
99.	2009	MANPADS program
100.	2009	Audit and oversight of hydrocarbon resources
101.	2009	Resolve remaining Saddam Hussein-era sovereign debts
102.	2009	Resolve remaining oil-for-food claims before the end of 2009
103.	2009	KRG – approval of election law[b]
104.	2009	Out of country voting
105.	2010	Orders on remaining 1,600 release-eligible detainees in US custody
106.	2010	Pass human trafficking legislation after March 7 election

Notes: [a] US requests made to both the KRG and the GOI. [b] US requests made to the KRG.

In order to paint a broad picture, we can divide US requests to Iraqi partners into six general subject categories, which reflect several key components of US counterinsurgency strategy. Similar to other interventions, specific issue areas of requests were not found to be significant in determining the likelihood that a request would be complied with or not as other structural factors often had a greater influence on compliance outcomes.

Issue Areas of US Requests to Iraq

1. *Development* – projects or activities intended to expand infrastructure. Includes projects to reconstruct damaged urban areas, microfinance strategies, and vocational training programs. 14/106 (13.2 percent) of US requests. There was occasional overlap with requests classified as economic reforms, such as pricing for electricity services, which is an economic policy, with growth and development goals.

2. *Economic reform* – activities intended to change economic policies. Includes reforms to state-run social services, banking and financial systems, WTO negotiations, pricing of public services, and hydrocarbon profit distribution legislation. 26/106 (24.5 percent) of US requests. There was some overlap with requests classified as development requests and one that could also be considered military reform: the Iraqi Army protecting the central bank. This is a security protocol issue, but also one addressing physical banking security and messaging about financial security.

3. *Military reform* – actions intended to change military policies and institutions. Includes increasing Iraqi security forces, increased authority for Iraqi military commanders, and command and control protocols. 9/106 (8.5 percent) of US requests.

4. *Military strategy* – actions intended to guide military forces in the execution of the war effort. Includes policies regarding the distribution and buying back of weapons, use and oversight of militias, and sending three additional Iraqi brigades into Baghdad. 5/106 (4.7 percent) of US requests.

5. *Political reform* – actions intended to change government policies and institutions. Includes electoral policy, law enforcement, governance, protections for minorities, bureaucratic protocols, constitutional reforms, sectarian issues, and reconciliation programs. 48/106 (45.2 percent) of US requests.

6. *Political-military counterinsurgency strategy* – actions intended to implement a particular counterinsurgency program. Includes pacification projects, negotiated ceasefire agreements, the provision of nonsecurity services in specific areas, and the Sons of Iraq program. There is occasional overlap with requests classified as political reform. 4/106 (3.8 percent) of US requests.

Development. As in other conflicts, compliance with development requests was slightly lower than the average across all requests: 4/14 (28.6 percent) compliance, 4/14 (28.6 percent) partial compliance, and 6/14 (42.9 percent) noncompliance. Development work is multifaceted and often takes significant institutional capacity that local allies rarely have at their disposal. Operating in a war zone often complicates progress even further. The insurgency in Iraq regularly targeted social service and infrastructure projects. As opposition grew against coalition forces in the years following the US invasion, development work became important to the goal of promoting the idea of an effective Iraqi state, but it also became increasingly difficult to accomplish.

Economic reform. Requests pertaining to economic reform had the lowest rate of compliance, a pattern unique to Iraq. In other conflicts including Vietnam and Sri Lanka economic-related requests from intervening forces produced one of the highest rates of compliance from local allies. High levels of local compliance might be expected for multiple reasons. First, allies share a strong interest in local economic stability. Second, intervening forces often have opportunities to build in budgetary controls over local financial policies absent in other policy areas less directly related to funding. And third, local allies often defer to intervening partners on financial or military matters out of consideration for their expertise. However, in Iraq, 13/26 (half) of all US requests related to economic reform resulted in Iraqi noncompliance, while 7/26 (26.9 percent) resulted in compliance and 6/26 (23.1 percent) in partial compliance. Iraqi defiance can be attributed to two major factors.

First, 20/26 (76.9 percent) of requests for economic reform in Iraq reported significant capacity failings, which is not surprising considering the tenuous economic infrastructure left behind by Saddam Hussein's regime. According to the RAND Corporation, before the American invasion "the information systems of the [Iraqi] Central Bank and commercial banks were antiquated and – in some cases – nonfunctioning. The Rafidain and Rasheed bank branches had no interbank voice and data communications ... The World Bank also noted that the Central Bank's supervisory capacity was largely nonexistent, and the bank suffered from the absence of any supervisory legislation."[61] US Under Secretary of the Treasury John Taylor was tasked with expanding Iraqi financial sector capacity, writing that after the US invasion, "the [Iraqi] banking system was in shambles. Electronic transfer of funds, widely made to people in developed countries was virtually non-existent, making Iraq's payment system the equivalent of a Model-T Ford."[62]

The second factor contributing to an unusually low rate of compliance with economic requests was that Iraq's historically statist emphasis on economic policies clashed with the free-market solutions advocated by American advisers. US officials in Iraq ran into cultural and institutional roadblocks when they tried to promote market policies. This resistance persisted even with seemingly straightforward economic programs. For example, the Provincial Reconstruction Team (PRT) in Babil was frustrated when six years after the invasion US officials were still unable to inspire Iraqi representatives to reform economic agricultural policies:

As the owner of around 70 percent agricultural land in the province, the government micromanages what farmers produce, guarantees the purchase of key crops, and, to a lesser extent, subsidizes agricultural inputs. These statist policies and other factors, from inadequate credit to the distortionary Public

Distribution System (PDS), have restricted private sector development in agriculture. Having come to rely on government largesse, it is little surprise that Babil farmers look to intervention and protection rather than open markets for their livelihoods.[63]

This type of disagreement between local and intervening forces with respect to economic philosophies was unique. The USA in Vietnam or the Soviets in Afghanistan, for example, shared a generally compatible viewpoint on economic growth with their local allies, which facilitated cooperation. In Iraq, however, local allies were threatened by US-designed market approaches and pushed back accordingly. One example is US attempts to compel the GOI to reform the PDS. The PDS was a ration system established in 1995 to provide Iraqis with staple household items. The system aimed to mitigate the effects of food shortages produced by sanctions. The program was run by the Iraqi Ministry of Trade (MOT) and depended on food imports (up to 480 tons per month), even though Iraq had historically been an agricultural exporter.[64] US officials commented that with sanctions and the PDS, Iraq went "from breadbasket to basket case."[65] The World Food Program (WFP) took over responsibility for the program after the disintegration of the Saddam-era Iraqi state.[66] Washington pressured the GOI to fix inefficiencies, market distortions, and the PDS burden on GOI finances. However, according to the US Embassy,

despite almost universal acknowledgement of its failures – its massive price tag, its distorting effects on domestic commodity markets, and its myriad opportunities for corruption – there have been no significant reform efforts. In 2007, the COR [Iraqi Council of Representatives] passed legislation that would have rendered households of government officials at the rank of Director General or higher ineligible for PDS rations. The GOI has not implemented this reform, however, and even the wealthiest Iraqis remain eligible for rations.[67]

The failure of Baghdad to reform the PDS can be attributed to a lack of willingness of Iraqi officials to risk modifying the popular social program, capacity problems hindering better administration, GOI reluctance to adopt market solutions more generally, and the US inability to pressure reform due to a lack of US budgetary oversight mechanisms and no unilateral ability with which to threaten to reform the PDS independently due to Iraqi legal sovereignty.

Military reform and military strategy. In both Iraq and Vietnam military-related requests produced the highest rate of local ally compliance. Saigon and Baghdad were more willing to take US advice on military actions out of deference to US military expertise. Attacking the enemy did not typically require potentially painful or complex internal adjustments, like many requests for political reform. Furthermore, it is less likely that low capacity would impact requests regarding military strategy, as the USA would only ask Saigon or Baghdad to undertake military operations Washington felt these small allies could handle.

Political reform. Similar to other counterinsurgency interventions, the majority of US requests to local Iraqi allies during intervention concerned political reform. After June 2004 and the handover of official legal sovereignty the USA was unable to undertake internal reforms for Baghdad, and therefore US forces were required to work through the Iraqis for political restructuring. Examples of political reforms requested include boycotting legislation to revoke immunities for American contractors, holding a constitutional referendum, accepting the transfer of inmates from Guantánamo Bay, and passing de-Ba'athification legislation.

Counterinsurgency projects. There were only a few requests pertaining to counterinsurgency strategy 4/106 (3.8 percent), and 75 percent resulted in partial compliance. This category is small because several other categories of issues also incorporate aspects of counterinsurgency strategy, including development or political reform. Since "counterinsurgency" policies can encompass a range of political and military actions, many demands potentially overlap with other issue categories. These requests include management of Concerned Local Citizens (CLCs), a US-designed COIN program to pay Sunnis to oppose al-Qaeda. Figure 4.1 compares rates of Iraqi compliance across categories of issues addressed in US requests.

Explaining the Compliance and Defiance of Local Allies in Iraq

As detailed in Chapter 2, I offer four variables to help explain the likelihood of Iraqi compliance with US policy requests. Namely, local capacity to implement the request, the interests of Washington and Baghdad or Erbil converging or diverging over the request, the dependency of US forces on Iraqi partners to implement the request, and a significant enemy offensive. However, the US war in Iraq did not experience a significant, coordinated acute enemy offensive similar to the Tet Offensive in Vietnam. Violence in Iraq was significant, but more diffuse, such that it did not serve as a sudden shock to local partners. However, similar to other interventions, I hypothesize that there is an important interaction effect between the interests and dependencies of allies producing a particular pattern of local compliance outcomes key to understanding when Iraqis partnered with the USA were likely to comply or

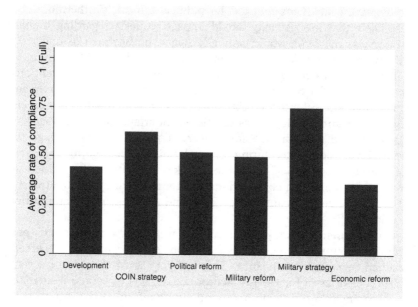

Figure 4.1 Rates of Iraqi compliance with US requests, by subject area, 2003–10

defy US requests. In this section I will address each variable in the context of US-Iraqi politics.

Local Capacity Limitations – Decreased Likelihood of Compliance

Consistent with other wars, failing Iraqi institutional capacity is strongly correlated with noncompliance with US policy requests. As in Vietnam and Afghanistan, the most common capacity issue affecting Iraqi ability to implement US requests came from deficient bureaucratic institutions. American dollars were readily available, but Iraqi officials nevertheless had to cope with bureaucratic machinery that was lacking in institutional capacity to execute policies. There are numerous examples of faltering Iraqi state structures affecting compliance outcomes. Consider the US concern regarding Iraq's failure to effectively spend allocated money. In 2007, Washington requested that Iraq execute its budgeted expenses, but Iraqi ministry spending was complicated by limited capacity to execute the specified projects, leaving allocated US money stagnant and ineffectual.[68] According to the US Government Accounting Office, Iraqi ministries faced staff shortages, weak accounting institutions, and security complications that severely limited the programs they could execute.[69]

Low capacity was cited as a problem throughout the Iraq conflict. Nearly half of all requests – 52/106 (49.1 percent) – made some mention of inadequate Iraqi ability contributing to policy outcomes. This is a higher rate than what was reported by intervening allies in other interventions. In the US wars in Vietnam and Afghanistan just over 25 percent of requests indicated that capacity was impacting the performance of local allies. Capacity was an even less-cited problem in Sri Lanka, where only 16.5 percent of Indian requests to Sri Lankan partners discussed capacity concerns. But these figures are not straightforward. Capacity failings being cited more frequently in Iraq than in Afghanistan does not mean Kabul had greater capabilities than Baghdad to cope with the insurgency.[70] In fact, the rates of substantive compliance for Iraq, Vietnam, and Afghanistan are strikingly similar – roughly one-third full compliance, one-third partial compliance, and one-third noncompliance.[71] If Baghdad were significantly less capable than Saigon and Kabul, a corresponding drop in rates of compliance with US requests would be expected since there is a significant correlation between low capacity and noncompliance. However, this was not the case.

Several factors contributed to the high US-reported rate of capacity failings by local Iraqi partners. First, the US directly governed Iraqi institutions through the CPA and became well versed in Baghdad's bureaucracy. This heightened sense of institutional awareness may have made US officials more able to recognize specific capacity failures in Iraq and more likely to discuss them in diplomatic correspondence than was the case in Vietnam and Afghanistan. Second, violence in Iraq quickly intensified between 2004 and 2007. The intensity of the rising insurgency largely caught US policymakers by surprise. This may have influenced the frequency with which capacity was cited as an issue in Iraqi compliance with US demands. Washington was desperate to ameliorate the situation and was making demands on Baghdad, even asking for reforms US policymakers likely understood would strain Baghdad's limited capacity. Moreover, a regime that would be able to handle certain challenges under normal circumstances may find itself incapacitated by insurgency. Once the insurgency became more subdued in 2008, Baghdad and Erbil were able to comply with US requests, as evidenced by the rise in compliance in 2008–9 illustrated in Figure 4.2.

Insufficient capacity to implement US policy requests could emerge from a number of sources, including the inherent limitations of weak Iraqi bureaucracies, internal disagreement within the Iraqi state, or obstacles to compliance produced by the war. Insurgent attacks in Iraq grew to 180 per day by June 2007, a state of insecurity that paralyzed projects and hampered development efforts.[72] Issues such as violence, combat operations, refugee influxes, and offenses by the enemy were cited as an issue affecting Iraqi compliance for 18/106 (17.0 percent) of requests. One example was observed in a 2007 US request to the GOI that it speed up the paperwork and processing of detainees, a task intended to move toward American withdrawal. However, as violence spiked in 2006–8, so did the number of insurgents captured by US forces. As the US Embassy observed, "Case processing is not improving fast enough to keep pace with GOI detainee intakes and the transfer of detainees from coalition forces in the coming months ... The GOI detention system will, under this scenario, become more of a warehouse than a detainee processing center for the courts."[73] The demands of the war outpaced Iraqi capacity, making Iraqi compliance with US policy requests increasingly difficult.

Furthermore, internal political complications can also affect the capacity of a regime to comply. The GOI and KRG are not unitary political entities. As in all governments, there was internal debates and conflict. For example, in 2006 the USA insisted the GOI hold members of the Iraqi MOI accountable for "killings, disappearances, and human rights abuses" committed by MOI officers.[74] The Iraqi Ministry of Human Rights (MOHR) and the Human Rights Directorate at the MOD investigated accusations of MOI misconduct; however, this effort fell short. The Minister of the Interior Jawad al-Bulani complied in part with requests from the MOD and MOHR by authorizing the arrest of multiple low-ranking MOI officials accused of human rights abuses.[75] However, high-ranking officials were largely excused, despite evidence against them. The commander of the 2nd National Police Division, for example, was believed by US officials "to have directly ordered torture and other abuse." [76] Yet Minister Bulani spared him from investigation, instead reassigning the commander to the Ministry's intelligence division. According to the US State Department, his "punishment for his alleged crimes has been the loss of four days of pay."[77] Sometimes certain elements of the GOI worked toward fulfilling US requests, while at the same time other officials worked in opposition, a tension that illustrates the difficulty of implementing coherent policy.

Interests and Dependencies

As in other interventions including Vietnam, Afghanistan (USA), and Sri Lanka, the prospect of potential short-term benefits for local allies was one of the highest predictors of compliance with US requests in Iraq.[78] If fulfilling the US request appeared likely to directly serve the interests of the regime or Iraqi policymakers, it had a higher likelihood of being adopted than if it couldn't immediately yield such benefits.

However, as hypothesized in Chapter 2, not all requests where the interests of local and intervening allies align over a proposed policy resulted in local compliance, because US dependencies on local allies also affected the ways interests translated into incentives for local allies to comply with or defy requests. Specifically I hypothesize that if the USA could act unilaterally and Iraqi and US interests converged, Iraqi allies were more likely *not* to comply, due to incentives to free ride. Conversely, if the USA could implement the request unilaterally, if necessary, and the interests of allies diverged, there was a higher probability of local compliance: in such cases Baghdad and Erbil were more willing to participate, despite their divergent interest, since the proposed policy could be implemented without them, and complying, at least in part, ensures they have a hand in the outcome. Under these conditions, going along with the request enables local allies to better affect the process of implementation and protect their interests.

I contend that the reverse is true if unilateral US action was not possible. If independent action was not possible and US and Iraqi interests converged, there was a higher likelihood of compliance, as Baghdad and Erbil were likely to adopt the requested actions to see the policy implemented and secure the benefit. If, however, the interests diverged and the USA could not implement the reform without Iraqi assistance, noncompliance by Iraqi allies was probable. This is because the GOI or KRG would have little motivation to comply and were unlikely to face immediate negative consequences for noncompliance. Below I provide several examples to illustrate this interaction and the impact on compliance outcomes.

Interests Diverge and the USA Is Dependent on Local Allies for Policy Implementation: Predicted Iraqi Noncompliance

If the interests of allies diverge and intervening forces are dependent on local allies to implement the policy noncompliance is likely. In Iraq this was observed in the November 2007 request from US Ambassador Ryan Crocker and US General David Petraeus to Iraqi Prime Minister Maliki asking that the GOI continue to offer legal immunity to US contractors in Iraq.[79] Public outcry in Iraq against alleged abuses by foreign contractors had hardened GOI positions against contractor immunity even though legal exemption had been afforded to contactors since the CPA. But US officials found they had little to compel the GOI to offer continued protection of contract personnel. The 2008 USA–Iraq Status of Forces Agreement thus specified: "Iraq shall have the primary right to exercise jurisdiction over United States contractors and United States contractor employees."[80] The USA depended on Baghdad to fulfill the request for extended immunity, as US personnel had no authority to determine Iraqi law in 2009. Baghdad chose not to comply, and in February 2011 a British contractor, Daniel Fitzsimons, was sentenced by Iraqi courts to twenty years in prison for the murder of two other foreign workers in 2009.[81] As predicted, under these conditions non-compliance was likely because the interests of allies diverge and intervening forces do not have unilateral ability to potentially threaten local allies into participation, a dynamic evident in the next example.

Interests Diverge and the USA Is Not Dependent on Local Allies for Policy Implementation: Predicted Iraqi Compliance

The USA made a controversial decision to fund the Sons of Iraq (SOI), or as it was initially called, Concerned Local Citizens (CLCs), an ad-hoc group of Sunnis that were willing to turn against al-Qaeda insurgents. The Anbar Awakening led by the SOI combined with the US troop surge reversed the course of the war, in favor of the counterinsurgency.[82] Nevertheless, despite its remarkable success undermining the insurgency the Shi'a-dominated regime in Baghdad did not welcome the SOI program. Some Shi'a officials viewed the program as a unilateral US decision to arm a rogue Sunni militia that would likely compete for power in Iraq challenging their regime.[83] Following the 2007–8 successes of the Sons of Iraq program against al-Qaeda, it was unclear what to do with SOI members. In late 2007 the USA asked Baghdad to match funds for the program and incorporate SOI elements into the government to reward SOI members for challenging al-Qaeda and to dilute Shi'a dominance in Baghdad.[84]

In January 2008, Baghdad agreed to match US funds for the program.[85] Although hostile to the idea of an armed, Sunni, nongovernment group controlling territory, the GOI recognized that if the USA continued to run the program without their input, the Sons of Iraq program would not only continue to exist on US funding, but would operate outside their influence. Matching funds and running programs to incorporate SOI elements into GOI institutions would give Baghdad more influence over the fate of the program. In 2008 Baghdad's cooperation on SOI funding was tenuous. The Office of SIGIR reported that in practice, GOI payments to the SOI were "consistently late." "Between October 2008 and December 2010, the GOI paid SOI salaries on time only about 42% of the time."[86] The USA had to provide additional payments to SOI entities due to Baghdad's tardiness or failure to pay.[87] Although US officials remained skeptical, Iraqi Prime Minister Nouri al-Maliki expressed support for SOI integration into the government, earmarking $300 million in matching USA–GOI funds for the SOI.[88] As American officials feared, as Baghdad gained authority over the SOI, reports emerged of Shi'a elements in the GOI targeting SOI figures.[89] According to National Public Radio (NPR), "in 2009, the fate of the Sons of Iraq was left in the hands of Iraq's Shiite-dominated coalition government, which agreed to pay the men and eventually either integrate them into the armed forces or give them civilian jobs. But scores have been arrested over the past year by the government."[90] After US withdrawal in 2011, Baghdad took advantage of its influence over SOI to largely dismantle the group, part of a larger pattern of sectarianism that isolated the Sunnis following US departure.

Therefore, albeit only partially and temporarily, Baghdad complied with US requests regarding supporting the SOI.[91] According to SIGIR, the GOI provided $196 million for reintegration programs. The US military reported that by July 2008 more than 14,000 SOI had been incorporated into the Iraq Security Forces; nonetheless it added, "developing broad support of the program has been a challenge."[92] Although Baghdad agreed to US demands that it provide matching funds to the SOI program in order to placate US policymakers, more importantly it did so in order to have greater influence over the program. US and GOI interests diverged in terms of the SOI, but US ability to run the program with or without Baghdad convinced Shi'a policymakers to participate in order to gain more influence over the program. As hypothesized in Chapter 2, coercion of local allies is possible under conditions where the interests of allies diverge and intervening allies are not dependent on local partners for implementation.

Interests Converge and the USA Is Dependent on Local Allies for Policy Implementation: Predicted Iraqi Compliance

When the interests of allies converge and free riding is not possible, local compliance is likely. In 2007 the USA asked Baghdad to issue an official letter requesting US assistance to remove 550 metric tons of yellowcake uranium, a starter material used to produce high-grade nuclear material from Tuwaitha, a nuclear research facility 18 km southeast of Baghdad. Significant US military presence in the country meant that the USA had the ability to remove the uranium, but depended on its Iraqi allies to issue an official letter from the GOI requesting US assistance in the process. The letter would address concerns of the International Atomic Energy Agency regarding which government had legal authority to transport uranium in Iraq.[93] Therefore, the USA issued two separate requests to Baghdad regarding the Tuwaitha yellowcake, one for an official letter regarding the removal, and one for Iraqi participation in the sale of substance. One request asked for a letter providing permission and required Baghdad's participation, while the other focused on removal, which could be implemented by the USA unilaterally.[94]

Baghdad had strong interest in having the yellowcake removed for various reasons. First, during this time Iraq was occupied by a foreign military sent to overthrow Saddam Hussein, in part to ensure that Iraq did not acquire nuclear weapons. The GOI was acutely aware of the potential consequences for Iraqi nuclear ambitions. Second, Baghdad exercised legal but not territorial sovereignty, and Iraqi forces, or US forces for that matter, did not have full control over Iraqi territory. There was a possibility that insurgents could get hold of the material and use it against Baghdad. Third, the USA presented Baghdad with a favorable offer that played to their interests. The US DOD would handle the difficult job of transporting this material, while the GOI could profit from the sale of the yellowcake. Fourth, by issuing the letter Baghdad could better influence the terms. It could lobby for a better price and legitimize itself as owner and legal authority. After negotiating the terms and joking with the USA that maybe Iraq could sell the yellowcake to Iran, officials in Baghdad complied and issued the letter.[95] The end result of the request was in Baghdad's interest and so long as the USA upheld Iraqi legal sovereignty, the task could not be accomplished without their authorization. Under such conditions, where the interests of allies converge and unilateral action by intervening forces is not possible, local ally compliance is likely.

Interests Converge and the USA Is Not Dependent on Local Allies for Policy Implementation: Predicted Iraqi Noncompliance

In 2006 the USA urged the GOI to invest in the reconstruction of Fallujah, a city that had been devastated by US military operations.[96] American policymakers felt it was important to have the GOI invest in rebuilding in order to increase the likelihood of suppressing the insurgency, and to better ensure lasting buy-in from Baghdad on reconstructing Sunni-dominated cities like Fallujah.[97]

Under US pressure the GOI pledged a few hundred million dollars to the effort, an amount described by Michael Schwartz as "meager" at best.[98] Additionally there were large discrepancies between what was promised and what was delivered, as at least 25–30 percent of what was eventually paid by the GOI was funneled into security operations instead of reconstruction.[99] By 2008 the USA had taken primary responsibility for rebuilding Fallujah. Entry into the city required a US-issued residency card, and the USA executed the majority of construction projects. As one US Marine said, "We were here, the battle part, last time. Now I'm back here fixing it, which is kind of ironic."[100] According to US NPR, "Americans are widely seen as the engine behind the rebuilding. Asked if he felt he was still being neglected by the Iraqi government, Sheikh Hamid al-Alwani, the head of the city council, didn't miss a beat. He says 'naam' or 'yes.' The Iraqi government has spent far less on rebuilding than the U.S. predicted."[101]

Baghdad failed to meet US expectations for investment in reconstruction of Fallujah because the USA had the resources to fund reconstruction unilaterally. Iraqi allies could free ride on US reconstruction efforts, benefiting while exerting minimal effort. In the illustrative case of a project to build a wastewater treatment system for Fallujah, Baghdad stalled payments to contractors for years. Furthermore, there was no plan to link houses to drainage; a necessary component of the project since without a link to homes, there would be no sewage to treat. In 2008 representatives from the US Army Corps of Engineers reported that although the Iraqi Ministry of Municipalities and Public Works had "verbally committed" to installing home connections, US engineers were "extremely concerned" that the ministry would not provide any services, instead insisting homeowners install their own connections. This ad-hoc solution would risk harming the system and undermining the entire effort. The Army Corps of Engineers "stated that if the Ministry cannot fund the house connections, the 'back up plan' is for the U.S. government to use Economic Support Funds."[102] As predicted, free riding by local allies is likely when the interests of allies converge, and the participation of local officials is not required.

Each of the 106 identified demands from the USA to its partners in the GOI and the KRG were coded to yield a picture of broad trends related to the proposed variables and the likelihood of Iraqi compliance. An ordered probit statistical model is provided in Table 4.3, accounting for capacity limitations and the proposed interaction effect between interests and dependencies in Iraq. Additional quantitative information is provided in Appendix C (Table C.1) specifying additional checks conducted, including fixed effects by year, and considering interests alone without dependencies as an alternative explanation. Additionally,

Table 4.3 *Ordered probit – local Iraqi compliance with US requests, 2003–10, interaction effect between US dependency on Baghdad and Erbil for policy and allied interests*

Local Iraqi capacity	US unilateral potential/ dependency	Allied interest convergence/ divergence[a]	Interaction with dependency and allied interest
-0.905^{**}	0.081	0.957^{**}	-1.118^{*}
(0.236)	(0.388)	(0.294)	(0.506)

Notes: $^{*}p < 0.05$; $^{**}p < 0.01$.
[a] The variable convergence/divergence of interests is produced by combining two variables related to local interests, "Private Benefit for Local Allies," and "Threat to Private Benefits for Local Allies." The variable was created in order to have a robust measure of Baghdad's interests based on costs and benefits, instead of substituting one (costs or benefits) for interest while excluding the other. For more on coding see Chapter 3.

the data was reanalyzed using robust standard errors clustered by the issue area of request, which provided a statistically and substantively comparable result to the results presented in Table 4.3.

Iraqi Compliance with US Requests over Time

Figure 4.2 charts the number of requests made by the USA each year, juxtaposed with the Iraqi rates of compliance with those demands.[103] Since complicated policies requested by the USA can take several years to implement, some years had more instances of compliance than new requests made, as Iraqi allies fulfilled US requests made earlier in the war. The number of requests made by Washington peaks in 2007, while Iraqi compliance reaches a high point in 2008–9. The latter peak can be attributed to (1) numerous requests made in 2007; (2) an expansion of Iraqi governance capacity; (3) the fulfillment of multiple complex requests that took years to implement; and (4) a marked decrease in violence in 2008–11 compared to 2007, allowing for greater Iraqi freedom of movement to implement programs.

The absence of requests in 2003–4 is unsurprising, given that the CPA was governing Iraq during this period. After the June 2004 transfer of sovereignty from the CPA to the Iraqi Interim Government there is a steady rise in US demands on Iraqi partners each year until 2007. This continual increase from 2004 to 2007 reflects the gradual nature of the transfer of authority from American caretaker organizations to Iraqi bureaucracies over the course of several years. If the June 28,

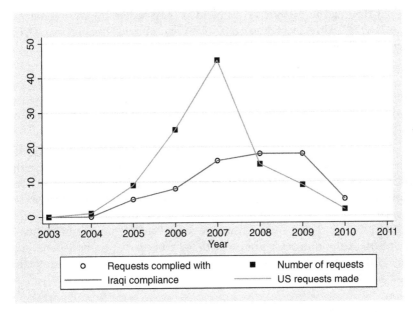

Figure 4.2 Rates of US requests and Iraqi compliance, 2003–11

2004 transfer had heralded an immediate Iraqi assumption of responsibility across the board, it would be reasonable to expect a spike in demands in 2004–5 (followed by a slow decline), similar to what was observed in Vietnam in 1964–6 (see Chapter 10, Figure 6.2).

Furthermore, before leaving office in June 2004, L. Paul Bremer's CPA promulgated orders at a frenzied pace. The Americans assumed it was easier to unilaterally implement policies under US authority in 2004, as opposed to negotiating policies through the GOI once Iraq was legally sovereign.[104] US decision makers may also have made fewer demands in the years immediately following CPA activity because they had largely pursued their agenda while in command a short while earlier.

Furthermore, the decision-making capacity of the GOI was intimately tied to the level of violence. As sectarian conflict and US demands increased in 2006–7, a corresponding rise in GOI rates of compliance (Figure 4.2) failed to materialize because Iraqi political leaders were frequently involved in sectarian violence, effectively failing to comply with US requests against such sectarian infighting. Increased violence in 2006–8 made opportunities for compromise between Sunnis, Shi'as, and Kurds increasingly scarce.[105] For example, in August 2006, the USA asked Baghdad to provide "even handed [non-sectarian] enforcement of the law."[106] But according to a US Government Accounting Office report in May 2007,

Iraq's Shi'a-dominated government bears responsibility for engaging in sectarian-based human rights violations ... death squads affiliated with the Ministry of Interior targeted Sunnis and conducted kidnapping raids in Baghdad and its environs, largely with impunity ... the Iraqi government and many Iraqi security force units are still applying the law on a sectarian basis.[107]

The US Embassy's 2008 assessment of the Iraqi parliament reflected similar sectarian concerns, commenting that the Council of Representatives' "problems are Iraq's problems: deep-seated sectarian, ethnic and personal animosities that inhibit and frequently prevent compromise, cooperation and agreement."[108] US anti-sectarian policy demands were more often than not ignored, as the USA had no unilateral ability to implement the policies without Baghdad's participation and Baghdad's divergent vision for Iraq's future regarding sectarian tensions motivated Shi'a defiance.

Conclusion

There were several unique features of alliance politics in the US intervention in Iraq in 2003–11. First, as in other counterinsurgency partnerships, there was an interaction between Washington's capability to independently undertake the policy US officials requested of local partners and the convergence or divergence of US and local ally interests. This interaction affected the likelihood of local Iraqi compliance with US requests. Second, the USA did not support one state government in Iraq, as is conventional in counterinsurgency interventions. Instead Washington maintained two distinct high-level diplomatic partnerships in Iraq – one with the Shi'a-dominated regime in Baghdad, and a second with Kurdish-held Erbil. Although the vast majority of US requests were directed at Baghdad, the independent US relationship with the autonomous Kurdish area reflected sectarian divisions in Iraq and, while practical in terms of negotiating specific policies, served to reinforced those divisions. Third, Baghdad was unique among local allies in resisting US requests for economic reforms, in part due to historical Iraqi reliance on statist economic solutions, which conflicted with Washington's insistence on market-based economic policies.

Additionally, two broad dynamics in Iraqi politics regularly affected alliance politics, namely (1) sectarian divisions between Sunni, Shi'a, and Kurdish blocs and (2) tensions over the division of authority and resources between federal and regional entities. These dynamics stand at the heart of Iraq's political future and were apparent in alliance relations throughout the US intervention.

In an effort to preserve Iraq's territorial integrity, at one level the USA promoted a strong agenda for reconciliation between sectarian groups. For example, in August 2006 the Bush administration required that Baghdad fulfill eighteen "benchmarks" in order to receive $1.5 billion in Economic Support Funds. Many of the benchmarks promoted Sunni–Shi'a reconciliation.[109] Roughly half of the benchmarks, such as establishing joint US–Iraqi security stations in Baghdad, were fulfilled, at least in part. Others, such as passing critical hydrocarbon legislation to clarify oil rights between competing regional and sectarian groups, were never completed. Baghdad received the supplemental funding anyway.[110]

Also, despite US requests that Baghdad pacify sectarian tensions, key features of the US occupation of Iraq hardened sectarian divisions. For example, the US reliance on Kurdish and Shi'a partners to combat the Sunni insurgency contributed to cycles of violent reprisals between sectarian militias. Even the US-promoted elections reinforced sectarian lines as candidates campaigned along such divisions.[111] Nevertheless, it is interesting that Iraqis allied with the USA complied with requests from US officials at roughly the same rates as local allies in other counterinsurgency interventions. At times, in fact, the USA leveraged sectarian groups against one another to incentivize compliance with US policies, including, for example, pressuring the KRG to compromise on election law and delay referendum on Article 140 regarding the status of Kirkuk.[112] The use sectarianism in Iraq as a way to coerce cooperation from local allies may at first seem like commendably potent coercive diplomacy. But such approaches actually illustrate a long-standing paradox in counterinsurgency interventions, namely that effective short-term tactics to accomplish particular tasks can undermine overarching strategic goals. In Iraq, leveraging sectarian divisions to pressure local allies to comply with US policies may have furthered particular US plans, while hardening the very sectarian divisions the USA was persistently demanding officials in Baghdad and Erbil set aside.

US officials lamenting the difficulties of counterinsurgency intervention have recognized this particular political complexity, at least in abstract terms. Former CIA official Chester Cooper noted in Vietnam, for example, "Americans should help, not substitute for, the government of our ally. To the extent that we Americans 'take charge,' we postpone (and may even jeopardize) the achievement of our ultimate objectives."[113] In other words, accomplishing certain tasks can provide the impression of progress for counterinsurgents, while in the long run contributing to dynamics that undermine instead of support the overall counterinsurgent cause. Leveraging local political dynamics that are harmful to the larger counterinsurgency mission to drive local allies toward compliance with the counterinsurgency protocol of intervening forces similarly may provide the appearance of success, yet undermines the COIN cause by hardening those divisions.

The 2006 *US Field Manual for Counterinsurgency* cautions, "tactical success guarantees nothing ... Tactical actions thus must be linked not only to strategic and operational military objectives but also to the host nation's essential political goals. Without those connections, lives and resources may be wasted for no real gain."[114] In managing its alliance in Iraq, this US COIN precept to have battleground tactics serve strategic goals was violated by multiple policies that ultimately reified sectarianism between Sunnis and Shi'as, against American strategic interests. Using divisions in Iraqi society to promote policy implementation undermined the likelihood that a cohesive, territorially sound Iraq would thrive intact after US withdrawal. In this context local compliance with US requests failed to serve the long-term strategic goals of the USA, a process that underscores the importance of investing in sound political strategies above pressing for more compliant proxies.

Notes

1 US requests for economic reform in Afghanistan were slightly lower than average when compared with rates of compliance across all subjects in Afghanistan, but were not nearly as low as the rates observed in Iraq.
2 Only 5/106 requests were made to the KRG, two of which were also made to Baghdad. This means in the data only 3/106 demands were made to the KRG alone, while the vast majority (103/106) of requests analyzed were for Baghdad.
3 These figures account for 105 of 106 requests. One request, a proposal for a MANPADS (Man-Portable Air-Defense Systems) Reduction Program made on October 11, 2009 (see US Department of State, US Embassy Baghdad, "Post Proposes Shoulder-Fired Missile Abatement Program for Iraq [MAN-PADS Reduction]," 09BAGHDAD2736, October 11, 2009) remains classified. Public sources do not make clear whether the Iraqi government adopted the program.
4 US Department of State, US Embassy Baghdad, "Summing Up Iraq's Year of Anti-corruption," 08BAGHDAD4058, December 29, 2008.
5 Donald P. Wright and Timothy R. Reese, *On Point II: Transition to the New Campaign: The United States Army in Operation Iraqi Freedom, May 2003–January 2005*, US Department of the Army, 1st ed. (Fort Leavenworth, KS: Combat Studies Institute Press, 2008), 156, https://history.army.mil/html/bookshelves/resmat/GWOT/OnPointII.pdf.

6 US Department of Defense, "Backgrounder on Reconstruction and Humanitarian Assistance in Post-War," News Transcript, Office of the Assistant Secretary of Defense (Public Affairs), March 11, 2003, www.au.af.mil/au/awc/awcgate/dod/t03122003_t0311bgd.htm.

7 Office of the Special Inspector General for Iraq Reconstruction (SIGIR), *Hard Lessons: The Iraq Reconstruction Experience*, by Stuart J. Bowen (Washington, DC: US Government Printing Office, 2009), 60, https://permanent.access.gpo.gov/lps108462/Hard_Lessons_Report.pdf;
Anthony H. Cordesman and Emma R. Davies, *Iraq's Insurgency and the Road to Civil Conflict*, Vol. 2, Center for Strategic and International Studies (Westport, CT: Praeger Security International, 2008), 740–1; Andrew Rathmell, "Planning Post-conflict Reconstruction in Iraq: What Can We Learn?," *International Affairs* 81, no. 5 (2004): 1023–4.

8 Rathmell, "Planning Post-Conflict Reconstruction in Iraq: What Can We Learn?," 1024.

9 US Department of Defense, Donald Rumsfeld Files, "Memorandum for the President, Subject: Iraqi Interim Authority," April, 1, 2003, 09-M-2634, Declassified July 7, 2009.

10 Office of the SIGIR, *Hard Lessons: Iraq Reconstruction Experience*, 61.

11 Lt. Gen. Jay Garner (Ret.), "The Lost Year in Iraq," Interview, *PBS Frontline*, October 17, 2006. See also Sharon Otterman, "Iraq: Interim Authority," Council on Foreign Relations, February 16, 2005, www.cfr.org/backgroun der/iraq-interim-authority.

12 James Dobbins et al., *Occupying Iraq: A History of the Coalition Provisional Authority* (Santa Monica, CA: Rand Corporation, 2009), 14, www.rand.org/pubs/monographs/MG847.html.

13 Charles 'Chuck' Costello, interviewed by Haven North, "Oral Histories: The Iraq Experience Project," United States Institute of Peace, October 14, 2004, 11, www.usip.org/sites/default/files/file/resources/collections/histories/iraq/costello.pdf [accessed April 4, 2012].

14 Dobbins et al., *Occupying Iraq*, xvii.

15 Ibid., 14.

16 L. Elaine Halchin, "The Coalition Provisional Authority (CPA): Origin, Characteristics, and Institutional Authorities," Text, UNT Digital Library, September 21, 2006, 8, http://digital.library.unt.edu/ark:/67531/metacrs10420/.

17 Dobbins et al., *Occupying Iraq*, 14.

18 Ibid., xx; Sharon Otterman, "Iraq: Iraq's Governing Council," Council on Foreign Relations, May 17, 2004, www.cfr.org/iraq/iraq-iraqs-governing-council/p7665.

19 John Ehrenberg et al., *The Iraq Papers*, 1st ed. (New York: Oxford University Press, 2010), 184–5; The Coalition Provisional Authority, Archived Web Page, Regulations, Orders, Memoranda and Public Notices, Order Number 1, www.iraqcoalition.org/regulations/.

20 Alissa J. Rubin and Mark Fineman, "Council Moves to Further 'De-Baathify' Iraq," *Los Angeles Times*, September 17, 2003, http://articles.latimes.com/2003/sep/17/world/fg-council17.

21 L. Paul Bremer, "The Lost Year in Iraq," Interview, *PBS Frontline*, October 17, 2006, www.pbs.org/wgbh/pages/frontline/yeariniraq/interviews/bremer.html.

22 Dobbins et al., *Occupying Iraq*, xxvi.

23 Ibid., xxi.

24 Otterman, "Iraq: Iraq's Governing Council"; Dobbins et al., *Occupying Iraq*, 23.

25 Ali A. Allawi, *The Occupation of Iraq: Winning the War, Losing the Peace*, 1st ed. (New Haven, CT: Yale University Press, 2007), 160.

26 James C. Scott, *Weapons of the Weak: Everyday Forms of Peasant Resistance* (New Haven, CT: Yale University Press, 1987), 32–5.

27 Office of the SIGIR, *Hard Lessons: The Iraq Reconstruction Experience*, 120. By March 2004, the CPA and Iraqi Governing Council developed the Transitional Administrative Law (TAL), a document intended to guide Iraq's interim authority structures during the transition period between the CPA and a popularly elected government. The UN had determined Iraq was too politically unstable for immediate elections.

28 Sharon Otterman, "Iraq: The June 28 Transfer of Power," Council on Foreign Relations, February 16, 2005, www.cfr.org/backgrounder/iraq-june-28-transfer-power [accessed April 11, 2012]; L. Paul Bremer and Malcolm McConnell, *My Year in Iraq: The Struggle to Build a Future of Hope*, 1st ed. (New York: Simon & Schuster, 2006), 392–5.

29 Otterman, "Iraq: The June 28 Transfer of Power."

30 Office of the SIGIR, *Hard Lessons: The Iraq Reconstruction Experience*, 157, 165.

31 PBS NewsHour, "Iraq's Transfer of Power," Transcript, June 28, 2004, www.pbs.org/newshour/bb/middle_east/jan-june04/sovereignty_6-28.html [accessed April 11, 2012]. The IIG was a temporary organization. It fulfilled its mandate by holding elections on January 30, 2005 for (1) the 275-seat transitional National Assembly, (2) provincial assemblies in Iraq's eighteen provinces, and (3) a regional assembly in the Kurdish region. This elected government, also known as the Iraqi Transitional Government (ITG) crafted and passed a new constitution on October 15, 2005, a critical step in Iraqi self-rule. Elections for a permanent government were held on December 15, 2005 for Iraq's primary legislative body, the Council of Representatives (COR). And by May 2006 executive branch cabinet officials were appointed and approved by the COR. US Government Accountability Office, "Stabilizing and Rebuilding Iraq – U.S. Ministry Capacity Development Efforts Need an Overall Integrated Strategy to Guide Efforts and Manage Risk," GAO-08-117, October 2007, 15, www.gao.gov/new.items/d08117.pdf [accessed April 11, 2012]. But as the Iraqi state was solidifying during this period, so was the insurgency challenging it.

32 *Voice of America*, "Iraqis Take Control of Last U.S. Prison in Iraq," July 14, 2010, www.voanews.com/a/iraqis-take-control-of-last-us-prison-in-iraq-98504894/172159.html; Tim Arango, "Transfer of Prison in Iraq Marks Another Milestone," *The New York Times*, July 14, 2010, sec. World / Middle East, www.nytimes.com/2010/07/15/world/middleeast/15iraq.html.

33 US Department of State, US Embassy Baghdad, "CODEL Inhofe Meets with TNA Members," 05BAGHDAD5051, December 19, 2005. US intelligence agencies were clamoring for information that could protect US

personnel, while Washington sought to prevent intelligence from being utilized further for sectarian feuding.

34 US Department of Defense, Donald Rumsfeld Files, "Iraq Policy: Proposal for the New Phase – Memo from Secretary Rumsfeld to President George W. Bush," December 8, 2006, MDR 08-M-1641, Special Collections, 2, www.dod.mil/pubs/foi/specialCollections/Rumsfeld/DocumentsReleasedTo SecretaryRumsfeldUnderMDR.pdf. Later that month the Pentagon decided to "mandate the following objectives for Iraq: … 3. Accelerate Iraqi assumption of political and security responsibilities." See US Department of Defense, "A Bold Shift in Iraq Policy – Accelerate the Transition, Sustain the Partnership, and Stabilize the Region," December 4, 2006, Declassified July 19, 2010.

35 American Forces Press Service, "Iraqi Ground Forces Command Assumes Command and Control of 8th Iraqi Army Division," September 1, 2006, www.usf-iraq.com/news/press-releases/iraqi-ground-forces-command-assumes-command-and-control-of-8th-iraqi-army-division [accessed April 13, 2012]; American Forces Press Service, "7th Iraqi Army Division Now Controlled by Iraqi Government," November 3, 2007, www.usf-iraq.com/news/press-releases/7th-iraqi-army-division-now-controlled-by-iraqi-government [accessed April 13, 2012].

36 US Department of Defense, "Subject: Iraq – Illustrative New Courses of Action," November 6, 2006, Declassified October 18, 2010, MDR 10-M-1231, http://library.rumsfeld.com/ [accessed January 7, 2020].

37 US Government Accounting Office, "Stabilizing and Rebuilding Iraq," 9.

38 Ibid. For an example of US influence within Iraqi ministries consider the following summary of progress in Iraqi banking reform from October 2005,

The Embassy (USAID and IRMO [Iraq Reconstruction Management Office]) are two months ahead of schedule in preparing the balance sheets of Iraqi state-owned banks as well as their executive summary assessment. Reform options for the banking sector will be presented to the Ministry of Finance by mid-November. USAID and Treasury are gathering financial information requested by the Minister of Finance to support the development of a state-bank reform program. We have begun to provide direction for the work of the Minister of Finance via a new Joint Task Force on Budget and Finance, led by the Treasury Attache [excised by author for privacy] and the Minister of Finance Allawi. The first meeting on September 13 focused on expenditures in the 2006 budget. Treasury's interagency meeting with Iraqi officials on September 22 on the margins of the World Bank/IMF Meeting also helped to set the agenda and activities for the next task force meeting in Baghdad.

US Department of State, US Embassy Baghdad, "JCRED – Progress Report," 05BAGHDAD4084, October 3, 2005.

39 Ibid.; see also US Government Accountability Office, "Stabilizing and Rebuilding Iraq."

40 US Government Accountability Office, "Stabilizing and Rebuilding Iraq," 11.

41 According to one US Department of State memo, total US personnel at the Iraqi Ministry of Oil was closer to thirty.

A wide range of mission elements provide training and advice in a variety of technical and policy fields. The core on-board Embassy/MNF-I Ministry of Oil engagement team consists of an oil section chief and three specialists in the economic section; the DoE section (2); USAID (5); ITAO (3); Energy Fusion Cell (5 with more enroute); Army Corps of Engineers (GRD) (4); MND-N (2); MNF-I (4). This core staff meets weekly to compare notes and to plan trips to Ministry facilities.

US Department of State, US Embassy Baghdad, "Building Capacity at the Ministry of Oil," 07BAGHDAD3837, November 25, 2007. See also Andrew Kramer, "U.S. Advised Iraqi Ministry on Oil Deals," *The New York Times*, June 30, 2008, sec. International / Middle East, www.nytimes.com/2008/06/30/world/middleeast/30contract.html.

42 Kramer, "U.S. Advised Iraqi Ministry on Oil Deals."

43 Efforts focused on "training and ministerial capacity building programs." US Department of State, US Embassy Baghdad, "Building Capacity at the Ministry of Oil."

44 Kramer, "U.S. Advised Iraqi Ministry on Oil Deals."

45 The 2006 Iraqi national budget made no mention of any costs for national intelligence services. This is highly unusual for a government in the midst of fighting an intense counterinsurgency war, highly dependent on actionable intelligence. Payment for Iraqi agents was not being funneled through Baghdad. US Department of State, US Embassy Baghdad, "CODEL Inhofe Meets with TNA Members."

46 US Department of State, US Embassy Baghdad, "USEB 018: Iraqi Government Signals Tougher Sentences for Criminals and Terrorists," 04BAGHDAD17, July 2, 2004.

47 US Department of State, US Embassy Baghdad, "Ministerial Capacity Surge Assessment," 08BAGHDAD1008, April 1, 2008.

48 Ibid.

49 US Department of State, US Embassy Baghdad, "Joint Reconstruction Efforts Support Baghdad Security Plan," 07BAGHDAD3045, September 11, 2007.

50 US Department of State, US Embassy Baghdad, "UK Ambassador Proposes Mediation for Iraq's Constitutional Review, Other Key Issues," 07BAGHDAD1756, May 27, 2007.

51 US Department of State, US Embassy Baghdad, "Iraqi Politics: Shifting Alliances and the Emergence of Issue-Based Coalitions," 08BAGHDAD3791, December 3, 2008.

52 Secretary Rumsfeld's documents are available at: www.dod.mil/pubs/foi/specialCollections/Rumsfeld/ and http://papers.rumsfeld.com/. Other sources of released documents included the Office of the Secretary and Joint Staff Freedom of Information Act Library, www.esd.whs.mil/FOIA/Reading-Room/Reading-Room-List_2/.

53 See Appendix B.

54 Special Inspector General for Iraq Reconstruction, www.sigir.mil/index.html.

55 United States House of Representatives, *Making Emergency Supplemental Appropriations for the Fiscal Year Ending September 30, 2005, and for Other Purposes, Conference Report* (To Accompany H.R. 1268), (109 H. Rpt. 72), www.congress.gov/congressional-report/109th-congress/house-report/72/1.

56 US Department of Defense, "Measuring Stability and Security in Iraq," Report to Congress, December 2009, www.defense.gov/Portals/1/Docu ments/pubs/Master_9204_29Jan10_FINAL_SIGNED.pdf.

57 The US intervention in Iraq dated from March 20, 2003 to December 15, 2011.

58 The last request analyzed in this study predates American withdrawal from Iraq by twenty-two months for two primary reasons. One, the last document in the *Cablegate* database is from February 28, 2010. Primary source documents currently available to the public for the last two years of the war are limited at best. Due to the recent nature of the Iraq conflict, this information is usually classified. Second, at least in other conflicts including Vietnam and Sri Lanka, the later years of the war bring fewer demands as the foreign state prepares to leave. It would be reasonable to expect the same in Iraq. It is likely the USA made fewer new requests from 2010 to 2011 compared to other years of the occupation.

59 US Department of State, US Embassy Baghdad, "The New Joint Campaign Plan for Iraq." For US communications on Article 140 with GOI see also: US Department of State, US Embassy Baghdad, "AMB, CG and PM Discuss SOFA, SOI, Ambassador's Trip to Erbil, GOI/KRG Relations and Election Law," 08BAGHDAD3031, September 21, 2008.

60 US Department of State, US Embassy Baghdad, "Barzani Agrees to Push for New GOI Article 140 Committee Chair, Invigorate Committee," 07BAGH-DAD2466, July 25, 2007.

61 Dobbins et al., *Occupying Iraq*, 207.

62 John B. Taylor, *Global Financial Warriors: The Untold Story of International Finance in the Post-9/11 World*, 1st ed. (New York: W. W. Norton & Company, 2007), 200.

63 US Department of State, US Regional Embassy Office Hillah, "Taking Steps to Uproot the Statist Legacy in Babil's Agriculture," 09HILLAH24, April 5, 2009.

64 Refugees International, "Iraq: Fix the Public Distribution System to Meet Needs of the Displaced," April 10, 2007, www.unhcr.org/refworld/docid/47a6eef311.html.

65 US Department of State, US Regional Embassy Office Hillah, "Taking Steps to Uproot the Statist Legacy in Babil's Agriculture."

66 Refugees International, "Iraq: Fix the Public Distribution System."

67 US Department of State, US Embassy Baghdad, "Reforming the Public Distribution System – Easier Said Than Done," 09BAGHDAD2621, September 9, 2009.

68 US Department of State, US Embassy Baghdad, "Meeting between Deputy Secretary Negroponte and Deputy Prime Minister Barham Salih," 07BAGH-DAD1991, June 17, 2007.

69 Anthony Cordesman and Adam Mausner, "How Soon Is Safe? Iraqi Force Development and 'Conditions-Based' US Withdrawals," Center for Strategic and International Studies, February 5, 2009, 78, www.ecoi.net/en/file/local/ 1305282/1002_1236977391_csis-iraq.pdf.
70 See Chapter 10. It is likely that Vietnam and Afghanistan (USA) have relatively lower rates of capacity cited as issues in compliance processes because Americans aren't asking those regimes for reforms they believe are beyond the abilities of the Vietnamese and Afghans. Whereas in Sri Lanka, it is very likely that Colombo had fewer capability constraints.
71 See Chapter 10, Table 10.1.
72 James Glanz and Eric Schmitt, "Iraq Attacks Lower, but Steady, New Figures Show," *The New York Times*, March 12, 2008. sec. International/Middle East. www.nytimes.com/2008/03/12/world/middleeast/12iraq.html.
73 US Department of State, US Embassy Baghdad, "Are the Iraqi Prisons Working Yet? – An Assessment of Ministry of Justice/Iraqi Corrections Service (ICS) Operations," 09BAGHDAD2384, September 4, 2009.
74 US Department of State, US Embassy Baghdad, "Badr Leader Agrees the Militia Should Demobilize," 06BAGHDAD930, March 21, 2006.
75 US Department of State, US Embassy Baghdad, "Demarche to Iraqi Interior Minister on Site 4," 06BAGHDAD2842, August 7, 2006.
76 US Department of State, US Embassy Baghdad, "Human Rights in the Interior Ministry; Director Complains of Marginalization, Threats," 07BAGHDAD1377, April 23, 2007.
77 Ibid.
78 See Appendix C.
79 US Department of State, US Embassy Baghdad, "Implementation of Recommendations on Personal Protective Services: Status Report Update #1."
80 "Iraq – U.S.A., Status of Forces Agreement (Nov 2008)," Article 12, Jurisdiction, www.state.gov/documents/organization/122074.pdf [accessed May 10, 2012].
81 Schmidt, "Immunity Gone, Contractor in Iraq Sentenced to Prison."
82 Stephen Biddle, Jeffrey A. Friedman, and Jacob N. Shapiro, "Testing the Surge: Why Did Violence Decline in Iraq in 2007?," *International Security* 37 no. 1 (2012): 7–40.
83 Hampton Stephens, "Analysts, U.S. Officials Differ on Maliki's Plans for Sons of Iraq," *World Politics Review*, September 11, 2008, www.worldpoliticsreview .com/articles/2651/analysts-u-s-officials-differ-on-malikis-plans-for-sons-of-iraq; Agence France-Presse (AFP), "US Buys 'Concerned Citizens' in Iraq, but at What Price?," October 16, 2007, http://afp.google.com/article/ALeq M5iMzKGlyT_ahqRjtyXrAUrKIQLncA [accessed May 10, 2012]. See also US Department of State, US Embassy Baghdad, "Maliki on Concerned Local Citizens, Strategic Partnership Declaration, and Large-Scale Detainee Amnesty," 07BAGHDAD3721, November 13, 2007.
84 US Department of State, US Embassy Baghdad, "CG and CDA Discuss Foreign Fighters and Syria, Turkey and the PKK, and UNSCR with PM."
85 US Department of State, US Embassy Baghdad, "Concerned Local Citizens Program: Securing Communities," 08BAGHDAD164, January 22, 2008.

86 Office of the SIGIR, "Sons of Iraq Program: Results Are Uncertain and Financial Controls Were Weak," 18.

87 Ibid.

88 US Department of State, US Embassy Baghdad, "Maliki on Cabinet Shake-Up, Return of Tawafaq, and Major Legislative Challenges," 08BAGH-DAD166, January 22, 2008.

89 Office of the SIGIR, "Quarterly Report to Congress," October 30, 2008, 49.

90 Lourdes Garcia-Navarro, "Bitterness Grows amid U.S.-Backed Sons of Iraq," *National Public Radio*, June 24, 2010, www.npr.org/templates/story/story.php?storyId=128084675.

91 US request that Baghdad match SOI/CLC funds coded as compliance based on SIGIR Reports, integration into security forces coded as partial compliance.

92 Office of the SIGIR, "Quarterly Report to Congress," July 2008, 96, www.globalsecurity.org/military/library/report/sigir/sigir-report-2008-07.htm.

93 Lauren Johnston, "'Dirty Bomb' Depot Dispute," *CBS News*, July 9, 2004, www.cbsnews.com/news/dirty-bomb-depot-dispute/; Alissa J. Rubin and Campbell Robertson, "U.S. Helps Remove Uranium from Iraq," *The New York Times*, July 7, 2008, www.nytimes.com/2008/07/07/world/middleeast/07iraq.html?_r=0.

94 US Department of State, US Embassy Baghdad, "Meeting with the Minister of Science and Technology Regarding the Tuwaitha Site," 07BAGH-DAD2028, June 20, 2007; US Department of State, US Embassy Baghdad, "Meetings with GOI Officials regarding the Tuwaitha Site," 07BAGH-DAD1960, June 14, 2007.

95 US Department of State, US Embassy Baghdad, "Tuwaitha Request Letter," 07BAGHDAD2924, August 31, 2007; US Department of State, US Embassy Baghdad, "Tuwaitha Update: Meeting at Most – GOI Letter to IAEA Signed," 08BAGHDAD36, January 4, 2008; US Department of State, US Embassy Baghdad, "CG and CDA Discuss Foreign Fighters and Syria, Turkey and the PKK, and UNSCR with PM."

96 US Department of State, US Embassy Baghdad, "Fallujans Mobilized for Election amid Increased Tension in City." But not all US policymakers backed the reconstruction of Fallujah. Two days before his November 2006 resignation, US Secretary of Defense Donald Rumsfeld wrote in a classified memo,

> Clearly, what U.S. forces are currently doing in Iraq is not working well enough or fast enough. Following is a range of options: Stop rewarding bad behavior, as was done in Fallujah when they pushed in reconstruction funds, and start rewarding good behavior. Put our reconstruction efforts in those parts of Iraq that are behaving, and invest and create havens of opportunity to reward them for their good behavior. As the old saying goes, "If you want more of something, reward it; if you want less of something, penalize it." No more reconstruction assistance in areas where there is violence.

> "Rumsfeld's Memo of Options for Iraq War," *The New York Times*, December 3, 2006, sec. International / Middle East, www.nytimes.com/2006/12/03/world/middleeast/03mtext.html.

97 Office of the SIGIR, "Information on Government of Iraq Contributions to Reconstruction Costs," SIGIR 09-018, April 29, 2009, www.dtic.mil/docs/citations/ADA508864.

98 Michael Schwartz, *War without End: The Iraq War in Context* (Chicago: Haymarket Books, 2008), 117.

99 US Department of State, US Embassy Baghdad, "The 2006 Iraq Budget," 06BAGHDAD955, March 23, 2006; Schwartz, *War without End*, 115–17.

100 Anne Garrels, "Long-Awaited Fallujah Rebuilding Shows Promise," *National Public Radio*, January 23, 2008, www.npr.org/templates/story/story.php?storyId=18319948.

101 Ibid.

102 Office of the SIGIR, "Falluja Waste Water Treatment System. Falluja, Iraq," SIGIR PA-08-144-08-148, October 27, 2008, vii, www.dtic.mil/docs/citations/ADA529001.

103 Compliance in the chart includes partial compliance and full compliance, from both Baghdad and Erbil.

104 Office of the SIGIR, *Hard Lessons: The Iraq Reconstruction Experience*, 155.

105 For example, US demands such as, "take measures to reduce sectarian violence" and "ensure the Baghdad Security Plan does not protect sectarianism," July 25, 2007, "GOI to cease sectarian appointments and politically motivated prosecutions," and seek the "prosecution of government and security officials who break the law" (ending protections based on sectarian affiliation). See US Department of State, US Embassy Baghdad, "The New Joint Campaign Plan for Iraq"; US Congress, Public Law 110–28, "U.S. Troop Readiness, Veterans' Care, Katrina Recovery, and Iraq Accountability Appropriations Act," May 25, 2007, 13, www.congress.gov/110/plaws/publ28/PLAW-110publ28.pdf. See also US Department of State, US Embassy Baghdad, "Allawi Back in Iraq; Seeks to Build Centrist Coalition," 07BAGHDAD612, February 20, 2007.

106 US Congress, "110th Congress Public Law 28, Conditioning of Future United States Strategy in Iraq on the Iraqi Government's Record of Performance on Its Benchmarks," May 25, 2007, 13, www.congress.gov/110/plaws/publ28/PLAW-110publ28.pdf.

107 US Government Accountability Office, "Securing, Stabilizing, and Rebuilding Iraq, Iraqi Government Has Not Met Most Legislative, Security, and Economic Benchmarks," GAO-07-1195, September 7, 2007, 44–5.

108 US Department of State, US Embassy Baghdad, "Iraq's Council of Representatives," 08BAGHDAD495, February 21, 2008.

109 Kenneth Katzman, "Iraq: Reconciliation and Benchmarks" (Congressional Research Service, RS21968, May 12, 2008).

110 US Government Accountability Office, "Securing, Stabilizing, and Rebuilding Iraq"; Kenneth Katzman, "Iraq: Politics, Governance, and Human Rights" (Congressional Research Service, RS21968, November 10, 2011); Kenneth Katzman, "Iraq: Reconciliation and Benchmarks" (Congressional Research Service, RS21968, June 5, 2008, August 4, 2008, and September 3, 2008).

111 Jeremy M. Sharp, "The Iraqi Security Forces: The Challenges of Sectarian and Ethnic Influence," Congressional Research Service, RS22093, January

18, 2007, www.cfr.org/iraq/crs-iraqi-security-forces-challenge-sectarian-ethnic-influences/p12616 [accessed May 15, 2012].

112 US Department of State, US Embassy Baghdad, "Delayed Gratification: Election Law Adopted," 09BAGHDAD3157, December 7, 2009; US Department of State, US Embassy Baghdad, "KRG Officials on Article 140 and Kirkuk," 10BAGHDAD64, January 11, 2010.

113 Cooper et al., "The American Experience with Pacification in Vietnam – Volume I – An Overview of Pacification (Unclassified)," 33.

114 US Army and Marine Corps, *Counterinsurgency, FM 3-24*, 1–28.

5   The USA in Afghanistan

A western diplomat said Karzai had given "two fingers" to the western donors who had pumped millions of dollars into establishing democratic elections in the country.   —Jon Boone, *The Guardian*

Alliances can be surprisingly fractious. Negotiating dissimilar interests and institutions inevitably creates friction between security partners. But the alliance between the USA and its partners in Kabul was at times exceptionally tense. Afghan President Hamid Karzai harshly blamed the USA for a myriad of Afghan problems, condemning the US mission for civilian deaths and sluggish development.[1] In exchange, US officials were known to label Karzai "paranoid" and "conspiratorial."[2] As US Ambassador to Afghanistan Karl Eikenberry reflected, Karzai was the kind of partner that required "frank ... (and perhaps, at times, confrontational) dialogue."[3] Peter Galbraith, the deputy chief of the UN mission to Afghanistan, found Karzai so difficult that he insinuated the Afghan president had a drug habit.[4]

Inimical relations between the USA and Afghan partners were not limited to President Karzai. In 2011 US Treasury advisers "described the working conditions at DAB [Afghanistan's National Bank] as 'hostile',," while US and International Security Assistance Force (ISAF) soldiers were being targeted and killed with alarming regularity by the Afghan forces they were training. According to the *Long War Journal*, local Afghan allies killed sixty-one ISAF personnel in 2012 alone, a figure that accounted for 15 percent of total coalition deaths in 2012.[5] Investigating the US-Afghan partnership, I explore these tensions and the structures that fostered or inhibited Afghan compliance with requests from Washington.

Summary Findings

The Government of the Islamic Republic of Afghanistan (GIRoA) complied with US policy requests about as often as local allies in other large-scale counterinsurgency interventions. Fifty of 148 (33.8 percent) of US

Table 5.1 *Timeline of the Afghan–US alliance during the US intervention*

2001	In response to the 9/11 attacks, the USA invades Afghanistan. In the months that follow, American and North Atlantic Treaty Organization (NATO) forces wage a successful campaign against the Taliban with the support of the Northern Alliance.
2002–3	The US-backed Northern Alliance gains ground against the Taliban.
2004	Hamid Karzai becomes the first democratically elected president of Afghanistan.
2005–6	Utilizing their support networks in the border regions of Pakistan, the Taliban increase their attacks. In October 2006, US troops in Afghanistan are transferred to the NATO ISAF.
2009	In February, in order to counter a resurgent Taliban, US President Barack Obama commits 17,000 more troops to Afghanistan. In November, President Karzai wins a second term in a controversial election.
2011	Al-Qaeda leader Osama Bin Laden is killed in Pakistan by US forces in May.
2014	In May, President Obama announces a timetable to withdraw all US forces from Afghanistan by 2016. In September, Ashraf Ghani becomes the new president of Afghanistan.
2015	In March, President Obama announces that troop withdrawal will be delayed at President Ghani's request. In May, Afghan officials hold informal peace talks with Taliban representatives in Qatar.
2019	US President Donald Trump presses US peace talks with the Taliban, refusing to set a timetable for US withdrawal but commenting, "great nations do not fight endless wars."

requests to Kabul were complied with, 42/148 (28.4 percent) resulted in partial compliance, and 56/148 (37.8 percent) went unfulfilled. This is similar to the rates of compliance observed in other large-scale counterinsurgency interventions: approximately equal parts full compliance, partial compliance, and noncompliance.

Furthermore, as in other counterinsurgency interventions, compliance by Kabul with US policy requests was affected by the interaction of converging or diverging interests between Washington and Kabul and US dependency on Afghan partners to implement the policy. Interestingly, in Afghanistan one aspect of this interaction effect did not hold up, and a new pattern, not observed in other interventions, emerged. Surprisingly, free riding by local allies in Afghanistan was relatively infrequent under the same conditions observed in other conflicts, including Iraq, Vietnam, and Sri Lanka. Hypothesis 3B in Chapter 2 proposes that free riding by local counterinsurgency (COIN) allies is expected when (1) the interests of local allies and intervening allies align over a proposed policy and (2) the intervening ally is capable of implementing the policy independently. But this hypothesis was not supported in Afghanistan. In Iraq, for example, the rate of full compliance with US requests in this category of interests and US dependency was 3/24 requests (12.5 percent). This figure is significantly lower than Iraq's overall rate of full compliance, 33.0 percent, a finding supporting Hypothesis 3B that predicts that compliance from local allies is unlikely under those conditions due to incentives to free ride. In Afghanistan, on the other hand, the rate of compliance with US requests under the same conditions was three times that of Baghdad with 15/40 (37.5 percent). This figure is in fact greater than the average rate of compliance for all US requests to Kabul from 2002 to 2010 of 50/148 (33.8 percent), which contradicts the predicted outcome of noncompliance.

Why were Afghans allied with the USA less likely to free ride under conditions predicted to foster free-riding behaviors? The answer may be potentially rooted in Afghanistan's notorious problems with corruption.[6] One reason that Kabul appears to have frequently chosen to comply, at least in part, with requests from US officials that US forces could implement independently, is because these proposed policies often offered funding opportunities that Kabul valued above the political or military outcomes produced by the USA implementing the request unilaterally. Partially complying enabled Kabul to influence which patronage networks were connected with these funds. Despite endemic corruption in Iraq and Vietnam, local allies tended to accept US forces unilaterally implementing programs where interests aligned. Such free riding enabled local partners in Iraq and Vietnam to reap the benefits of policy implementation without incurring the costs. However, leaders in Kabul appeared to have a different calculation and did not reliably free ride under those conditions. The GIRoA made decisions that prioritized opportunities to access cash instead of allowing the USA to implement policies that would provide political or security benefits. In this study, I unsurprisingly find that all local allies are profit seeking, but Kabul consistently prioritized access to money above other interests. If given the choice between free riding on US efforts to implement a policy aiming to strengthen Kabul's hold on power, or to participate in the policy and ensure particular networks were on the payroll, Kabul tended to opt for the latter, whereas local allies in other interventions tended to opt for the former and free ride on policies implemented unilaterally by intervening patrons. This suggests that US allies in Afghanistan emphasized the USA as a source of funding, but Kabul was less apt to rely on American patrons as a source of political-military support. The tendency of Kabul to emphasize cash relates to the palpable hostility between Washington and Kabul, suggesting that leaders in Kabul did not trust the USA to reliably implement policies that would benefit their regime, instead likely viewing money as the most dependable way to use the US intervention for their benefit. These dynamics only deepened animosities as US policymakers lamented the patronage networks that defined local Afghan politics.

Noting the magnitude of corruption in Afghanistan is not meant to imply that other local proxies in other interventions were not corrupt. Graft and nepotism are endemic in virtually all large-scale military interventions, and are typically deepened by the processes of intervention. The injection of vast amounts of money by large, wealthy allies into relatively small economic systems creates massive disruptions, producing widespread incentives for nepotism and graft. Consider, for example, how one US State Department official summarized the situation in Vietnam:

Corruption pervades all aspects of Vietnamese life, and it is brazenly practiced. For example ... we now have given South Vietnam $30 million a year for refugee relief. In my many conversations with the hard-pressed American refugee personnel, it was estimated that only half of the supplies ever reach the refugee. The officials of the Government of South Vietnam, and the Province Chiefs supported by them, have the keys to the warehouses, and they keep much of the goods for themselves. Each refugee is supposed to receive the equivalent of $45 for resettlement. It was estimated to me by a U.S. official adviser to the refugee program that 75 per cent of this amount is siphoned off before it reaches these people.[7]

Corruption was rampant in the US wars in Vietnam and Iraq. Yet in Afghanistan bribery was so ubiquitous that one Afghan prosecutor described that in his district corrupt practices are "so rampant that 'it is more like looting.' People are losing patience, he added, and this makes it easier for the Taliban to convince people to work against the government."[8] As observed in this chapter, indeed in Afghanistan corruption was so prolific that it provided a different logic to local allies, uniquely affecting their decisions to comply with or defy certain requests from US allies, and interestingly, inspiring them to opt for a potential cash payout instead of free riding. The other proposed hypotheses and the interaction between interests and dependencies were significant and consistent with other interventions.

Constructing Kabul: Hamid Karzai and the Evolution of the US–Afghan Alliance

The previous chapter on Iraq discussed authority in Baghdad and the steady transition from US domination of the Iraqi state to increasing Iraqi sovereignty. The following section on Afghanistan similarly discusses the role of US advisers in Afghan bureaucracies in order to clarify how much legal authority Kabul exercised within its institutions. As cautioned by the US Senate Committee on Foreign Relations, Kabul "cannot be held accountable for processes over which it has little or no control."[9] However, there are interesting differences between Iraq and Afghanistan regarding local autonomy. In Iraq, for example, Americans directly administered local bureaucracies under the Coalition Provisional Authority (CPA). In Afghanistan, however, while Americans ran or supported institutions that competed with local bureaucracies, they never dominated the Afghan state in a fashion similar to the CPA. The Afghans at the start of the US intervention had more legal sovereignty than the Iraqis, but were at times marginalized by US funding being channeled around, rather than through Afghan state institutions.

At the onset of the Afghan intervention the USA instituted conflicting directives. On the one hand, US military operations aiming to eliminate al-Qaeda made Washington a major player in shaping Afghanistan's political future. On the other hand, the George W. Bush administration was wary of getting bogged down in an Afghan quagmire, similar to the British or Soviet experience. In order to expel the Taliban while limiting US commitment, US policymakers aimed to initially rely on the Northern Alliance as a readymade Afghan alternative to the Taliban. US officials then hoped the United Nations would lead post-Taliban development efforts. As Secretary of Defense Donald Rumsfeld wrote in October 2001, "The USG [US government] should not agonize for post-Taliban arrangements ... the USG should begin discussing international arrangements for the administration of Kabul."[10]

The December 2001 conference in Bonn established the Afghanistan Interim Administration (AIA), appointing Hamid Karzai as chair with relative consensus.[11] The AIA convened an "Emergency Loya Jirga" (Emergency Grand Council) in June 2002 to extend Karzai's term by two years, as well as renaming the AIA the Transitional Islamic State of Afghanistan (TISA), selecting 1,500 Afghan representatives.[12] In January 2004, a 500+ person Loya Jirga passed a constitution, and presidential elections held in October 2004 affirmed Karzai's presidency.

During this lengthy transition (2001–4) the Afghan government, such as it was, did not govern much of Afghanistan. Furthermore, according to Barnett Rubin, "even within Kabul, Karzai had only limited control over his own government, many of whose top officials led militias that had fought or were still fighting against the Taliban with U.S. support."[13] The varied regional, ethnic, and militia leaders that had been cobbled together to form the post-Taliban Afghan state vied to maintain control of their territories, resisting federal consolidation. Regional commanders (often also referred to as Afghanistan's warlords) pressed Kabul for concessions.[14] Even in December 2012, eleven years after the Taliban were driven from Kabul, Ismail Khan, a key anti-Taliban regional commander, called on independent militias to fight the Taliban, a direct challenge to Kabul's state security forces that had been organized to unite anti-Taliban forces and protect Afghan territory.[15] Due to Kabul's tenuous grasp of authority, US officials frequently negotiated directly with the regional warlords. Bypassing the Karzai administration was more efficient, but further weakened Kabul's reach and legal authority.[16]

In Afghanistan, the USA and its local allies were remarkably uncoordinated partners in the initial years of the intervention. Divisions between US, international, and Afghan political organizations planted the seeds for tense relations. Six months after the USA sent forces into Afghanistan, Rumsfeld wrote to his deputies: "I may be impatient. In fact I know I'm a bit impatient. But the fact that Iran and Russia have plans for Afghanistan and we don't concerns me."[17] This was not just sarcasm. Peter Tomsen, US Special Envoy to the Afghan resistance from 1989 to 1992, similarly noted a fundamental failure to coordinate US strategy and approaches to the Afghan state during the intervention following September 11, 2001.

> Inside Afghanistan, the U.S. embassy, the U.S. military, and the CIA often operated in separate stovepipes and at cross purposes. There was no integrated U.S. policy enforced by the White House to coordinate all U.S. agency efforts. Sometimes one agency backed Afghans who competed with Afghan rivals supported by other agencies, a problem that the Soviet leadership in the 1980s constantly struggled with but never resolved ... While the CIA and U.S. military were boosting warlords

the US Department of State backed the regime in Kabul.[18]

The lack of US strategic and institutional coherence regarding Afghanistan took a toll on the counterinsurgency effort. Afghanistan's 70 percent illiteracy rate and lack of state administrative traditions made it difficult to establish bureaucracies. And low institutional capacity motivated US officials to circumvent Kabul altogether. According to the US Senate Committee on Foreign Relations,

> Most international donors, including the United States, channel much of their aid "off-budget," meaning it does not go through the Afghan Government ... the U.S. Government is working to meet its Kabul Conference commitment to fund up to 50 percent of our aid "on-budget" by FY 2012 from approximately 21 percent in FY 2009, 35 percent in FY 2010, and 37-45 percent in FY 2011.[19]

This tendency to fund projects by circumventing Kabul was also a response to Kabul's endemic corruption. Consider, for example, US Agency for International Development (USAID)'s decision to use

Catholic Relief Services instead of Afghan government agents to distribute funds for a cash-for-work program in Ghor Province because "villagers reported that some officials involved with government-run programs ... diverted commodities to their own pockets. It is unclear if this is blatant corruption or whether officials felt food they took was an expected return for their involvement in the organizing the activities."[20]

Although efficient in the short term, the practice of excluding Kabul from the aid process institutionalized ad hoc practices, hindered local institutional development, and reinforced Kabul's reputation for ineffectiveness. From 2001 to 2009 less than 20 percent of aid was channeled through Afghan government institutions.[21] This meant parallel governance institutions were competing with Kabul for political influence. For example, while many considered Provincial Reconstruction Teams (PRTs) helpful for embedding the coalition military presence in particular locations, others, including Hamid Karzai, critiqued the PRT system for perpetuating reliance on governance programs run by entities outside the Afghan government. As Kenneth Katzman summarized:

President Karzai criticized the PRTs as holding back Afghan capacity-building and called for their abolition as "parallel governing structures" ... The Afghan government and some outside organizations have long argued that the PRTs have hampered Afghan government efforts to acquire the skills and resources to secure and develop Afghanistan on its own. USAID observers say there has been little Afghan input, either into development project decision making or as contractors for facility and other construction.[22]

The problem of Kabul's tentative authority was compounded, according to the Special Inspector General for Afghanistan Reconstruction (SIGAR), by a chronic shortage of educated Afghans, which reportedly created "donor practices of hiring Afghans at inflated salaries, [drawing] otherwise qualified civil servants away from the Afghan Government."[23] These salaries could be "10 to 20 times the amount of base government salaries ... Moreover donors provide salary support outside the Afghan planning and budgeting process, thereby hindering the GIRoA's ability to assume responsibility for managing its civil service."[24] Furthermore, while corruption often motivated US decisions to circumvent Kabul, several important early decisions by US officials opened the floodgates for corrupt practices in Kabul, including advocating that the coffers of the state be positioned in the hands of the president and a few ministries without significant checks and balances, or failing to institutionalize systematic audits or budget oversights. Corruption made Kabul an unwieldy ally at times, but US policymakers contributed to these unsavory characteristics by failing to institutionalize practices to guard against overt graft early on in the intervention.[25]

Nevertheless, elements of the US mission understood that long-term US success in Afghanistan hinged on the survivability of the regime in Kabul and worked to support policies connected to Kabul despite the difficulties. As Peter Tomsen noted, even though the US Department of Defense and Central Intelligence Agency (CIA) readily bypassed Kabul, the US Department of State was dedicated to bolstering Kabul. This is supported by the data regarding US requests to Kabul, indicating that US officials not only asked the GIRoA to help when their participation was required, but also encouraged Kabul's involvement in tasks the USA could implement otherwise. The majority of US requests required GIRoA participation – 93/148 (62.8 percent). This is a figure similar to Iraq – 71/106 (67.0 percent) and Sri Lanka – 51/79 (64.6 percent).[26] This indicates that even when the USA could implement certain policies without Kabul, the US State Department would nevertheless often include Kabul to bolster local capacity and build legitimacy.

The 2009 strategic review deepened US investment in Afghan institutional development through a "civilian surge" that increased advisers to the GIRoA from 400 in 2009 to 1,300 by 2012.[27] The role of US advisers varied widely depending on conditions in the Afghan office. The adviser guide to the Afghan Ministry of Interior (MOI), overseeing the Afghan police, rated MOI offices according to competency. In May 2011 the average office ranked on the level of "Ministry organization cannot accomplish the function without significant coalition assistance,"[28] and no MOI office was yet ranked "self-reliant."[29] By April 2012 that rating had not improved.[30] The Ministry of Defense performed a little better. By April 2012 the MoD "was assessed as requiring some coalition assistance to accomplish its mission."[31]

Furthermore, after more than a decade of intervention the United States maintained full responsibility for training and arming the Afghan military through the NATO Training Mission and Combined Security Transition Command – Afghanistan (NTM-A/CSTC-A). US personnel were integrated in Afghan security institutions, but unlike in Iraq under the CPA, or Afghanistan under Soviet control, US advising in Afghanistan did not deny Afghan legal sovereignty. In fact, US documents regularly detail the complaints of US officials struggling to persuade Afghan partners to follow US advice. Preparing for a February 2010 meeting on electoral reform, US Ambassador Eikenberry wrote, "Karzai's willingness or unwillingness to consider our perspectives and serious efforts to build a sustainable system of representative governance will be a good indication of his willingness to partner with us in the year ahead."[32] US officials could not readily dictate Kabul's governance policies due to policies to respect Afghan legal sovereignty, yet American advisers observed Kabul struggling year after year with basic bureaucratic functions.

In order to gather data on the US-Afghan alliance, over 1,000 US primary source documents (2,500+ pages) were analyzed to identify 148 unique demands from the US government to Kabul from 2001 to 2010. As in the investigation of Iraq, these documents were found in three locations. First, I examined declassified documents located on US government agency websites, including Secretary Rumsfeld's documents in the Special Collections Library.[33] A second source of material came from declassified documents released to the National Security Archive Afghanistan/Pakistan project. Last, the largest source of data for this chapter was US State Department documents published in the *Cablegate* database released by *Wikileaks*.[34] Its database contains 2,961 cables from the US Embassy, detailing correspondence between US and Afghan allies.[35]

In addition, reports published by SIGAR often clarified the outcome of US requests.[36] Since the USA usually provided funding for programs it asked the GIRoA to adopt, SIGAR materials provide reliable accounts of Afghan performance. Another helpful resource was a set of reports the US Department of Defense was mandated to produce, "Report on Progress toward Security and Stability in Afghanistan." These reports monitored progress, growth, and developments in Afghan governance.[37]

Documents analyzed date from September 11, 2001 to September 11, 2012, which fails to cover a large portion of the war after 2012. The 148 US requests to Afghan partners date from January 10, 2002 to February 26, 2010, providing coverage of the war until early 2010. The last request analyzed predates American withdrawal by more than nine years for two primary reasons. One, as of the date of this analysis, the war is still ongoing. Second, the last document contained in the *Cablegate* database is dated February 28, 2010. Primary source documents from the later years of the war are limited since public, high-level US-Afghan correspondence is currently unavailable.

What Kinds of Policies Did the USA Ask for in Afghanistan?

The USA proposed a wide variety of requests to the regime in Kabul, including legislation on women's rights, counternarcotics protocols, anti-corruption measures, passing legislation to protect forests, postponing elections, and engaging in military operations. Table 5.2 provides a summary of US requests to Afghan partners identified in US Department of State records.

Table 5.2 *Summary of US requests to Afghan allies*

	Year	US request to Afghan allies
1.	2002	Take the lead providing national security, developing national security forces
2.	2002	Military factions to come under the control of interim administration
3.	2002	Include women and minorities in Loya Jirga planning
4.	2002	Assist US forces in finding Taliban leaders
5.	2002	Work toward poppy elimination
6.	2002	Public statements against opium
7.	2002	Programs to encourage women in government
8.	2002	Provide adequate funding for daily functioning of Ministry of Justice (MOJ) and Ministry of the Interior (MOI)
9.	2002	US funds should not fall to drug traffickers
10.	2002	Authorize use of Darulaman site for Afghan police or reimburse the USG
11.	2002	Require anti-narcotics certification for trainees
12.	2002	Retain trained personnel for 2+ years
13.	2002	Convert currency at highest rate
14.	2002	Exempt imports and USG-designated personnel working on police construction from taxes
15.	2002	Accept US personnel for police expansion and grant immunities
16.	2002	Prepare a new constitution
17.	2002	Elections by June 2004
18.	2003	Ministry of Rural Rehabilitation and Development (MRRD) – Recommendations for Alternative Livelihood Projects
19.	2003	Written USG approval for projects over $25K
20.	2003	Ministry of Finance to follow specific currency transfer protocol for allocating grant funds
21.	2003	Responsibility for low-level enemy combatants
22.	2003	Enforce poppy ban
23.	2003	Drug intelligence, investigative, and interdiction units
24.	2003	Autonomous and independent drug enforcement units in MOI
25.	2003	Review and strengthening of narcotics laws
26.	2003	Training programs for police, prosecutors, and judges for counternarcotics
27.	2003	Anti-corruption, narcotics-related programs across government levels
28.	2003	Transparency in the MOI for counternarcotics
29.	2003	Establish long-term development programs in post-eradication areas
30.	2003	Legislation on anti-money laundering and forfeiting
31.	2003	Drug treatment and rehabilitation programs – Ministry of Public Health
32.	2003	Arrest, charge, and incarcerate increasing numbers of drug traffickers
33.	2003	Services and vocational training for former child laborers and soldiers
34.	2003	Pay Senior Police Advisor (SPA) from SPA funds transferred from USG to MOI
35.	2003	Authorize access to locations in major cities for police training facilities
36.	2004	Senior advisor for police reform
37.	2005	Sign US Afghanistan Charter of Partnership
38.	2005	Customs reform
39.	2005	Increased customs revenue
40.	2005	References to eradication in Afghanistan Compact

Table 5.2 (*cont.*)

	Year	US request to Afghan allies
41.	2006	Ministry of Counternarcotics (MCN) assign personnel to Poppy Elimination Program (PEP) teams
42.	2006	Operate the Counter Narcotics Justice Center (CNJC)
43.	2006	Authorize access for training facility in Bamiyan
44.	2006	Reduce land area contaminated by mines
45.	2006	Restructure government ministries
46.	2006	National appointments mechanism
47.	2006	Annual performance-based reviews
48.	2006	Ratify UN Convention Against Corruption
49.	2006	Census and statistical capabilities
50.	2006	GIRoA to contribute to cost of elections
51.	2006	Single national identity document
52.	2006	Implementation of the National Action Plan for Women
53.	2006	Expanded judicial capacity for each province
54.	2006	Separate prison facilities for women and juveniles
55.	2006	Land disputes
56.	2006	Increased education, reporting, and monitoring on human rights
57.	2006	Pollution control – regulations
58.	2006	Curriculum overhaul
59.	2006	Human resource study
60.	2006	Disaster response program
61.	2006	Assistance to the disabled
62.	2006	Assistance to refugees
63.	2006	Strategy for privatization
64.	2006	Restructure state-owned commercial banks
65.	2006	Provide Afghanistan National Development Strategy
66.	2006	Provide reports on benchmark progress
67.	2006	Reforms toward World Trade Organization (WTO) accession
68.	2006	Border management initiative
69.	2006	Prepare and submit to council key commercial laws
70.	2006	Reform business/trade licensing and customer protections
71.	2006	Technical regulations inspectorate
72.	2006	Budget execution
73.	2006	Work with Islamabad on formula for cross-border Jirgas
74.	2007	Respond to detainee transfer proposal within two weeks
75.	2007	Investigate bank robberies and prosecute guilty parties
76.	2007	Reforms at the MOI – Police
77.	2007	Electricity development projects
78.	2007	Reform tax administration
79.	2007	Attorney General Sabit to follow legal procedures for anti-corruption initiatives
80.	2007	Transparent anti-corruption review process
81.	2007	Engage religious figures in public messaging
82.	2007	Year-round poppy elimination program
83.	2007	Focus poppy elimination on Helmand

Table 5.2 (*cont.*)

	Year	US request to Afghan allies
84.	2007	Counternarcotics trust fund – disseminate development money faster
85.	2007	Give Pak border issue to Organization of the Islamic Cooperation (OIC) to study
86.	2007	Public information campaigns against poppy, narcotics, and heroin
87.	2007	MCN to pay PEP team salaries after six months
88.	2007	Provide office space and law enforcement personnel
89.	2008	Domestic law enforcement agencies assigned to Financial Transactions and Reports Analysis Center of Afghanistan (FinTRACA)
90.	2008	Hawala registration and oversight
91.	2008	Separate Hawala regulatory division in the Central Bank's financial supervision department
92.	2008	Open an Afghan customs academy
93.	2008	Do not include inflammatory language in the Joint Coordination and Monitoring Board (JCMB)'s Plan
94.	2008	Transfer remaining Afghan nationals at US Naval Station Guantánamo Bay (GTMO) to Afghanistan
95.	2008	Resettle Uighurs from GTMO
96.	2008	Cluster Munitions Convention (CCM) – broadly interpret Article 21, Ratification
97.	2009	Karzai – stop publicly lambasting the USA for civilian casualties
98.	2009	Suspend assessment of fees for security companies on US contracts
99.	2009	Change varied fees for private security companies
100.	2009	Lift 500-person employee ceiling on private security companies
101.	2009	Rollback Shia Family Law provisions regarding women
102.	2009	Coordinated and longer hours at Af-Pak border crossing points
103.	2009	Operational independence for FinTRACA
104.	2009	Appoint director general to FinTRACA
105.	2009	Ratify anti-money laundering law
106.	2009	MOU on transit trade with Pakistan
107.	2009	Issue statements on Taliban abuses
108.	2009	More transparent licensing procedures for private security companies
109.	2009	Stop insisting USA supports Abdullah in the election
110.	2009	Meetings with US officials – productive US-Afghan engagement
111.	2009	Do not return Dostum to Afghanistan
112.	2009	Provide international community compact
113.	2009	Afghan anti-corruption commission with international participation
114.	2009	Discourage violence over election results
115.	2009	Affirm support for election process, do not interfere with Independent Electoral Commission (IEC) or Electoral Complaints Commission (ECC)
116.	2009	Merit-based appointments of key ministers – reject Ishmael Khan
117.	2009	GIRoA "Compact" with Afghan people
118.	2009	Deliver early results on "Compact"
119.	2009	Host ministerial conference on Afghanistan in Kabul
120.	2009	No immediate revision of Status of Forces Agreement (SOFA)

Table 5.2 (*cont.*)

	Year	US request to Afghan allies
121.	2009	Expand mobile money services – MOI and other Ministries
122.	2009	Approve Community Defense Initiative (CDI) program
123.	2009	Pass forest law
124.	2009	Allow for legal lumber sale
125.	2009	Postpone elections to 2011
126.	2009	Fund and expand the CDI
127.	2009	Satisfy the mining trigger required for Heavily Indebted Poor Countries (HIPC) completion
128.	2009	Conclude Afghanistan–Pakistan Transit Trade Agreement (APTTA)
129.	2009	Prosecute Paktya Governor Juma Khan Hamdard
130.	2009	Engage in Operations in Marja – governance
131.	2010	Announce integration strategy before London Conference
132.	2010	Reintegration paper edits
133.	2010	President Karzai to create a political party
134.	2010	Reforms to better formulate influential political parties
135.	2010	Jirgas on peace and integration policy – increased positions for women and minorities
136.	2010	Increased district governance in Spin Boldak
137.	2010	Torkham Gate transfer Minister of Commerce and Industries (MOCI) land
138.	2010	Accept US package approach for District Delivery Program (DDP)
139.	2010	Lessons learned from DDP
140.	2010	Remove Specified Governors – Abobaker
141.	2010	Remove Specified Governors – Salangi
142.	2010	ECC to include two international members
143.	2010	Da Afghanistan Breshna Sherkat (DABS) as commercialized national electricity utility
144.	2010	Strengthen oversight of NGOs and charities
145.	2010	Withdraw "sweeping" amendments to the 2005 Electoral Law
146.	2010	Change in IEC leadership
147.	2010	Retain internationals from the United Nations Assistance Mission in Afghanistan (UNAMA) and ISAF on the Candidate Board (Disbandment of Illegal Armed Groups, or DIAG)
148.	2010	Election reform announcements and action

In order to paint a broad picture, we can divide US requests to Afghan partners into six general categories of subjects, which reflect several key components of US counterinsurgency strategy. Similar to other interventions, issue areas of requests were not found to be significant in determining the likelihood that a request would be complied with or not as other structural factors had a greater influence on compliance outcomes than the general subject of the proposed policy.

Issue Areas of US Requests to Afghanistan

1. *Development* – projects or activities intended to support economic growth and expand infrastructure. Includes projects to reduce areas contaminated by landmines and allow for the legal sale of lumber. 27/148 (18.2 percent) of US requests. There is potential overlap with requests classified as economic reforms, such as pricing for electricity services, an economic policy, with growth and development goals.
2. *Economic reform* – actions intended to change economic policies. Includes reforms regarding currency conversion policies, import policies, and customs reforms. 25/148 (16.9 percent) of US requests.
3. *Military reform* – actions intended to change military policies and institutions. Includes increasing security forces, and command and control protocols. 3/148 (2.0 percent) of US requests.[38]
4. *Military strategy* – actions intended to guide military forces in the execution of the war effort. Includes policies regarding assisting US forces and weapons employed in the field. 4/148 (2.7 percent) of US requests.
5. *Political reform* – actions intended to change government policies and institutions. Includes law enforcement, governance, bureaucratic protocols, constitutional reforms, sectarian issues, and reconciliation programs. 89/148 (60.2 percent) of US requests.
6. *Political-military counterinsurgency strategy* – actions intended to implement counterinsurgency strategy. No requests fell into this category for the US war in Afghanistan in the available sources.

Figure 5.1 compares rates of Afghan compliance across categories of issues addressed in US requests.

Explaining the Compliance and Defiance of Local Allies in Afghanistan

As detailed in Chapter 2, I propose four variables to help explain the likelihood of Afghan compliance with US policy requests. Namely, local capacity to implement the request, the interests of Washington and Kabul converging or diverging over the request, the dependency of US forces on Afghan partners to implement the request, and a significant enemy offensive. However, the US war in Afghanistan has not (yet) experienced a significant, coordinated acute enemy offensive similar to the Tet Offensive in Vietnam. The violence thus far in Afghanistan has been significant, but more diffuse, such that it did not serve as a critical shock to local partners. However, similar to other interventions,

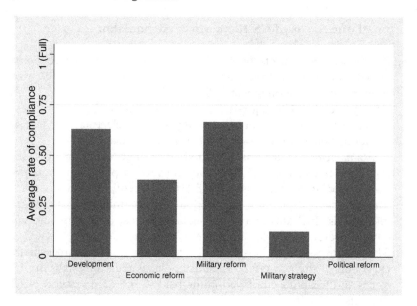

Figure 5.1 Rates of Afghan compliance with US requests, by subject area, 2002–10

I hypothesize that there is an interaction effect between interests and dependencies producing a particular pattern of compliance outcomes that is key to understanding local free-riding behaviors and to specify when intervening forces are likely to have coercive leverage over local clients. In this section I will address each variable in the context of US-Afghan politics.

Local Capacity Limitations – Decreased Likelihood of Compliance

As expected, a lack of local governance capacity is strongly correlated with a lower likelihood of compliance with US requests in Afghanistan. Kabul experienced significant bureaucratic shortcomings, as 26.4 percent of US requests (39/148) to Kabul ran up against a specific capacity limitation that affected the compliance process. The 26.4 percent frequency of capacity being cited as an issue was comparable to Vietnam, where 27.6 percent of requests discussed capacity shortcomings. US officials asked Kabul for essential reforms, such as issuing a federal identity document, creating separate prisons for women and children, or establishing rudimentary bookkeeping protocols in government agencies. As one high-level US official summarized, "Governance is the most conceptually difficult area of the [US-Afghan] Strategic Partnership."[39] Coping with issues in failing governance required tackling the toughest of Afghanistan's political problems. US efforts to promote reforms confronted fundamental challenges in Afghanistan such as widespread illiteracy. The request, for example, that Kabul require anti-narcotics certification for new Afghan police recruits was complicated by the fact that 70 percent of new recruits were illiterate and could not meaningfully comprehend the certifications presented to them.[40]

Additional Factors Related to Afghan Capacity – Wartime Environment and Local Internal Politics

Since Kabul was making decisions regarding compliance while facing a serious military and political threat from the Taliban insurgency, it is important to recognize that Afghan capacity was also affected at times by this wartime environment. Factors such as combat operations, refugee influxes, and enemy military activities were cited as an influence in 11/148 (7.4 percent) of US requests. This included, for example, the 2006 demand for Kabul to organize a national census to provide much-needed information for aid distribution, which was canceled by the Afghans in 2008 due to security concerns.[41]

Local internal politics also affected capacity as illustrated by a 2006 US request for increased Afghan private sector development and privatization of the national airline, Ariana. Efforts at privatization failed, in part, due to competing domestic interests, including parliamentarian resistance.[42] Similarly, influence of local bureaucratic tensions in determining outcomes of requests became evident to US policymakers managing Afghanistan's borders. The US Embassy reported in 2005, "When the [Afghan] Border Police worked by themselves, they generated serious disagreements with the Ministry of Finance (that controls the Customs Service). This disagreement has blocked further progress of the unilateral border plan designed earlier."[43] Two years later, the relevant Afghan interagency situation had only deteriorated, further stalling border management reforms requested by Washington. The US State Department reported in 2007, "At issue during this meeting were the roles of [Government of Afghanistan] ministries at the border and in part the lack of mutual support and misunderstandings between MoF and MoI [Ministries of Finance and Interior]."[44]

Afghan interagency squabbling affected the process of complying with US-requested reforms in 24.5 percent of requests. This was a smaller percentage than in Iraq, where 37.7 percent of requests had a record of local domestic political complications, but more than the 9.5 percent in

Vietnam, and 11.4 percent in Sri Lanka. However, local politics had varied effects on compliance outcomes. For example, against US advice, Afghan President Hamid Karzai "Afghanized" the Electoral Complaints Commission (ECC), a watchdog group intended to deal with fraud in Afghanistan's shaky electoral process. The US demanded the ECC contain at least two international members in order to prevent the watchdog organization from becoming a rubber stamp for pro-Karzai election fraud. In this instance, however, President Karzai was thwarted by the Afghan upper house of Parliament, which rejected Karzai's plan to "Afghanize" the institution. The GIRoA effectively complied with US demands, despite the president's efforts to do otherwise.[45]

Interests and Dependency

Kabul's compliance with US policy requests was influenced by the interaction of US dependency on Kabul to implement the proposed policy and whether or not US and allied Afghan interests converged or diverged over the policy. If the USA could act unilaterally and Afghan and US interests diverged, there was a higher probability of compliance. In such cases Afghan officials were more willing to agree, since the proposed policy could be implemented without them, leaving Kabul isolated or undermined. A reverse relationship was observed if US unilateral action was not possible and US and Afghan interests converged. There was a higher likelihood of compliance under such circumstances, as Kabul was likely to adopt the requested actions for the sake of its own interests. If the two parties' interests diverged and the USA could not implement the reform without Afghan assistance, there were higher rates of noncompliance. The GIRoA had no interest motivating compliance and would likely suffer no immediate consequences for failing to comply.

Interests Diverge and the USA Is Dependent on Local Allies for Policy Implementation: Predicted Afghan Noncompliance

If the interests of Washington and Kabul diverged and the USA was dependent on Kabul to implement the request, noncompliance was likely. This finding was robust in Afghanistan. There are dozens of examples of the Afghans not complying with demands they determined were costly if US forces did not have an independent capacity to implement the policy. The 2002 request that US personnel be exempted from Afghan income tax is one example. The US Status of Forces Agreement signed between the USA and Afghanistan stipulates that US "military and civilian personnel, contractors, and contractor personnel shall not be liable to pay any tax or similar charge assessed within Afghanistan."[46] However, GIRoA officials held that this did not cover income gained through foreign or aid contracts and pressed US personnel in Afghanistan to pay income tax. Said Mubin Shah, Deputy Minister of Finance for customs and revenue, explained with some irony, "The international community should be happy we are implementing the rule of law ... We should work together to solve this problem and impose the rule of law, because a lot of foreign contractors are evading their taxes."[47]

Interests Diverge and the USA Is Not Dependent on Local Allies for Policy Implementation: Predicted Afghan Compliance

As in other large-scale counterinsurgency interventions, Washington was able to coerce its Afghan partners by threatening to unilaterally implement reforms that went against the immediate interests of the GIRoA. Under those circumstances the Afghans often participated in order to protect their interests to the greatest extent possible, rather than have US allies implement the request unilaterally and impose potentially higher costs. These dynamics can be illustrated with an additional example of US requests relating to corruption reform.

By 2009, Washington had grown frustrated with faltering Afghan efforts to curb corruption, and aimed to establish a new anti-corruption commission. Washington was insisting on an organization with direct international participation, instead of previous models that relied ultimately on Afghans.[48] By late 2010, Kabul complied and established the Monitoring and Evaluation Committee (MEC), a board consisting of three Afghans and three international appointees. President Karzai was careful to appoint to the MEC Mohammed Yasin Osmani, a Karzai loyalist and former chief of the High Office of Oversight (HOO or HOOAC), despite international resistance to Osmani. *The Guardian* newspaper reported that under Osmani HOOAC "got nowhere."[49] According to one former official, Osmani's inclusion in the MEC "was a real kick in the teeth for internationals and signaled Karzai had no intention of going after those who were corrupt."[50]

However, the Americans were able to get a limited level of compliance from Kabul regarding the MEC and international participation because the Chief of the Monitoring Group (MEC) was being established by the United Nations (with the international and Afghan committee designating the MEC chair), not the Afghan government. US officials appear to have grown weary of anti-corruption organizations made meaningless because they were controlled by Kabul. The international community signaled to President Karzai that they were determined to establish a group other than the HOO to investigate corruption in Afghanistan. Because the UN was taking a large role in establishing the MEC, the USA could threaten, at least in part, to establish an anti-corruption organization in Afghanistan to publicize corruption in the GIRoA that would be outside the influence of regime elites.[51]

Kabul had more to lose from failing to participate than from joining and influencing the MEC. SIGAR has reported that the GIRoA has attempted to interfere with certain MEC tasks. According to SIGAR, the Department of State lauded

the progress the MEC has made in a difficult environment and noted that its work is receiving proper attention from the Afghan government ... [Nevertheless] the HOO has also continued to interfere with the MEC's work, although with diminishing effect ... Although President Karzai had attempted to limit the scope of the MEC in February 2012, the Afghan government did not prevent the MEC from conducting its work this quarter.[52]

Interests Converge and the USA Is Dependent on Local Allies for Policy Implementation: Predicted Afghan Compliance

The scenario in which US and Afghan interests converge, but the USA does not have unilateral capability to implement a policy, predicts compliance because local allies must participate in order to benefit from the policy. Since local participation is required by the nature of the task at hand, there is no option to free ride. This logic was evident in Afghanistan. To use an example that once more details Afghanistan's serious problems with corruption, in 2006 Kabul was asked by Washington to move forward to ratify the UN Convention Against Corruption (UNCAC). The GIRoA had signed the UN convention in 2004, but it had not yet been ratified by the legislature. By August 2008 both houses of parliament in Afghanistan completed the ratification process, effectively complying with the US demand.[53]

There were several reasons why there were more benefits for Kabul than costs for signing and ratifying the convention, something that only they (not the Americans) could do in Afghanistan. First, with 140 signatories, the UNCAC is a popular convention. States such as North Korea, Somalia, and Chad that have refused to sign the agreement have risked deepening their reputations for corruption. Second, for individual lawmakers in Afghanistan, taking a public stance against signing or ratifying the anti-corruption agreement would be tantamount to publicly endorsing Afghanistan's system of corruption, a prevalent and poorly concealed system, but nevertheless a despised part of Afghan political life. Third, there are aspects of the convention that can potentially benefit Kabul, or the elites in office; these include asset recovery, where signatories are able to pursue government monies that have been hidden in private bank accounts in other countries. This could potentially be used as a tool to punish particular individuals, while allowing others to act with impunity.

Fourth, the institutions tied to the convention, including the HOO or HOOAC, were established under the guidance of Afghan President Karzai. In 2010, SIGAR stated that HOO

has not been able to do more because it suffers from significant gaps in capacity: It lacks sufficient independence ... in contravention of generally accepted standards and ethical codes for oversight organizations, both the Director General and Deputy Director General of the HOO are also employed as presidential advisors within the Office of the President ... holding two government positions simultaneously can, and in this case does, create a conflict of interest.[54]

Lastly, policymakers in Afghanistan could balance pressure by ratifying the convention, but neglecting to follow up on promised reforms. This was a regular occurrence in Afghanistan (and Iraq; see Chapter 4). As the State Department summarized in an investment risk summary in 2012:

Based on the Penal Code, corruption is a serious criminal act; articles 260 to 267 state that anyone accepting or giving a bribe can be charged with criminal acts. While these anti-corruption laws exist, enforcement has been very limited. President Karzai created the High Office of Oversight for the Implementation of Anti-Corruption Strategy ("HOO") to coordinate anti-corruption measures for the government; this office, however, does not control penalties and fines and has been largely ineffective. Afghanistan acceded to the United Nations Convention Against Corruption (UNCAC) in August 2008, but is not a party to the OECD Convention on Combating Bribery of Foreign Public Officials. The early 2011 establishment of the Independent Monitoring and Evaluation Committee (MEC) for Anti-corruption should assist the Afghan Government in assessing its compliance with UNCAC. However, questionable Afghan Government commitment to supporting the MEC and early administrative challenges plague the new organization.[55]

Signing the UN Convention against Anti-Corruption would placate international donors by depicting Kabul as doing "something" about Afghanistan's rampant corruption, yet GIRoA officials could evade meaningful implementation. Since Afghan participation was required for the policy to be implemented, and since the GIRoA demonstrated interest in having the convention ratified, Kabul complied with the request to sign.

Interests Converge and the USA Is Not Dependent on Local Allies for Policy Implementation: Predicted Afghan Noncompliance

When the interests of intervening and local forces converge over a given policy, and the participation of local allies is not required to fulfill the policy, local partners are likely to fail to comply because there is an opportunity to free ride. Local partners declining to comply can push intervening forces to implement the policy unilaterally, enabling local allies to reap the benefits of the policy without paying any associated costs. Indeed, there are many examples of Afghans free riding off American efforts in Afghanistan.[56] US requests to Kabul regarding investment in energy infrastructure and the issuance of a single national identity card are two examples. Kabul opted to push the USA to invest in energy infrastructure, and has relied on US iris scanning and biometric data to compile a national identity database.[57] Kabul could benefit from the US efforts without making significant investment.

However, compared to other large-scale counterinsurgency interventions, under these conditions there was significantly less free riding by local allies in Afghanistan. Consider, for example, that Saigon's and Baghdad's rates of full compliance when free riding is expected, namely when the interests of allies converge and the USA is able to unilaterally implement the policy, were 10/35 requests (28.6 percent) and 3/24 requests (12.5 percent) respectively. This is significantly lower than rates of full compliance with US requests in Vietnam and Iraq, namely 43.8 percent and 33.3 percent respectively. In contrast, the Afghan government complied with US requests expected to inspire Afghan free riding at a rate of 37.5 percent, a greater frequency than the overall rate of full compliance for all requests in the Afghan conflict, 33.8 percent.

Why were the Afghans different than Vietnamese, Sri Lankan, and Iraqi allies by failing to capitalize on these opportunities to free ride on intervening partners in these instances? The answer appears entrenched in Afghanistan's chronic issues with corruption. Kabul frequently chose to comply with requests that the USA could implement independently because these policies created opportunities to access US funding and to influence which Afghan actors would benefit from these policies. Complying, at least in part, with these requests allowed Afghan administrators to access cash and augment their influence over decision-making processes.

US demands in 2002 that Kabul adopt a strong position against poppy cultivation provides one example. The Karzai administration made opium cultivation and trafficking illegal, published eradication goals, and created multiple counter-drug enforcement organizations.[58] The creation of the Afghan-controlled Counternarcotics Directorate (CND) and Poppy Eradication Force (PEF) allowed Afghans to influence which organizations would be shut down and which could continue to operate illicitly. According to the Congressional Research Service, "Many Afghan government officials are believed to profit from the drug trade ... Corrupt practices range from facilitating drug activities to benefiting from revenue streams that the drug trade produces."[59] If counternarcotics decisions were left to the US Drug Enforcement Agency (DEA), Afghans connected to the GIRoA that profit from the drug trade would risk their lucrative arrangements. However, by exerting greater influence over the law enforcement process, at the request of the Americans, Afghans in charge could decide who gets punished, as well as potentially also turning these positions of influence into new opportunities to profit.

This logic was evident across multiple drug-related US policy requests to Kabul, including a request to enforce the poppy ban.[60] Several Afghan officials took advantage of this opportunity because it allowed them to choose which opium operations could be shut down and which would thrive.[61] This selective enforcement of the ban and its potential profits for GIRoA officials was particularly well documented in Nangarhar Province from 2005 to 2007.[62] Afghan officials in Nangarhar were able to solicit payments and shut down competitors; additionally, limited opium production boosted prices (and profits), while putting Nangarhar on the US list of "good performers" due to the net decrease of drug production. According to one report, this increased foreign aid, "which was all too easily siphoned off" by GIRoA officials.[63] Similar situations emerged in other drug initiatives encouraged by the USA, including requests that the Ministry of Rural Rehabilitation and Development take responsibility for alternative livelihood projects,[64] and the establishment of Drug Intelligence, Investigative and Interdiction Units.[65]

Aside from drug issues, other types of policy requests were complied with, at least partially, by Afghan officials preferring to cooperate with US programs in order to access cash. In October 2006, for example, the USA sought to transfer responsibility for border management to the GIRoA.[66] There were capacity limitations, but the GIRoA had strong motivation to control and profit from border customs revenues. In Herat, for example, the US Embassy reported "tremendous leakage in imported cargo, much of which either bypasses the customs house via other transport routes or else is sold into the local market before reaching the customs house."[67] The Inspector General reported that Afghan customs organizations were "rife with corruption," with "up to 70 percent of potential border revenue lost because of corruption."[68] High-level officials from the Ministry of Finance (Customs) and Ministry of the Interior (Border Police) had strong motivation to be involved to augment both licit and illicit profits, which led to bureaucratic infighting for control over border management. In 2009, border management was still dominated by US forces, but the Afghan Ministries of Finance and Interior were playing an increasingly important role. The Afghan bureaucrats profited by partially complying with the US request, rather than free riding and letting the Americans carry the entire load.[69]

While customs collection and counternarcotics may be prone to corrupt practices, even issue areas that are not traditionally associated with corrupt networks, such as natural disaster relief, became sources of illicit profits in Afghanistan.[70] Kabul's disaster management bureaucracies relied on "provincial governors and warlords" to distribute aid to rural areas. This gave local officials "tremendous power to control allocations of aid and its disbursement."[71] According to the US State Department, GIRoA provincial administrators were documented as conducting "wholesale robbery" of humanitarian assistance intended for flood victims.[72] Washington sought to channel disaster relief aid through the Afghan government in order to promote popular support for the regime, a strategic goal in the counterinsurgency campaign. However, GIRoA "compliance" with an American request for aid distribution likely harmed instead of strengthened Kabul's legitimacy. The US Ambassador commented that unless these corrupt officials were removed, "efforts in the area of humanitarian assistance meant to gain the support of the local populace will more likely serve to underline local corruption and further erode local support."[73]

Afghan compliance in situations of potential free riding, where local allies in other interventions tended to fail to comply, is at least partially motivated by incentives to protect and augment predatory behavior and corrupt practices. In Afghanistan, the motivation for profit was more influential than motivations to free ride on US efforts or to reap political or military benefits from US unilateral action. Free riding risked forfeiting lucrative illicit sources of income and political power.

Again, Afghans were not unique among small partners in seeking to protect corrupt institutions, but Kabul appears exceptional in the severity and depth of corrupt institutions. Although the influx of cash that accompanies foreign troop deployment makes corruption an enduring problem in any counterinsurgency intervention, in Afghanistan, it reached unprecedented proportions. As one US DEA agent noted, "The big problem in this country is criminality and corruption. It's huge. It's just rampant. It's rife. It's beyond anything we've seen in Colombia or Mexico or any place else."[74] According to the United Nations Office on Drugs and Crime, in 2009, Afghans paid approximately $2.5 billion in bribes, the equivalent of 23 percent of Afghanistan's GDP.[75] In fact, according to the UN, the two greatest sources of revenue in Afghanistan are drugs and bribes.[76] When ranked, Afghanistan has continued to be found more corrupt than Iraq, and is listed second to notoriously corrupt states such as Somalia and North Korea.[77]

Furthermore, reports exchanged between the US Embassy in Afghanistan and policymakers in Washington also document Afghanistan's pervasive corruption. For an example, one December 2009 report detailed a US-backed pilot program to pay Afghan police using mobile-money, an electronic, cell phone-based fund dispersal method intended to prevent superiors from siphoning cash intended to pay subordinates. Superiors accustomed to skimming off a percentage of their subordinates' salaries were unhappy about the program. In fact, the Afghan National Police commander in Jalrez in Wardak Province called the company providing technical assistance for the mobile-money program, and directly asked for "a cut of his subordinates' salaries. After his request was refused, he ordered his subordinates to give him their phones and PIN numbers. On November 22, the commander collected 45 phones and demanded payment of their salaries."[78] Other instances include the "numerous occasions" where the Attorney General and President Karzai authorized the release of detainees based on personal connections and nepotism detailed in US State Department records:

In April, President Karzai pardoned five border policemen who were caught with 124 kilograms of heroin in their border police vehicle. The policemen, who have come to be known as the Zahir Five, were tried, convicted and sentenced to terms of 16 to 18 years each at the Central Narcotics Tribunal. But President Karzai pardoned all five of them on the grounds that they were distantly related to two individuals who had been martyred during the civil war … Separately, President Karzai tampered with the narcotics case of Haji Amanullah, whose father is a wealthy businessman and one of his supporters. Without any constitutional authority, Karzai ordered the police to conduct a second investigation which resulted in the conclusion that the defendant had been framed.[79]

Each of the 148 identified requests from the USA to its partners in Afghanistan were coded to yield a picture of broad trends related to the proposed variables and the likelihood of Afghan compliance. An ordered probit statistical model is provided in Table 5.3, accounting for capacity limitations and the proposed interaction effect between interests and dependencies in Afghanistan. Additional information is provided in Appendix C that specifies additional checks conducted, including fixed effects by year, and considers interests alone without dependencies as an alternative explanation. Additionally, the data was reanalyzed using robust standard errors clustered by the issue area of request, which

Table 5.3 *Ordered probit – local Afghan compliance with US requests, 2002–10, interaction effect between US dependency on Kabul for policy and interests*

Local Afghan capacity	US unilateral potential / dependency	Allied interest convergence/ divergence[a]	Interaction with dependencies and allied interest
-1.092^{**}	1.245^{**}	2.192^{**}	-1.880^{*}
(0.240)	(0.348)	(0.286)	(0.441)

Notes: $^{*}p < 0.05$; $^{**}p < 0.01$.

[a] The convergence/divergence of interests is produced by combining two variables related to interests, "Private Benefit for Local Allies," and "Threat to Private Benefits for Local Allies." The variable was created in order to have a robust measure of Kabul's interests based on costs and benefits, instead of substituting one (costs or benefits) for interest while excluding the other. Since the USA is making the request, the model assumes Washington is interested in the request. Therefore convergence/divergence of interests between allies is based on the interests of the local ally (Kabul). Interest is determined by taking into account costs (estimated by the variable, "Threat to Private Benefits for Local Allies") and benefits (estimated by the variable "Private Benefits for Local Allies") for Kabul. These are measured by using documentary evidence. For more on coding see Chapter 3.

provided a statistically and substantively comparable result to the results presented in Table 5.3.

Afghan Compliance with US Requests over Time

For 50/148 (33.8 percent) of US policy requests the regime in Kabul complied in full, and for 42/148 (28.4 percent) it complied in part; 56/148 (37.8 percent) of US requests were not fulfilled. If partial compliance and full compliance are combined, Afghans were compliant for 92/148 (62.2 percent) of US demands. These are similar percentages to the rates observed in other large-scale counterinsurgency interventions.[80]

Figure 5.2 charts the number of requests made by the USA by year, juxtaposed with compliance from Kabul with those requests.[81] For 2004, 2005, and 2010, Kabul's rate of compliance exceeds new US requests, as Afghans fulfilled requests made by US officials in previous years. The choppy nature of the number of US requests and the level of Afghan compliance over the course of the war in Afghanistan (see Figure 5.1) compared with Iraq's steady rise and fall reflects differences in the nature of the two American interventions.

There is a significant drop in the number of US requests from 2004 to 2005. There are several potential reasons why. First, as previously noted,

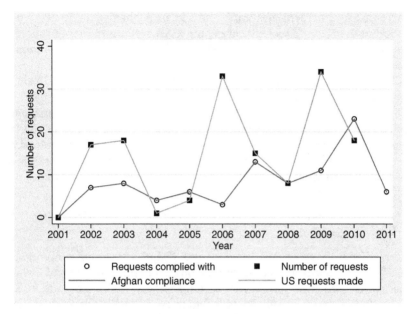

Figure 5.2 Rates of US requests and Afghan compliance, 2001–11

Kabul's low institutional capacity motivated US agents to circumvent the regime and implement projects without the Afghan government's participation.[82] Second, as Iraq grew increasingly violent during this period, Afghanistan became less of a priority. Although US military officials refuted claims that the mission in Afghanistan was compromised because of Iraq, other officials argued that Afghanistan languished in Iraq's shadow.[83] Representative Bill Delahunt stated in 2006, "We became too focused on Iraq, and we forgot about Afghanistan."[84] Ambassador Peter Tomsen too argued that Afghanistan undeniably took a backseat to the conflagration in Iraq:

Air assets and sophisticated intelligence equipment were moved to the Iraq theater. Even the large U.S. Army generators at the Kandahar Airport were taken out of Afghanistan. A U.S. CentCom official explained why [to Tomsen]: "We're simply in a world of limited resources, and those resources are in Iraq … Anyone who tells you differently is blowing smoke."[85]

The spike in US demands noted in 2006 is largely a product of "The Afghanistan Compact," a document requiring Kabul to implement a long list of reforms, benchmarks, and priorities. The 2006 increase in US requests is also a response to the resurgence of the Taliban and an increase in US troops. Between 2005 and 2006 Taliban suicide attacks increased 400 percent and armed attacks in Afghanistan climbed from 1,558 to 4,542.[86] As violence grew, the USA submitted more proposals to its Afghan partners to compel them to reform in an effort to counteract losing ground in the war.

The decrease in US requests to Afghanistan in 2008 was likely a symptom of the change in US administrations, while the spike in 2009 is a consequence of President Obama's strategic review, the start of the surge, and an American reinvestment. The approximately 30,000 American troops in Afghanistan at the end of 2008 more than doubled to 63,000 by 2010.[87] As US expenditures in Afghanistan rose, the number of demands Americans put on Afghan partners similarly increased.

Conclusion – The Curious Pattern of Free Riding in Afghanistan

The alliance between Washington and Kabul was often contentious, providing a shaky foundation for any joint strategic effort. David Galula warned how such disjointed efforts could severely hamper counterinsurgency.[88] US efforts in Afghanistan to augment the US-allied Afghan state were less substantial and less clearly defined than in the wars in Vietnam or Iraq. After ten years of war, for example, in his 2011 congressional testimony, retired Lt. Gen. David W. Barno noted the "overall U.S. advisory effort in Afghanistan today is fragmented, non-standard, decentralized and largely lacking any bureaucratic power or centralized senior leadership."[89] This confused investment in Afghan institutions began at the very outset of the war, with Bush administration officials wavering between a limited counterterrorism mission against al-Qaeda and an extended post-Taliban counterinsurgent state-building endeavor – a debate that carried into the Obama and Trump presidencies as well. While markedly increasing US advising and assistance to Kabul, the Obama administration remained wary of too much investment that might risk trapping the USA in perpetual support of Kabul and corruption. At the same time it recognized that a limited US mission would almost certainly leave Kabul enfeebled, open to Pakistani and Taliban machinations, and fertile ground for spreading anti-American terrorism.[90]

US Counterinsurgency Field Manual 3-24 claims local allies as fundamental, stating, "U.S. forces committed to a COIN effort are there to assist a HN [Host Nation] government. The long-term goal is to leave a government able to stand by itself. In the end, the host nation has to win on its own. Achieving this requires development of viable local leaders and institutions."[91] Yet US institutional policies such as excluding

Karzai to avoid inefficiencies and corruption counteracted this long-term goal. Persistent circumvention of Kabul led to the emergence of competing, mostly non-Afghan, organizations providing goods and services, which added to the political fragmentation of Afghanistan. The political spaces created by absent and uneven governance in Afghanistan enable the Taliban to assume control and expand insurgent authority. The Taliban's ability to provide seminal order and basic services, like judicial hearings, in contrast to the local government's failing plays significantly in the Taliban's favor. As reporter Robert Fisk summarized regarding the political situation, "Nobody supports the Taliban, but people hate the government."[92]

A regime such as Kabul, with low capacity and high corruption, can be expected to be particularly prone to free riding – the common practice of small allies aiming to capitalize on a partnership by shifting shared burdens to larger partners. However, Kabul demonstrated a disinclination for free riding under conditions that inspired other local COIN partners to free ride, namely when the interests of local and intervening allies converge, and intervening forces can potentially unilaterally implement the policy, local allies in Afghanistan usually did not take advantage of those opportunities, instead, opting to participate in order to tap into potential sources of revenue. Such full or partial compliance with US policy requests enabled Afghan officials to access additional funding and ensure that their favored networks were also connected to lucrative projects. With time these corrupt practices hardened into institutions, becoming increasingly difficult to dislodge. Like many components of counterinsurgency, there is inertia and feedback to these political dynamics. Corruption in Kabul motivated US policymakers to circumvent the GIRoA, in effect convincing GIRoA officials that the USA would not reliably back the regime, confirming choices in Kabul to emphasize the USA as a funding source.

This helped produce a weak, profiteering regime in Kabul and exceptionally tense relations between local and foreign allies. US military efforts to target the Taliban are failing to address the political fissures in Afghanistan that create space for the Taliban. As historian Vanda Felbab-Brown noted,

the persistent inability to establish good governance, even in areas repeatedly cleared by ISAF and the ANSF [Afghan National Security Forces], has often made any security gains highly ephemeral. The state's manifestation, though meager, has often been outright malign[ant] from the perspective of many Afghans. It has been characterized by rapaciousness, nepotism, corruption, tribal discrimination, and predatory behavior from government officials and power brokers closely aligned with the state.[93]

The politics of alliance management in counterinsurgency is complex, but critical work, as turning a blind eye to corrupt practices within local regimes can undermine all other counterinsurgency efforts and produce opportunities for insurgents to assert themselves politically as an alternative to a failing system.

Notes

1 Tahir Khan, "Karzai's Anti-US Comments Win Little Support," *The Express Tribune*, March 16, 2013, http://tribune.com.pk/story/521471/karzais-anti-us-comments-win-little-support/; Matthew Rosenberg, "False Claims in Afghan Accusations on U.S. Raid Add to Doubts on Karzai," *The New York Times*, January 25, 2014; Alissa J. Rubin and Rod Nordland, "U.S. General Puts Troops on Security Alert after Karzai Remarks," *The New York Times*, March 13, 2013, www.nytimes.com/2013/03/14/world/asia/karzais-remarks-draw-us-troop-alert.html.

2 BBC, "Wikileaks Cables Say Afghan President Karzai 'Paranoid'," December 3, 2010, www.bbc.com/news/world-south-asia-11906216; Elizabeth Rubin, "Karzai in His Labyrinth," *New York Times Magazine*, August 4, 2009, www.nytimes.com/2009/08/09/magazine/09Karzai-t.html.

3 US Department of State, US Embassy Kabul, "Karzai on the State of U.S.-Afghan Relations."

4 Fox News, "White House Slams Karzai for Latest Anti-American Outburst," April 5, 2010, www.foxnews.com/politics/2010/04/05/white-house-slams-karzai-latest-anti-american-outburst/; Jon Boone, "US-Afghan Relations Sink Further As Hamid Karzai Accused of Drug Abuse," *The Guardian*, April 7, 2010, www.guardian.co.uk/world/2010/apr/07/hamid-karzai-galbraith-substance-abuse; US Department of State, US Embassy Kabul, "Karzai on ANSF, Cabinet, and 2010 Elections."

5 Office of the Special Inspector General for Afghanistan Reconstruction (SIGAR), "Limited Interagency Coordination and Insufficient Controls over U.S. Funds in Afghanistan Hamper U.S. Efforts to Develop the Afghan Financial Sector and Safeguard U.S. Cash," July 20, 2011, 7, www.sigar.mil/pdf/audits/2011-07-20audit-11-13.pdf; Bill Roggio and Lisa Lundquist, "Green-on-Blue Attacks in Afghanistan: The Data," *The Long War Journal*, August 23, 2012, www.longwarjournal.org/archives/2012/08/green-on-blue_attack.php.

6 United Nations Office on Drugs and Crime (UNDOC), "Corruption in Afghanistan: Bribery As Reported by Victims," January 19, 2010, www.unodc.org/documents/data-and-analysis/Afghanistan/Afghanistan-corruption-survey2010-Eng.pdf; "Corruption Index 2011 from Transparency International: Find Out How Countries Compare," *The Guardian*, December 1, 2011, sec. News, www.guardian.co.uk/news/datablog/2011/dec/01/corruption-index-2011-transparency-international; Sarah Chayes, *Thieves of State: Why Corruption Threatens Global Security*, 1st ed. (New York: W. W. Norton & Company, 2016).

7 US Department of State, "Sen. Edward Kennedy's Report on His Recent Trip to Vietnam. For Komer and MacDonard from Grant."

8 US Department of State, US Embassy Kabul, "PRT/Lashkar GAH-Naw Zad Prosecutor Says Official Corruption Causes Alienation," 06KABUL874, March 1, 2006.

9 US Senate Foreign Relations, "Evaluating U.S. Foreign Assistance to Afghanistan: A Majority Staff Report Prepared for the Use of the Committee on Foreign Relations," 120th Congress, 1st Session, 2011, S. Prt. 112-21, 19, www.gpo .gov/fdsys/pkg/CPRT-112SPRT66591/pdf/CPRT-112SPRT66591.pdf.

10 US Department of Defense, "Strategy," Attachment, "U.S. Strategy in Afghanistan," Donald Rumsfeld to Douglas Feith, National Security Council, October 16, 2001, 7:42 a.m., Secret/Close Hold/ Draft for Discussion, 3, http://nsarchive.gwu.edu/NSAEBB/NSAEBB358a/doc18.pdf.

11 Barnett R. Rubin, "Crafting a Constitution for Afghanistan," *Journal of Democracy* 15, no. 3 (2004): 8; James Dobbins, conversation with the author, May 2012. According to the agreement, the AIA cabinet would act as both executive and legislative authority, with Karzai as head of state.

12 Larry P. Goodson, "Afghanistan's Long Road to Reconstruction," *Journal of Democracy* 14, no. 1 (2003): 82; Rubin, "Crafting a Constitution for Afghanistan," 9.

13 Rubin, "Crafting a Constitution for Afghanistan," 9.

14 Barnett R. Rubin, *Afghanistan in the Post-Cold War Era* (New York: Oxford University Press, 2013), 143; William Maley and Susanne Schmeidl (eds.), *Reconstructing Afghanistan: Civil-Military Experiences in Comparative Perspective* (New York: Routledge, 2014), 13.

15 Graham Bowley, "Afghan Warlord Ismail Khan's Call to Arms Rattles Kabul," *The New York Times*, November 12, 2012 sec. World/Asia Pacific, www.nytimes.com/2012/11/13/world/asia/ismail-khan-powerful-afghan-stokes-concern-in-kabul.html.

16 Seth G. Jones, *In the Graveyard of Empires: America's War in Afghanistan* (New York: W. W. Norton & Company, 2009), 110; Kimberly Marten, *Warlords: Strong-Arm Brokers in Weak States* (Ithaca, NY: Cornell University Press, 2012), 2; Tomsen, *The Wars of Afghanistan*, 632; Ahmed Rashid, *Descent into Chaos: The U.S. and the Disaster in Pakistan, Afghanistan and Central Asia*, revised ed. (London: Penguin Books, 2009).

17 US Department of Defense, Donald Rumsfeld Files, "Subject: Afghanistan," April 17, 2002. See also Donald C. Bolduc, *Bureaucracies at War: Organizing for Strategic Success in Afghanistan* (Carlisle Barracks, PA: US Army War College, 2009).

18 Tomsen, *The Wars of Afghanistan*, 632.

19 US Senate Foreign Relations, "Evaluating U.S. Foreign Assistance to Afghanistan," 18–19.

20 US Department of State, US Embassy Kabul, "Precariously Perched between Crisis and Recovery, Ghor Residents Hope for a Plentiful 2009 Wheat Harvest," 09KABUL1000, April 21, 2009.

21 US Senate Foreign Relations, "Evaluating U.S. Foreign Assistance to Afghanistan," 18–21.

22 Kenneth Katzman, "Afghanistan: Post-Taliban Governance, Security, and U.S. Policy" (Congressional Research Service, January 4, 2013), 41.

23 US Senate Foreign Relations, "Evaluating U.S. Foreign Assistance to Afghanistan," 3.

24 Office of the SIGIR, "Quarterly Report to Congress," October 30, 2010, 11.

25 Anthony H. Cordesman, *How America Corrupted Afghanistan: Time to Look in the Mirror* (Washington, DC: Center for Strategic and International Studies, September 9, 2010).

26 The US intervention in Vietnam is more of an outlier. For 49/105 (46.7 percent) the USA was dependent on Saigon for assistance, roughly 15–20 percent less often than what was observed in Iraq, Afghanistan, and Sri Lanka. This indicates the USA was more likely in Vietnam to ask for tasks it could accomplish unilaterally than in the other wars.

27 The White House, Office of the Press Secretary, "Remarks by the President in Address to the Nation on the Way Forward in Afghanistan and Pakistan," Eisenhower Hall Theatre, United States Military Academy at West Point, West Point, New York, December 1, 2009, www.whitehouse.gov/the-press-office/remarks-president-address-nation-way-forward-afghanistan-and-pakistan [accessed October 4, 2012]; Patrick Cronin, *Civilian Surge: Key to Complex Operations* (Washington, DC: National Defense University, 2009).

28 NATO Training Mission – Afghanistan, Combined Security Transition Command – Afghanistan, "Afghan Ministry of Interior (MoI) Advisor Guide," (May 2011, Version 1), 2–7, 2–8, http://publicintelligence.net/isaf-afghan-ministry-of-interior-advisor-guide/.

29 Ibid., 2–5.

30 US Department of Defense, "Report on Progress toward Security and Stability in Afghanistan," April 12, 2012, 16, www.defense.gov/pubs/pdfs/Report_Final_SecDef_04_27_12.pdf.

31 Ibid., 14.

32 US Department of State, US Embassy Kabul, "GIRoA Appears to Retreat on Electoral Reform," 10KABUL577, February 15, 2010.

33 Secretary Rumsfeld's documents are available at: www.dod.mil/pubs/foi/specialCollections/Rumsfeld/. Other sources of released documents included the Office of the Secretary and Joint Staff Freedom of Information Act Library, www.esd.whs.mil/FOIA/Reading-Room/Reading-Room-List_2/.

34 See Appendix B.

35 A discussion of the *Cablegate* database can be found in Appendix B.

36 Office of the SIGAR, www.sigar.mil/index.html.

37 Reports on Progress toward Security and Stability in Afghanistan. See www.defense.gov for individual reports.

38 There are likely more demands made by the USA regarding military reform and military strategy that are not yet captured in the dataset examined because much of the material from the US Department of Defense is currently unavailable to the public.

39 US Department of State, US Embassy Kabul, "Scenesetter: U.S.-Afghan Strategic Partnership Talks in Kabul – March 13."

40 US Department of State, "Letter of Agreement on Police and Justice Projects"; US Department of State, Department of Defense Inspectors General,

"Interagency Assessment of Afghan Police Training and Readiness, Department of State Report No. ISP-IQO-07-07, Department of Defense Report No. IE-2007-001."

41 Pia Heikkila, "Afghan Census Cancelled Due to Security Fears," *The Guardian*, June 11, 2008, www.guardian.co.uk/world/2008/jun/11/afghanistan.internationalaidanddevelopment.

42 US Department of State, US Embassy Kabul, "Charge's Initial Call on New Afghan Finance Minister," 09KABUL558, March 12, 2009.

43 US Department of State, US Embassy Kabul, "Urgent Resource Request to Support Afghan Border Management Initiative (BMI)," 05KABUL518, December 20, 2005.

44 US Department of State, US Embassy Kabul, "EXBS Afghanistan Advisor Monthly Border Management Initiative Reporting Cable – April 2007."

45 Sardar Ahmad, "Karzai Appoints New Elections Chief for Afghanistan," *Agence France-Presse*, April 17, 2010; Jon Boone, "Hamid Karzai Takes Control of Afghanistan Election Watchdog," *The Guardian*, February 22, 2010, www.guardian.co.uk/world/2010/feb/22/karzai-afghanistan-electoral-complaints-commission.

46 US Department of Defense DoD/OGC, "Fact Sheet: Tax Exceptions Accorded U.S. Contractors and U.S. Contractor Personnel under the Agreement Regarding the Status of United States Military and Civilian Personnel of the U.S. Department of Defense Present in Afghanistan in Connection with Cooperative Efforts in Response to Terrorism, Humanitarian and Civic Assistance, Military Training and Exercises, and Other Activities (U.S.-Afghanistan Status of Forces Agreement (SOFA))," March 28, 2011, www.acq.osd.mil/dpap/ops/docs/TAB_A_-_Incoming%5B1%5D.pdf.

47 Rod Nordland, "Afghan Tax on Foreign Contractors Hits Resistance," *The New York Times*, January 17, 2011, sec. World / Asia Pacific, www.nytimes.com/2011/01/18/world/asia/18afghan.html.

48 US Department of State, US Embassy Kabul, "Election Preparations in RC-North: A Message to Governor Atta," 09KABUL2425, August 19, 2009.

49 Lianne Gutcher, "Afghanistan's Anti-corruption Efforts Thwarted at Every Turn," *The Guardian*, July 19, 2011, www.guardian.co.uk/world/2011/jul/19/afghanistan-anti-corruption-efforts-thwarted.

50 Maria Abi-Habib, "Karzai Picks Ally to Fight Graft," *The Wall Street Journal*, December 30, 2010, http://online.wsj.com/article/SB10001424052748704543004576051692092314726; Gutcher, "Afghanistan's Anti-corruption Efforts Thwarted at Every Turn."

51 United Nations Mission in Afghanistan, *Second Six Month Report: Independent Joint Anti-corruption Monitoring and Evaluation Committee* (July 25, 2012), unama.unmissions.org. See also http://unjoblist.org/vacancy/?249686 for listing of the Chief of MEC position, duties, and qualifications.

52 Office of the SIGAR, "Quarterly Report – Special Inspector General for Afghanistan Reconstruction," July 30, 2012, 102, www.sigar.mil/pdf/quarterlyreports/2012-07-30qr.pdf.

53 United Nations Convention Against Corruption Signature and Ratification Status, www.unodc.org/unodc/en/treaties/CAC/signatories.html.

54 Arnold Fields, "Special Inspector General for Afghanistan Reconstruction (SIGAR): Quarterly Report to the United States Congress" (SIGAR, 2010), 14–15.

55 US Department of State, Bureau of Public Affairs, "2012 Investment Climate Statement – Afghanistan," Report US Department of State, June 7, 2012, www.state.gov/e/eb/rls/othr/ics/2012/191093.html.

56 For more see Elias, "The Likelihood of Local Allies Free-Riding."

57 Michelle Nichols, "Afghanistan Plans National Electronic ID Cards," *Reuters*, December 12, 2010, www.reuters.com/article/2010/12/12/us-afghani stan-identification-cards-idUSTRE6BB0P720101212.

58 John A. Glaze, "Opium and Afghanistan: Reassessing U.S. Counternarcotics Strategy" (US Army War College, Strategic Studies Institute, October 2007), www.strategicstudiesinstitute.army.mil/pdffiles/pub804.pdf.

59 US Department of State, Bureau for International Narcotics and Law Enforcement Affairs, "International Narcotics Control Strategy Report, Volume I, Drug and Chemical Control," March 2010, 98, www.state.gov/documents/organization/137411.pdf.

60 US Department of State, "Objectives for Certification to the Government of Afghanistan."

61 Mankin, "Rotten to the Core."

62 Michael Nicoletti, "Opium Production and Distribution: Poppies, Profits and Power in Afghanistan" (Theses and Dissertations. Paper 74. DePaul University, March 2011), http://via.library.depaul.edu/etd/74, 38; Mankin, "Rotten to the Core"; "Afghanistan's Opium Poppies: No Quick Fixes," *The Economist*, June 19, 2008, www.economist.com/node/11591396.

63 Mankin, "Rotten to the Core."

64 US Department of State, "Authorization to Sign Implementing Agreement for Immediate-Impact Counternarcotics Project," 2003STATE068522, March 14, 2003, The National Security Archive, FOIA Number 2001105588; US Department of State, "Objectives for Certification to the Government of Afghanistan."

65 US Department of State, "Objectives for Certification to the Government of Afghanistan."

66 US Department of State, US Embassy Kabul, "Afghan Commerce Minister Visit – Opportunity to Stress Our Priorities," 06KABUL5238, October 29, 2006.

67 US Department of State, US Embassy Kabul, "Urgent Resource Request to Support Afghan Border Management Initiative (BMI)."

68 Ibid.; Office of the SIGAR, "Inspectors General Fiscal Year 2013 Joint Strategic Oversight Plan for Afghanistan Reconstruction," July 2012, 21, www.sigar.mil/pdf/strategicoversightplans/fy-2013.pdf.

69 US Department of Defense, *United States Plan for Sustaining the Afghanistan National Security Forces*, June 2008, Report to Congress in Accordance with the 2008 National Defense Authorization Act (Section 1231, Public Law 110-181), 31, https://archive.defense.gov/pubs/United_States_Plan_for_Sus taining_the_Afghanistan_National_Security_Forces_1231.pdf; US Department of State, US Embassy Kabul, "Afghanistan Terror Finance – Disrupting External Financing to the Taliban," 07KABUL1555, May 8, 2007; US

Department of State, US Embassy Kabul, "EXBS Afghanistan Advisor Monthly Border Management Initiative Reporting Cable – April 2007"; US Department of State, US Embassy Kabul, "Urgent Resource Request to Support Afghan Border Management Initiative (BMI)."

70 The London Conference on Afghanistan, "The Afghanistan Compact," February 2006, www.nato.int/isaf/docu/epub/pdf/afghanistan_compact.pdf.

71 Kevin Savage et al., "Corruption Perceptions and Risks in Humanitarian Assistance: An Afghanistan Case," HPG Working Paper (Humanitarian Policy Group, July 2007), 12, www.odi.org/sites/odi.org.uk/files/odi-assets/publications-opinion-files/4162.pdf.

72 US Department of State, US Embassy Kabul, "PRT/Badghis: Provincial Officials Steal Humanitarian Aid," 07KABUL1861, June 5, 2007.

73 Ibid.

74 Nic Jenzen-Jones, "Chasing the Dragon: Afghanistan's National Interdiction Unit," Small Wars Journal, September 5, 2011, http://smallwarsjournal.com/blog/chasing-the-dragon-afghanistan's-national-interdiction-unit.

75 UNDOC, "Corruption in Afghanistan"; Associated Press, "Afghanistan: U.N. Counts Cost of Bribes," The New York Times, January 20, 2010, sec. International/Asia Pacific, www.nytimes.com/2010/01/20/world/asia/20briefs-Afghanistan.html.

76 UNDOC, "Corruption in Afghanistan."

77 Daniel Schulman, "Corruption in Afghanistan: It's Even Worse Than You Think," Mother Jones, January 21, 2010, http://motherjones.com/politics/2010/01/corruption-afghanistan-its-even-worse-you-think; "Corruption Index 2011 from Transparency International."

78 US Department of State, US Embassy Kabul, "Corruption Threatens Mobile Money Pilot Program for Police," 07KABUL1861, December 3, 2009.

79 US Department of State, US Embassy Kabul, "Complaints to GIRoA on Pre-trial Releases and Pardons of Narco-Traffickers," 09KABUL2246, August 6, 2009, "WikiLeaks Archive – A Selection from the Cache of Diplomatic Dispatches," The New York Times, June 19, 2011, www.nytimes.com/interactive/2010/11/28/world/20101128-cables-viewer.html#report/corruption-09KABUL2246 [accessed January 8, 2013].

80 Because at the time of writing the war was ongoing, not every US request currently designated as unfulfilled is certain to remain unfulfilled. Consider for example the 2009 US request that Kabul allow legal lumber sales. In 2006 Kabul banned timber sales in order to protect what remained of Afghanistan's forests. US Department of State, US Embassy Kabul, "Kunar's Timber Industry and Smuggling: Solutions Await a New Cabinet," 09KABUL3792, November 28, 2009. The ban created the opportunity for substantial profit for the Taliban from illicit lumber trafficking. The Islamic State as well reportedly engaged in lumber smuggling on the Afghan border with Pakistan. As of 2016 Kabul's prohibition of timber sales had yet to be overturned; however, that status could change before full US withdrawal. See Zia-U-Rahman Hasrat, "IS Runs Timber Smuggling in Business in Afghanistan, Officials Say," Voice of America, February 8, 2016, www.voanews.com/content/islamic-state-timber-smuggling-afghanistan/3182282.html; Kenneth Katzman, "Afghanistan: Post-Taliban Governance, Security, and U.S.

Policy" (Congressional Research Service, RL30588, February 17, 2016), 62. The request is coded as noncompliance, but is flagged for review. Two other requests similarly fall into this category.

81 Compliance here includes partial compliance and full compliance.

82 US Senate Foreign Relations, "Evaluating U.S. Foreign Assistance to Afghanistan," 18–19. Although more efficient, eliminating the Afghan government from decision-making processes on development not only inhibits the growth of institutional capacity, it also likely limits the number of demands made on those partners. In the initial years following September 11, the USA may have asked less of Kabul because it was increasingly running independent assistance programs in theater.

83 National Commission on Terrorist Attacks, *The 9/11 Commission Report: Final Report of the National Commission on Terrorist Attacks upon the United States* (2004), 563.

84 Congressional Record, "Proceedings and Debates of the 109th Congress," 2nd Session, Vol. 152, No. 99, July 25, 2006, H5841, www.congress.gov/crec/2006/07/25/modified/CREC-2006-07-25-pt1-PgH5837.htm.

85 Tomsen, *The Wars of Afghanistan*, 633; quoting David Rohde and David E. Sanger, "How a 'Good War' in Afghanistan Went Bad," *The New York Times*, August 12, 2007, sec. International / Asia Pacific, www.nytimes.com/2007/08/12/world/asia/12afghan.html.

86 Tomsen, *The Wars of Afghanistan*, 624.

87 Ian S. Livingston, Heather L. Messera, and Michael O'Hanlon, "Afghanistan Index: Tracking Variables of Reconstruction & Security in Post 9/11 Afghanistan" (Washington, DC: The Brookings Institution, February 28, 2011), 4; Amy Belasco, "Troop Levels in the Afghan and Iraq Wars, FY 2001–FY 2012: Cost and Other Potential Issues" (Congressional Research Service, July 2, 2009).

88 Galula, *Counterinsurgency Warfare*, 61.

89 David Barno (Lt. Gen). *2014 and Beyond: U.S. Policy towards Afghanistan and Pakistan*, Testimony before the House of Foreign Affairs Committee Subcommittee on the Middle East and South Asia, Center for a New American Security, November 3, 2011, www.cnas.org/files/documents/publications/CNAS%20Testimony%20Barno%20110311.pdf.

90 Vanda Felbab-Brown, *Aspiration and Ambivalence: Strategies and Realities of Counterinsurgency and State-Building in Afghanistan* (Washington, DC: Brookings Institution Press, 2012), 24–5.

91 US Army and Marine Corps, *Counterinsurgency Field Manual, FM 3-24*, 1–26.

92 Robert Fisk, "Robert Fisk: 'Nobody Supports the Taliban, but People Hate the Government'," *The Independent*, November 26, 2008, www.independent.co.uk/voices/commentators/fisk/robert-fisk-nobody-supports-the-taliban-but-people-hate-the-government-1036905.html.

93 Felbab-Brown, *Aspiration and Ambivalence*, 53.

6   The USA in Vietnam

> The Vietnamese in the street is firmly convinced that the U.S. totally dominates the GVN [Government of South Vietnam] and dictates exactly what course shall be followed. However, the bitter and tragic truth is that the U.S. has been kept at such a distance from GVN circles and power that in joint councils or plans our views may be heard, some portions of our logic may be endorsed but with confrontations or matters that represent any truly revolutionary departure from existing GVN practices etc., we are light weights and presently do not possess the leverage or power to carry the day.   —The Pentagon Papers

Prior to Afghanistan and Iraq, Vietnam was criticized as a bothersome aberration in US military history – a long, miscalculated war against an impoverished but tenacious enemy.[1] In contrast to WWII, a war Kurt Vonnegut pronounced the US "fought for near-holy motives," the justification for Vietnam was more ambiguous.[2] As one US veteran confided, "I want it to have been worth something, and I can't make myself believe that it was."[3] When pressed for a justification for the US commitment in Vietnam, President Johnson often cited the geopolitical and moral obligation to contain communism and demonstrate US resolve to rescue an enfeebled anti-communist ally. In April 1965, Johnson claimed, "our objective is the independence of South Viet-Nam and its freedom from attack. We want nothing for ourselves – only that the people of South Viet-Nam be allowed to guide their own country in their own way."[4] But over the course of the expansive US intervention, Washington insistently pressed Saigon to reform, issuing over 100 specific high-level policy requests. In this chapter I analyze these requests and the key factors that affected the likelihood that South Vietnam would comply.

In particular, I offer three primary findings explaining patterns of compliance with US requests by the GVN. First, as hypothesized in Chapter 2, a specific pattern of interaction between interests and dependencies created powerful incentives for Saigon to collaborate with certain kinds of American requests, and to disregard others. Second, the shock of the 1968 Tet Offensive led to a sharp increase in local compliance, suggesting that significant enemy activity can motivate clients to cooperate with the demands of their larger patrons.[5] Third, as American forces withdrew from the war, there was an increase in GVN compliance, suggesting that a decrease in commitment can increase coercive leverage for patrons, perhaps because they are more readily able to credibly issue threats against local clients. However, as cautioned by the US experience in Vietnam, extracting concessions from a local ally by issuing threats and withdrawing can undermine the very counterinsurgency mission that motivated the intervention in the first place as local allies may be pressured into complying with requests that facilitate foreign withdrawal, but undermine their stability.

In order to test the theory offered in the context of the war in Vietnam, I analyzed over 2,500 US primary source documents (10,000+ pages) identifying 105 policy requests from the US government to its Vietnamese allies from 1964 to 1973. Each request was found in the primary source documents contained in the twelve volumes of the *Foreign Relations of the United States* (FRUS) on Vietnam published by the US Department of State Office of the Historian.[6] The outcome of each of these requests was determined by using FRUS volumes, as well as the Pentagon Papers, declassified US documents in the Digital National Security Archive (DNSA) and the Declassified Document Reference System (DDRS) as well as online US government document repositories including the US Agency for International Development (USAID) and the Defense Technical Information Center (DTIC).

Summary Findings

Contrary to the recurrent reports of defiant, uncooperative allies in Saigon, South Vietnam was more likely than not to comply with US requests, either fully or partially. Of the requests examined, 46/105 (43.8 percent) resulted in GVN compliance, 25/105 (23.8 percent) in partial compliance, and 34/105 (32.4 percent) in noncompliance.[7] This is a higher rate of compliance than Washington was able to secure in either Iraq or Afghanistan. As detailed in Chapter 10, Table 10.1, comparing compliance across counterinsurgency interventions, South Vietnam has the lowest level of noncompliance. Furthermore, the rate of full compliance with US requests in Vietnam was relatively high, with only the Soviets in Afghanistan securing greater rates of full compliance. Although US forces in Vietnam were undoubtedly frustrated that one-third of all American requests were not carried out, this figure is consistent across large-scale third-party counterinsurgency interventions and is likely related to local regimes coping with significant capacity limitations

Table 6.1 *Timeline of the Vietnam–US alliance during the US intervention*

1961	Committed to the Diem regime, the USA pledges military assistance to Saigon in October and the first US operational units are deployed to South Vietnam in December.
1963	There is a Buddhist rebellion in South Vietnam against Prime Minister Diem. With US acquiescence, General Duong Van "Big" Minh leads a successful coup against Diem in November.
1964	A series of coups increases instability in South Vietnam. On August 2, the US destroyer *Maddox* in the Gulf of Tonkin claims to have been hit by North Vietnamese torpedoes. Lyndon Johnson orders air strikes against North Vietnamese naval and oil targets, marking the beginning of a long US bombing campaign. On August 7, Congress adopts the Gulf of Tonkin Resolution, a joint resolution of support of South Vietnam.
1965	Nguyen Cao Ky takes power in South Vietnam in June in another coup and the South Vietnamese Army suffers setbacks. Major US ground combat operations in Vietnam begin in March.
1968	The Tet Offensive in January shocks US forces and the American public. During Tet, the National Liberation Front launches the first phase of a countrywide offensive.
1971	In July the demilitarized zone separating North and South Vietnam is turned over to the South Vietnamese and by August all ground combat offenses are assigned to the Army of the Republic of Vietnam (ARVN). US troops in Vietnam assume a more defensive role.
1972	The last US combat battalion is withdrawn in August, but the US Air Force and Navy continue to provide support to the ARVN. North Vietnam launches the Easter Offensive against South Vietnam in March. US bombing continues.
1973	On March 29 the last American servicemen are withdrawn from South Vietnam.
1974	On December 13 the North Vietnamese Army launches the last campaign of the war by striking Phuc Long Province.
1975	Fighting expands as territory falls to the North Vietnamese Army. Saigon surrenders in April, uniting Vietnam under the communist north.

as well as dealing with competing pressures to combat insurgents, neutralize rivals, and placate foreign patrons.

Complying with reforms proposed by foreign patrons can be complicated. Even straightforward requests can become impressively problematic. The seemingly banal 1964 request that the GVN "carry out a sanitary clean-up of Saigon," provides one illustrative example.[8] American officials had received numerous complaints of uncollected garbage creating health issues. The South Vietnamese government was short on capacity and infrastructure to handle the problem, increasingly relying on the USA to expand local capacity by providing aid and converting US transport trucks into municipal garbage collection vehicles.[9]

These measures were perennially insufficient as Saigon's trash crisis nevertheless deepened after 1964 with the influx of US troops. Even though US bases were built to handle their waste, the arrival of thousands of American troops had unforeseen effects on services in South Vietnam, including the capital city. According to Neil Sheehan, in 1965, "sanitation services collapsed in Saigon, because the workers quit en masse and rushed away to labor at the base construction sites for much higher salaries than the municipality could pay."[10] The problem worsened as military operations in the countryside intensified, causing the cities to flood with refugees. Before the US buildup in South Vietnam, 80–85 percent of the population was located in rural communities.[11] Military operations by the National Liberation Front (NLF), GVN, and the USA however pushed hundreds of thousands of rural tenants into urban centers. Between 1960 and 1970 Saigon alone reported a 45 percent increase in population.[12] American officials reported on the migration into the cities:

Vietnam has become an urban society ... Except for some efforts by AID in the area of public works (water, electricity and road building) and the Vietnamese government's concentration on security measures, the cities of Vietnam have been residual claimants on the time, energy and resources of pacification officials. While such questions as poverty, pollution, sanitation, housing, traffic congestion, noise, and crime are not, strictly speaking, insurgency related, they do bear heavily on the government's ability to enlist the positive support of the people in its capital.[13]

Sanitation conditions in Saigon were so dire that in 1967 the bubonic plague resurged to "alarming proportions" decades after the concerted French effort to reduce its incidence throughout the region.[14] Saigon made an effort to expand sanitation, but efforts by the GVN were always lagging behind expanding demand. US frustration with the garbage issue caused American military units to periodically take over, or hire US contractors to organize Korean or Vietnamese workers to do the job independently. Nevertheless, USAID and Department of State officials felt it was important that the GVN municipality provide public services and promoted GVN participation.[15] Saigon's garbage problem illustrates the complications of policy implementation while combating an insurgency, how readily local capacity can be strained, and how structural factors related to the war are often underpinning compliance outcomes, dynamics detailed in this chapter on USA–South Vietnam relations.[16]

What Kinds of Policies Did the USA Ask for in Vietnam?

The USA proposed a wide variety of requests to Saigon, including building schools, holding elections, reorganizing military units, addressing minority grievances, providing refugee services and military operations in Laos and Cambodia. Table 6.2 provides a summary of US

Table 6.2 *Summary of US requests to Vietnamese allies*

	Year	US request to South Vietnamese allies
1.	1964	Rural development – land reform
2.	1964	Increase police forces
3.	1964	Increased GVN civil forces
4.	1964	Amnesty, reconciliation programs
5.	1964	Reorganization – political forces
6.	1964	Strategic Hamlet Program
7.	1964	Operation Hop Tac
8.	1964	Price stability – rice
9.	1964	Program for national mobilization
10.	1964	Reorganization – military forces
11.	1964	Implement national mobilization – draft
12.	1964	Standardize and clarify police procedures
13.	1964	Increase the Republic of Vietnam Armed Forces (RVNAF) military forces – by at least 50,000 men
14.	1964	Accepting increased US involvement in GVN military
15.	1964	Program for the improvement of the port of Saigon
16.	1964	Saigon sanitation
17.	1964	Increased compensation – military
18.	1964	Increased compensation – political
19.	1964	Offensive guerilla force
20.	1965	Centralize pacification program
21.	1965	Reroute funds directly to province chief
22.	1965	GVN officials visit provinces
23.	1965	Unify cadre system
24.	1965	GVN survey and research capacity
25.	1965	Increase funds for youth programs
26.	1965	Improve transportation systems – coastal water transportation
27.	1965	GVN press center – daily press briefings
28.	1965	Increase pay – rural teachers
29.	1965	Train broadcasters
30.	1965	Agree to bombing halt
31.	1965	Comprehensive system of rewards
32.	1965	Revitalize interreligious council
33.	1965	Improve image abroad
34.	1965	Announce program for social and political reconstruction
35.	1965	Reduce draft age [from 20 to 17 or 18]
36.	1965	Support anti-communist labor parties
37.	1965	GVN take first strikes against North Vietnam (NVN)
38.	1965	Montagnard grievances
39.	1966	Enlarge customs force
40.	1966	Simplify import procedures
41.	1966	Appoint port director
42.	1966	Advisory council – represent minorities
43.	1966	Advisory council – establish electoral laws
44.	1966	Establish security stockpiles
45.	1966	Build elementary schools

Table 6.2 (*cont.*)

	Year	US request to South Vietnamese allies
46.	1966	Train health personnel
47.	1966	Vocational training for graduating students
48.	1966	Distribute new varieties of seed to farmers
49.	1966	Advisory council to draft constitution
50.	1966	Expand rural electrification
51.	1966	Increased lottery sales
52.	1966	Eliminate ninety-day grace period on payment of duties
53.	1966	Require advanced deposit by importers
54.	1966	Use foreign exchange to purchase imports – reduce inflation
55.	1966	Move forward with elected village executive councils
56.	1966	Raising exchange rates on import duties
57.	1966	Special currency exchange rate
58.	1966	Increased GVN revenues
59.	1966	Assistance programs for refugees
60.	1966	Employment for refugees
61.	1966	Schools for refugee children
62.	1966	Vocational training for refugees
63.	1966	Expand credit to farmers
64.	1966	Limit spending to certain amount to prevent inflation
65.	1967	Centralize rural development in remaining hamlets
66.	1967	Regular meetings, weekly or regular lunches between GVN officials and their US counterparts
67.	1967	Associate a program with a national hero
68.	1967	Transfer land authority to villages and rental law enforcement
69.	1967	Project Take-Off
70.	1967	Public sacrifice campaign
71.	1967	Consult US government (USG) before executing prisoners
72.	1967	Make the province chief the "key" figure in pacification efforts
73.	1967	Reaffirm military reforms – promotion based on merit and ARVN efficiency
74.	1967	Appoint prominent civilians to new government, including opposition leaders
75.	1967	Bring various religious and political groups into government
76.	1967	Promote changing the draft laws
77.	1967	Veterans' benefits
78.	1967	Build prisons and holding facilities
79.	1967	Reorient pacification efforts – mobility, offensiveness, and dedication
80.	1968	Ad valorem taxes
81.	1968	Increase taxes on petroleum
82.	1968	Thieu and Ky – cooperation
83.	1968	Add provision for run-off election
84.	1968	Move collectively with the USA to Paris Peace talks
85.	1968	Reappraisal of the bureaucracy
86.	1968	Pass political party law that brings anti-communist political forces together
87.	1968	Dismiss officials US Military Assistance Command, Vietnam (MACV) indicates are corrupt or ineffective

Table 6.2 (*cont.*)

	Year	US request to South Vietnamese allies
88.	1968	Paris Peace Negotiations
89.	1968	Invest in Lien Minh
90.	1969	Accept the NLF as a party in talks
91.	1969	Compromise on the shape of the table
92.	1969	Offer to talk to the NLF
93.	1969	Accept US plans for troop withdrawal
94.	1970	Drop demand for mutual withdrawal in peace framework
95.	1970	ARVN units to cross into Cambodia
96.	1971	Thieu to allow Ky to run in elections
97.	1971	Address rising drug problem
98.	1971	Send ARVN to Laos
99.	1971	Controlled pull-out of Laos
100.	1972	Vietnam reunification to be determined after peace accords
101.	1972	Concurrence on US proposal for Paris – October 1972
102.	1972	Drop insistence on "except for purely defensive purposes and on a temporary basis" at the end of final paragraph of proposal
103.	1972	Accept tripartite electoral commission
104.	1973	Attend international conference on Vietnam
105.	1973	Concurrence on US proposal for Paris – January 1973

requests to South Vietnamese partners identified in US Department of State records.

In order to paint a broad picture, we can divide US requests to Saigon into six categories of subjects, which reflect US counterinsurgency approaches to win the war by weakening the insurgency, expanding local government services, reforming local bureaucracies, expanding the economy, and pacifying restive areas.

Issue Areas of US Requests to Vietnam

1. *Development* – projects or activities intended to support economic growth and provide social services. Includes land reform, school construction, and assistance to refugees and veterans. 27/105 (25.7 percent) of US requests.
2. *Economic reform* – actions intended to change economic policies. Includes raising exchange rates and increased taxes. 12/105 (11.4 percent) of US requests.
3. *Military strategy* – actions intended to guide military forces in the execution of the war effort. Includes putting forces into Laos,

invading Cambodia, and striking North Vietnam. 5/105 (4.8 percent) of US requests.

4. *Military reform* – actions intended to change military policies and institutions. Includes military promotion policies, enemy POW treatment, reducing the draft age, and compensation levels. 9/105 (8.6 percent) of US requests.
5. *Political reform* – actions intended to change government policies and institutions. Includes policies toward opposition parties, protocol for administering funds, and legal and constitutional issues. 43/105 (41.0 percent) of US requests.
6. *Political-military counterinsurgency strategy* – actions intended to implement US counterinsurgency (COIN) strategy. Includes COIN projects (Project Take-Off, Operation Hop Tac), public sacrifice campaigns, and pacification activities. 9/105 (8.6 percent) of US requests.

Development. Interestingly, US policy requests pertaining to development had relatively low rates of compliance. This may at first seem surprising considering Saigon's interest in development activities and the vast American funds available for projects. However, large development projects require time, logistical coordination, and institutional know-how. They are often difficult to implement, and as a result, partial compliance was almost twice as likely as full compliance, as development projects were initiated but became bogged down and often left uncompleted. More than half of development requests were impacted by Saigon's faltering capacity, 10 percent more than other requests. Development is hard, but it is even more difficult during combat where capacity can be further strained. As an example, in 1966, the USA asked the GVN to work with USAID providing schools for refugees. The task proved gradually more difficult as military activities throughout Vietnam produced more refugees.[17]

Economic reform. There were notably high levels of compliance with US requests for economic reform. Because Washington was directly funding much of South Vietnam's budget, it was able to put specific oversight mechanisms in place on economic policies, providing Washington opportunities to dictate terms and conditions. USAID's Commodity Import Program (CIP) is one such example. CIP was a complex aid program, aiming to balance varying fiscal and economic pressures. From 1954 to 1975 under CIP the USA provided over $4 billion worth of goods and materials to Vietnamese importers, an amount that totaled 70 percent of nonmilitary aid to South Vietnam.[18] US budget officials conducted reviews of GVN military budgets, and recommended programs to be funded from the CIP counterpart fund. Payments made from this fund were verified and US officials approved additional spending on a monthly basis. In support of principal agent models advocating for aid to be provided conditionally, USAID reported, "on occasion, withholding of monthly releases for military budget support has been used to achieve broad political objectives."[19] Therefore, high levels of compliance with economic reforms were likely affected by direct US oversight mechanisms over South Vietnamese economic programs. Aid conditionality in this context was successful, but highly contextualized and limited to items tied to budgetary policy. Additionally, note that despite these mechanisms, one-third of all US economic requests nevertheless resulted in noncompliance, as additional factors, such as capacity limitations and the interaction of interests and dependencies, affected compliance outcomes, as detailed later in the chapter.

Military strategy. Rates of compliance with US requests related to military strategy were exceptionally high. Saigon was less likely to perceive requests regarding military matters as threats to their regime because these policies targeted the enemy. Furthermore, it is improbable that limited capacity would have as significant of an impact on these requests compared to other issues, because the USA would only ask Saigon to undertake military operations well within its abilities. Otherwise, the USA would likely opt to rely on US forces.

Military reform. Similar to economic policies, US oversight of the GVN's military budget likely contributed to increased compliance with US demands for military reforms, such as force reorganization. Yet, despite the supervision, not every structural demand the USA asked of the ARVN was implemented. For example, US officials became increasingly frustrated with Saigon's execution of communist prisoners of war and asked the ARVN to consult American officials before shooting prisoners. American military officials feared Hanoi would retaliate against US prisoners of war. Ignoring US requests, GVN military courts continued to order executions.[20] Reform was not forthcoming and the USA could not coerce the GVN.

Political reform. Similar to other interventions, requests to Saigon for political reforms were the predominant type of requests made by the USA – totaling to 41.0 percent of all requests in Vietnam. As an intervening military, the USA was unable to undertake internal reforms without GVN participation. Furthermore, compliance was lower with political reforms than other subjects, which is unsurprising since these requests included adding legal provisions for run-off elections or dismissing specific officials whom US military officials deemed corrupt.

Additionally, unlike the budgeting process, the USA did not have much direct oversight over the GVN civilian government, a testament to the political importance of local sovereignty, even if partial and incomplete. As a congressional report noted when comparing the civilian and military budgets, "the control exercised by the United States over expenditure of the 3 billion piasters attributed to the civil sector of the GVN budget is, however, virtually nonexistent."[21] There are several reasons why the USA did not have greater supervision of Saigon's civilian budget. First, the State Department did not want to undermine the autonomy of the South Vietnamese administration. The establishment of an effective independent noncommunist state administration was an essential component of US efforts in Vietnam. US policymakers feared that if the majority of internal decisions in Vietnam were made in Washington, the GVN would grow overly reliant on the USA, making US withdrawal increasingly difficult and would undermine Saigon's credibility in South Vietnam. Second, GVN officials fiercely resisted direct US political control. A long French colonial legacy, along with Saigon's motivation to avoid being rendered an American "puppet," meant that Saigon was wary of any significant US administrative control.

Counterinsurgency projects. These included US requests specifically pertaining to counterinsurgency policies, such as pacification campaigns that combined political, military, and socioeconomic efforts to undermine the political causes motivating insurgents.[22] These US policy requests had the lowest rate of compliance. This is because many of these requests required the adoption of US-designed programs that Saigon did not support. Operation Hop-Tac, an early project intended to pacify contested areas near Saigon, is one example. As the Pentagon Papers summarized,

While pacification received a low emphasis during troubled 1964–1965, there was one important exception: the Hop Tac program, designed to put "whatever resources are required" into the area surrounding Saigon to pacify it ... General Taylor and General Westmoreland began Hop Tac, setting up a new and additional headquarters in Saigon which was supposed to tie together the overlapping and quarrelsome commands in the Saigon area. The Vietnamese set up a parallel, "counterpart" organization, although critics of Hop Tac were to point out that the Vietnamese Hop Tac headquarters had virtually no authority or influence, and seemed primarily designed to satisfy the Americans. (Ironically, Hop Tac is the Vietnamese word for "cooperation," which turned out to be just what Hop Tac lacked.)[23]

The plan failed, in part due to strategic flaws, and in part because of the lack of cooperation from Saigon. Again, the Pentagon Papers commented on how "the Vietnamese were cynical about Hop Tac; it was something, speculation ran, that General Khanh had to do to keep the Americans happy, but it was clearly an American show, clearly run by the United States, and the Vietnamese were reluctant to give it meaningful support."[24]

Furthermore, pacification is laborious and Saigon was reluctant to commit troops. In mid-June 1967 Robert "Blowtorch Bob" Komer designed "Project Take-Off" a pacification effort that Komer headed as deputy commander of MACV's Civil Operations and Revolutionary Development Support (CORDS) program. The South Vietnamese were asked to participate, as US Ambassador Ellsworth Bunker expressed, "as is often the case ... GVN performance remains the crucial factor" in the pacification program's success.[25] But the timing of Take-Off worked against the USA. Focused on the upcoming elections, President Thieu and Prime Minister Ky were wary of new initiatives and their potential for political risk. According to historian Richard A. Hunt:

In the absence of an express top-level South Vietnamese direction, CORDS was forced to act on its own, bidding advisers and program managers in August to coordinate Takeoff programs with their counterparts. In other words, CORDS issued guidance to the South Vietnamese through its advisory network hoping its advisers would be able to get local officials committed to Takeoff. As an exercise in leverage, Takeoff was inauspicious since CORDS failed to convince the government to issue orders to its own officials. Without obvious, high-level South Vietnamese endorsement, Takeoff could be viewed only as an American effort.[26]

Yet, not all US-designed counterinsurgency programs were sunk by a lack of cooperation from Saigon. Project Take-Off was soon reorganized under the Phuong Hoang or Phoenix program, a brutal campaign against "VCI," Viet Cong (NLF) infrastructure. The Phoenix program aimed to "neutralize," via capturing, interrogating, or killing communist NLF shadow government operators active in contested areas of South Vietnam. It was extraordinarily controversial, frequently labeled as a Central Intelligence Agency (CIA) assassination campaign. George Jacobson, Deputy for the Office of Civil Operations and Rural Development Support (DepCORDS), quipped in 1972, "I sometimes think we would have gotten better publicity for molesting children."[27] Nevertheless, President Thieu, in the aftermath of the Tet Offensive, complied with US requests to support the program, signing a decree in July 1968. The GVN provided operators, interrogators, intelligence officers, and infiltrators. Despite widespread criticism of Saigon's implementation efforts, after 1968, high-level GVN support helped make the Phoenix program devastatingly efficient.[28] After the war, a senior NVA officer, Colonel Bui Tin, admitted the program neutralized thousands of communist cadres, while communist authorities later said Phoenix was "the single most effective program" anti-communist forces used in the entire war.[29]

The experience of US pacification programs in Vietnam suggests that such programs are unlikely to succeed without leadership, or at a minimum, significant buy-in from local allies. This is an uncontroversial point considering pacification is a project to remake local politics and the governance reforms comprising pacification strategies directly depend on the performance of local officials. In the case of Vietnam, as Richard Hunt noted, "pacification encompassed both military efforts to provide security and programs of economic and social reform and required both the U.S. Army and a number of U.S. civilian agencies to support the South Vietnamese."[30] The South Vietnamese managed the program, with US support. Yet, it is tempting for intervening powers, such as the USA, to attempt to insert local allies into US-designed pacification plans, instead of accepting inevitable compromises required for local allies to design and lead such efforts. Figure 6.1 compares rates of Vietnamese compliance across categories of issues addressed in US requests.

Although the likelihood of compliance varies by subject, the general topic of requests was not itself significant in explaining patterns of South

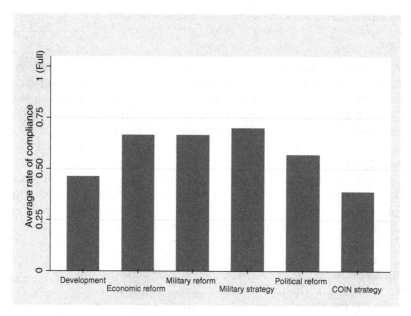

Figure 6.1 Rates of Vietnamese compliance with US requests, by subject area, 1964–73

Vietnamese compliance with US requests. Other factors such as interests, dependencies, capacity, and Tet as a significant enemy offensive had a significantly more evident impact on the likelihood of cooperation.

Explaining the Compliance and Defiance of Local Allies in Vietnam

As detailed in Chapter 2, I propose four variables to help explain the likelihood of South Vietnamese compliance with US policy requests. Namely, the GVN's capacity to implement the request, the interests of Washington and Saigon converging or diverging over the request, the dependency of US forces on Vietnamese partners to implement the request, and a significant enemy offensive. I hypothesize that there is an interaction effect between interests and dependencies producing a particular pattern of compliance outcomes that is key to understanding free-riding behaviors and when intervening forces are likely to have coercive leverage over local clients. In this section I will address each variable in the context of alliance politics between Washington and Saigon.

Local Capacity Limitations – Decreased Likelihood of Compliance

When institutional capacity is lacking, compliance by a local ally, such as Saigon, with requests from larger partners is less likely. The most common capacity issue affecting South Vietnam's ability to implement demands came from deficient bureaucratic institutions. American dollars were readily available to invest in reforms, but GVN bureaucracies were at best underdeveloped, and at worst inept at carrying out the tasks required. According to American documents, two processes steadily fostered second-rate performance in the GVN. First, GVN administrators based promotions on loyalty instead of effectiveness. Second, the insurgent NLF methodically assassinated capable GVN administrators, leaving ineffective GVN personnel in office because they supported NLF messages that American allies were corrupt and ineffectual.

Faltering bureaucratic capacity impacted compliance with American demands through a number of pathways. In anticipation of greater military efforts and US involvement, in 1964 Washington demanded that the GVN "increase the armed forces [RVNAF] (regular plus paramilitary) by at least 50,000 men."[31] The GVN complied by raising force levels, but increases did not materialize until late 1965. The Pentagon Papers reported that Saigon's "failure to provide funds was blamed as a major reason for these military manpower deficiencies."[32] This puzzled American policymakers. As Secretary of Defense McNamara declared, "there is an unlimited appropriation available for the financing of aid to Vietnam. Under no circumstances is a lack of money to stand in the way of aid to that nation."[33] However, according to Ambassador Taylor, the problem was that US funds were not being distributed to local GVN recruiters to enlist troops.

Washington also expressed that administration of the ARVN was a major problem. By 1968 the average desertion rate in the ARVN was 17.7 per 1,000 soldiers, one of the highest in modern history.[34] Inadequate housing and leave allowances motivated desertions, while inaccurate reporting that did not account for soldiers who left their units to join others closer to home led to misleading statistics of force capacity. This dynamic demonstrates how the institutional capacity of local allies is a complex variable. Limited institutional capacity, such as GVN failure to provide ample veterans' benefits, led to practices of inflating operational force figures as wounded or missing ARVN were reported as active duty in absentia. This increased the probability that units in the field would perform poorly, which increased ARVN reliance on US personnel, and hindered ARVN institutional growth. Multifaceted capacity limitations reinforce one another, a process that cripples the collective counterinsurgency effort and the likelihood that local forces would succeed in independent operations. As Vietnam historian George Herring noted, "Much time and money were spent training and equipping the South Vietnamese from 1965–1967," yet, the ARVN "became more than ever dependent on the United States and was ill prepared to assume the burden of the fighting at some later, unspecified date."[35]

Furthermore, as the US forces in Vietnam were acutely aware, coercion and interalliance bargaining with the GVN took place amidst a sizable insurgency, adding a layer of complexity to accomplishing the task of building a durable noncommunist regime. Unlike a significant enemy offensive, classic insurgency low-intensity tactics such as harassment, subversion, and assassinations over time decrease the likelihood of local ally compliance as diffuse violence and its effects complicate tasks.

Interests and Dependencies

In addition to capacity, interests and dependencies can also affect the likelihood of local compliance. At times only one ally can carry out a given task; while in other circumstances either ally is able to do so. For certain US requests, the USA could act unilaterally and take over tasks if Saigon failed to perform. US forces did so on multiple occasions, including assuming control of efforts to expand the port of Saigon, upgrading coastal transportation systems or funding youth programs. Other times the USA could not independently tackle the issue, leaving Washington reliant on its local allies. For example, US forces found they had to rely on the GVN to dismiss inept or corrupt officials, amend the national constitution, or lower the draft age. Whether or not the USA could potentially adopt the requested activity on its own, or if it would fundamentally require GVN participation, was a significant factor determining the likelihood of compliance. US dependency on Saigon to implement requests interacted with the interests of the USA and Saigon converging or diverging to produce a pattern of incentives motivating compliance or defiance.

If the USA could not act independently and the allies' interests diverged, rates of noncompliance were higher. Conversely if the USA could not act independently, and interests of the allies converged, there was a higher likelihood of compliance. The reverse pattern was observed if unilateral US action was possible. If the USA could act unilaterally and the allies' interests diverged, there was a higher probability of compliance, as the GVN agreed in order to protect its interests since the action would likely be implemented with or without them. Lastly, if the USA could act unilaterally and the interests of the two allies converged, the GVN was more likely to fail to comply because there were incentives to free ride.

Interests Diverge and US Unilateral Ability Not Possible: Predicted Vietnamese Noncompliance

When the USA was unable to independently implement the policy requested of its local allies in Vietnam, and US interests diverged with Saigon's, compliance by the South Vietnamese was unlikely. Saigon did not share the interests of its American allies, and the USA, unable to implement the policy without local collaboration, found itself powerless to compel reform. Consider a 1968 US request for a substantial increase in GVN consumer taxes on oil and petroleum, designed to increase GVN revenues by 3–4 billion piasters.[36]

Even though the GVN profited from taxing the extraction and distribution of petroleum, the government continued to provide sizable tax breaks for public oil consumption.[37] As a report from the US Department of Defense summarized,

ostensibly, the GVN has failed to act on raising petroleum taxes for the following reasons: Petroleum is thought to be a price leader; thus all prices will be raised if the price of petroleum is increased; it would be unpopular with the people; and it would be unpopular with the legislature which would have to approve a rise in petroleum taxes. In short, it is frequently claimed that in the interest of political stability the price of petroleum must not be allowed to rise.[38]

Saigon feared the political fallout that such price increases might invoke, and failed to implement the US-requested policy. Divergent interests and dependency on local allies to implement requests increased the likelihood of local noncompliance.

Interests Diverge and US Unilateral Ability Possible: Predicted Vietnamese Compliance

Conversely, if Saigon was threatened by a policy requested by the USA, yet the Americans were able to undertake the action unilaterally, there was a motivation for Saigon to comply in order to better protect its interests. If Washington was likely to undertake the action regardless of Saigon, South Vietnamese leaders rationalized that they were better off taking the lead and shaping the policy as best they could, as opposed to risk being isolated and undermined.

This logic was evident in peace negotiations with North Vietnam. With US public opinion of the war plummeting, Washington was under intense pressure to withdraw. US officials were pushing the GVN to agree to concessions facilitating a negotiated solution with communist forces in the north and south. President Thieu was appalled by US demands that Saigon accept the NLF as a negotiating party. North Vietnam was insisting on "four-party talks," while the GVN advocated "two-sided" negotiations in order to avoid recognizing the NLF. Saigon stalled negotiations for months by refusing to sit at a "square" table with four-party negotiations.[39] The issue was solved by agreeing to call negotiations whatever attendees chose (four-party or two-sided) and by adopting a Soviet-proposed compromise of one large round table and two rectangular ones.[40] The GVN reluctantly accepted the NLF as a party in negotiations, and later even acquiesced to US demands that it offer to directly negotiate with the NLF. Washington's ability to implement policies with or without the GVN, and GVN discomfort with the US position, convinced GVN officials to acquiesce to Washington's demands. This combination of allied interests and dependencies create a condition where intervening forces can coerce local allies. If the interests of allies diverge, but the foreign power has the ability to independently implement the requested policy, it can coerce compliance by threatening to execute the policy and exclude local allies altogether. Counterintuitively, such coercion is only possible if local allies are opposed to the request. If the local regime is interested in the policy, it has little incentive to participate.

Interests Converge and US Unilateral Ability Not Possible: Predicted Vietnamese Compliance

Under conditions where interests between allies converge, and intervening forces depend on local allies for enforcement, compliance is likely. For example, Saigon complied with US demands to place more taxes on an *ad valorem* basis, in order to allow revenues to "expand automatically with price increases."[41] This reform was important since inflation was causing a substantial decrease in revenue among taxes collected by a set fee per item. Although reforms were slow, the GVN promulgated a series of updated income and excise taxes from 1972 to 1974, "basing most of the new taxes on an ad valorem basis."[42] Note that this series of laws was passed when the USA was withdrawing forces and the GVN faced the prospect of fighting the war independently, which required independent funding and was not contingent on US oversight mechanisms. As hypothesized, because Saigon's participation was required to implement the request and the policy proposed served GVN interests, increased compliance was observed.

Interests Converge and US Unilateral Ability Possible: Predicted Vietnamese Noncompliance

Lastly, US unilateral capability to implement a request, if necessary, incentivized Saigon to not comply with US requests if the GVN anticipated it would benefit from the policy being requested. Saigon could free ride by waiting for the USA to undertake the activity, and thus profit from the policy, without incurring direct costs. For instance, in 1966 Washington requested that the GVN train more health personnel and expand health services. Yet, despite US officials seeing local provision of services as a key to increasing Saigon's legitimacy, a year later Americans had taken over the task.[43] Since the USA could substitute for the GVN, and Saigon was not threatened by the policy, Saigon could choose to opt out, yet still benefit from US efforts to expand health services. Under conditions where allied interests converge and intervening forces have the unilateral capability to implement the policy in question, free riding is likely.

Each of the 105 identified requests from the USA to its partners in Vietnam were coded to yield a picture of broad trends related to the proposed variables and the likelihood of Vietnamese compliance. An ordered probit statistical model is provided in Table 6.3, accounting for capacity limitations and the proposed interaction effect between interests and dependencies in Vietnam. Additional quantitative information is provided in Appendix C (Table C.3).

Table 6.3 *Ordered probit – local Vietnamese compliance with US requests, 1964–73, interaction effect between US unilateral capacity and allied interests*

Local Vietnamese capacity	US unilateral potential	Allied interest convergence/ divergence[a]	Interaction with unilateral ability and allied interest
−1.167*	0.539	1.161*	−1.117*
(0.285)	(0.347)	(0.367)	(0.491)

Notes: *p < 0.05.

[a] Regardless of approximation for interest, the interaction effect between unilateral capacity and interests outlined in Table 6.3 remains significant. The convergence/divergence of interests variable is produced by combining two interest variables, "Private Benefit for Local Allies," and "Threat to Private Benefits for Local Allies." The variable was created in order to have a robust measure of Saigon's interests based on costs and benefits, instead of substituting one (costs or benefits) for interest while excluding the other. Interest is determined by taking into account costs (estimated by the variable Threat to Private Benefits for Local Allies) and benefits (estimated by the variable Private Benefits for Local Allies). For more on coding see Chapter 3.

Significant Enemy Threats – Increasing the Likelihood of Local Compliance

Saigon was also appreciably more likely to comply with US policy requests directly following the 1968 Tet Offensive where insurgents attacked urban centers and undermined confidence that the USA and Saigon could defeat communist forces in the long term. The proximity of the enemy drove local and intervening allies closer. Although some instances of compliance in this period can be attributed to progression, as complex programs requested by the USA years earlier were starting to materialize in 1968, the majority of this upward trend in compliance is due to the shock of Tet. Consider, for example, the shift of high-level GVN opinions on draft procedures for South Vietnam. Starting in 1964, the USA began to press the GVN to implement a comprehensive program of national mobilization, including a national service law that would enforce a lower draft age. The Pentagon Papers describe how events related to this demand unfolded in 1964: "General Khanh signed a mobilization decree on April 4; at the time the decree satisfied the USG as meeting McNamara's recommendation on the subject. However, Khanh delayed signing implementing decrees for the mobilization decree indefinitely; and it has never become clear what it would have meant, if implemented."[44]

With the national draft age effectively set at 20, the USA continued asking the GVN to adopt programs lowering the draft age to 18 or 17 in order to add to ARVN strength.[45] This was a particularly vexing issue for US advisers since Americans were drafted at 18. Furthermore, according to US estimates, the NLF insurgents drafted all males from contested areas between 15 and 45.[46] Saigon was concerned that lowering the draft age would impact public opinion and motivate an American drawdown as fewer Americans would be needed to compensate for ARVN strength. Its official reasons for avoiding enforcement focused on social stability. As one historian summarized, "prior to Tet, the GVN had cited social and religious mores to resist American demands to lower the draft age from 21 to 18."[47]

However, the immediacy of the threat posed by the Tet Offensive motivated Saigon to comply. According to General William Westmoreland,

In March [1968] 19-year olds became eligible for the draft, followed on 1 May by 18-year-olds. The false starts at mobilizing manpower in previous years were partly due to the weak and unstable nature of the central government. But in this critical time, the Tet offensive had further crystallized support for an already strengthened and stabilized government. This solidifying effect was, in my estimation, the single development which enabled the mobilization program to be successful.[48]

In June 1968 President Thieu secured authority "to conscript all men between 18 and 38 into the regular forces, and those over 16 and between 39 and 50 into the Self-Defense Corps."[49] The conscription policies advocated by the USA since 1964 were implemented as US after-action reports noted:

Under the mobilization decrees and later laws, the strength of the Vietnamese Armed Forces rapidly increased. In the first six months of 1968 total strength rose by 122,000 men. The upsurge in volunteers was mainly attributable to the mobilization, effective enforcement of the draft, and, in the wake of the *Tet* offensive, a noticeably greater allegiance to the central government on the part of the people as a whole. This kind of growth of the South Vietnamese Army had been our goal for years.[50]

In this case, Tet motivated compliance because it not only inspired high-level GVN officials to issue orders, but also motivated the bureaucracy within Saigon to set aside infighting and focus on the war effort. As Robert Komer noted in his influential 1972 RAND study:

Most experienced observers on the scene have noted a marked improvement in overall GVN administrative performance beginning with Tet 1968. In part this is attributable to increased U.S. advisory influence and, occasionally pressure. In part it is simply that the earlier efforts, of 1965–1967, began to bear more fruit over time. But even greater influences on GVN behavior were the twin shocks of Hanoi's Tet and post-Tet offensives and the resultant clear beginning of U.S. de-escalation.[51]

However, as influential as the external shock of the Tet Offensive was on GVN compliance, not every reform US advisers requested was adopted in 1968. In particular, specific local and mid-level bureaucratic reorganization measures went unfulfilled. In August 1967, US Ambassador Bunker outlined two critical administrative reforms: (1) giving the province chief operational control over military forces engaged in province pacification, resources for development programs, and control of technical cadres, including agricultural, engineering, education, public health and public works officials and (2) centralizing rural development planning and coordination efforts in non-Revolutionary Development hamlets.[52]

Regarding the first reform, province chiefs had difficulty coordinating pacification efforts, including managing military forces and development cadres, in part because they lacked the authority to issue orders. Although most chiefs were ex-military commanders themselves, ARVN officers operating in the region were reluctant to surrender any authority to individuals outside the military chain of command. Similarly, the various development bureaucracies in Saigon refused to make their personnel accountable to local province authorities. The Tet Offensive did little to alter these calculations. As a 1969 US Department of Defense report summarized, "province chiefs have had considerable difficulty at times in getting cooperation and action out of a [sic] regular forces supporting them."[53]

The same report tracked marked failure in Bunker's second reform, centralizing rural development efforts in the areas requiring better coordination.[54] A lack of interest, as well as inadequate capacity on behalf of Saigon's ruling policymakers to impose changes on entrenched GVN bureaucratic organizations at the provincial and district levels, left these interests in power to maintain the status quo. The Tet Offensive did not impact compliance in these areas. This may be due to the strength of entrenched bureaucratic interests, as well the potential distance of these entities from the 1968 offensive. Tet was an effective campaign because it shocked Vietnam's urban areas and US public opinion. It was the first battle of the US war in Vietnam fought in urban settings, and proved the enemy could coordinate a sizable offensive in South Vietnam. GVN officials operating in more rural areas, some of whom were already accustomed to communist assaults, were either not under attack during Tet, or not nearly as shocked by the violence as Saigon. Therefore the threat exposed during the Tet Offensive was instrumental in motivating GVN compliance with US demands, but it was not a cure-all for Saigon's bureaucratic issues

South Vietnamese Compliance over Time

GVN compliance varied over time depending on the conditions of the war. Figure 6.2 graphs trends in the frequency of US requests alongside rates of GVN compliance. Interestingly, due to the threat demonstrated by the Tet Offensive, the number of requests Saigon complied with in 1968 rises above the number of new requests made by the USA in 1968, as Saigon catches up with US requests made in previous years.

Patterns in the direction, intersection, and slope of the data show five discreet sections of the graph, 1964–6, 1967, 1968, 1969–70, and 1971–3. Each is a distinctive segment of the war where changing dynamics impacted the number of new US demands and GVN rates of compliance. Note that two trends are particularly significant. First, the shock of the Tet Offensive motivates Saigon to cooperate to an unprecedented level in 1968. Second, both requests and compliance drop off after 1969 as the Nixon administration, weary of the war, largely gives up on asking the GVN for reforms.

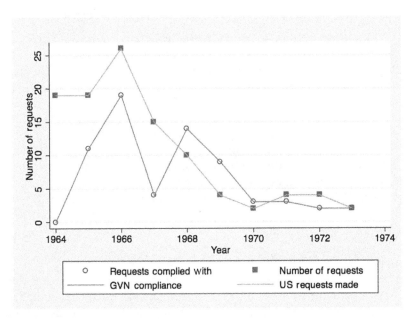

Figure 6.2 Rates of US requests and GVN compliance, 1964–73

1964–1966 – Rates of US Requests and Compliance by Saigon

At the start of the US intervention in Vietnam there was an upward trend in both the number of requests made by the USA and compliance by the GVN. Washington sharply increased aid to Saigon, and requested specific reforms in return. This funding combined with the American takeover of military operations freed the GVN to comply with relatively simple US demands, including requests that called for a larger customs force, higher exchange rates, and the creation of secured stockpiles of weapons and rice to insulate Saigon against attack. Throughout this period the GVN initiated more time-intensive projects, such as land reform.

1967 – Rates of US Requests and Compliance by Saigon

There was a sharp decrease in requests made and an even more dramatic drop in compliance during 1967. After a few years of US military intervention, the GVN had already completed simple reforms requested, and was resisting additional changes.

1968 – Rates of US Requests and Compliance by Saigon

The Tet Offensive of January 1968 is often cited as a turning point in Vietnam, and as illustrated by Figure 6.2, it also strongly affected rates of GVN compliance. As discussed later in the chapter, this spike in compliance in 1968 is a response to the NLF attacks on Saigon and other GVN urban strongholds, which shocked the GVN into recognizing the external threat posed by the enemy. There are multiple examples of similar heightened senses of emergency inspiring compliance with US requests. From the data in the graph a dramatic shift is evident between 1967 and 1968, where post-Tet the GVN was complying with more requests than the USA was making as it caught up on long-standing demands made between 1964 and 1967 in an effort to improve its position against the enemy.

1969–1970 – Rates of US Requests and Compliance by Saigon

Following the Tet Offensive and the initiation of numerous reforms that were stalled prior to 1968, this period contains more requests complied with than new requests. This degree of compliance is in part due to the completion of several reforms initiated in earlier years that took time to implement, as well as the immediate adoption of multiple new requests made by the USA during this period, including offering concessions to North Vietnam in order to initiate peace negotiations.

1971–1973 – Rates of US Requests and Compliance by Saigon

A drop in both the rate of new requests made and the rate of requests complied with illustrates the stalemate of the last years of direct American military involvement in Vietnam. The majority of new American demands pertained to compromises offered during peace talks. US officials had largely stopped asking for new substantive domestic reforms, and old requests, such as specific anti-corruption measures, continued to be unfulfilled by Saigon.

The trends observed in Figure 6.2 also indicate that new requests proposed by intervening forces drop significantly over time. There are several factors contributing to this trend. First, easy reforms are adopted early on, causing a spike in compliance near the beginning of intervention. Second, Washington makes a large number of requests early in the war, as expectations for reform are high. Third, over time, standard operating procedures in GVN bureaucracies become institutionalized and increasingly resistant to change.[55] As Saigon grew increasingly accustomed to US financial and military support, it became harder for Washington to use those factors to compel change.

Coercing South Vietnam through Withdrawal

Following Nixon's election in November 1968, the USA began to make fewer demands on local allies in Vietnam. Washington had accepted that in certain areas, such as corruption, Saigon was unlikely to reform. Furthermore, US officials under Nixon focused primarily on securing a peace agreement with North Vietnam that had eluded US policymakers. As American attention fixated on this task, issues related to the internal workings of the GVN became secondary.

According to State Department records, there were just sixteen new policy requests from the USA to South Vietnam between January 1969 and January 1973. This represents a notable reduction from previous years. There were, for example, eighteen requests in 1965 alone. Despite the limited data available for this period, there is a notably higher rate of full compliance, 75 percent compared to the rate of full compliance made on average throughout the war, 43.8 percent. Rates of partial compliance are lower at 18.75 percent, compared to the 23.8 percent observed across the intervention, including the period in question

(1969–73), and the rate of noncompliance, 6 percent, is much lower than the 32.4 percent average.

This higher rate of compliance at the end of the war is largely due to two factors: first, the kinds of requests asked by Washington, and second, the withdrawal of US troops, along with threats made to exclude Saigon. In the later stages of the war, US officials were only requesting new reforms they believed were either highly likely to be complied with by the GVN, or were critical to US withdrawal. The USA had leverage over Saigon because the USA had the ability to sign agreements with North Vietnam without the GVN's concurrence. As the primary military power opposing the NLF and Hanoi, the USA could agree to compromises at the negotiating table with, or without, its local allies. This depicts how the ability to act unilaterally places large allies in an advantageous bargaining position against their smaller allies.

The ability of the USA to coerce South Vietnam was evident from 1969 to 1973, when Washington pressured the GVN to comply with difficult measures or likely risk even greater losses by being shut out of negotiations entirely. According to historian Larry Berman, as National Security Adviser, Henry Kissinger initiated "a pattern of exclusion for the next four years – Kissinger negotiating an American troop disengagement with the North Vietnamese while informing Thieu only after the fact."[56] The USA was under intense pressure at home to end the war. Spurred on by the upcoming US election, in 1972, Kissinger decided that the USA–GVN delegation was going to have to make substantial concessions to the communists in order to get the USA out of the war. The GVN had little interest in conceding to the North's demands, since these allowances would only leave the South more vulnerable following US withdrawal, and tried to stall negotiations. In 1969, Saigon delayed talks for two months with the "famously stupid" debate over the shape of the negotiating table.[57] The negotiation process frustrated both the GVN and the USA. According to John Prados, Thieu did not appreciate being forced to compromise. "He did not want to be pushed from one position to another, as was the case with the 'shape-of-the-table' issue."[58] Similar to the trajectory of the USA–Karzai relationship during the US intervention in Afghanistan, what had once been a relatively cordial partnership between Washington and Saigon had devolved over the course of a counterinsurgency into an acrimonious and deteriorating union.

Due to the difficulties of getting the Vietnamese to compromise, Kissinger held secret negotiations without Saigon, allowing the USA to drop a key GVN demand for mutual US and North Vietnam withdrawal. President Thieu was informed of this concession only once the USA and North Vietnam had agreed.[59] Thieu could choose between accepting the compromise or walking out and risking the loss of post-withdrawal American aid. In the absence of a viable alternative, Saigon reluctantly complied with Kissinger's demand to acquiesce on the mutual withdrawal issue. This process was replayed in this period of the war on multiple issues, as 7/16 (43.8 percent) of requests fully complied with after 1969 were affected by the impending US withdrawal, and subsequent US activities to push forward on a peace compromise.

Instances of noncompliance during this period appeared under one of two conditions. First, as predicted, when Saigon's interests diverged and the USA was dependent on the GVN to implement the policy. For example, the USA pressured Thieu to allow Vice President Nguyen Cao Ky to run in the 1971 presidential elections. Thieu refused and the USA was helpless to change that ruling, as American officials could not implement electoral policy unilaterally.[60] Second, Saigon failed to comply when Americans had the unilateral ability to implement the policy, if necessary, but were unwilling to engage in coercive diplomacy to sufficiently motivate Saigon. In October 1972, on the eve of the US election, for instance, Kissinger attempted to force President Thieu into signing an agreement that would effectively end the war right before the US election. Historian George Herring writes:

In his haste to get an agreement Kissinger badly miscalculated Thieu's willingness to do what the United States told him and Nixon's willingness to support Thieu. Kissinger spent five days in Saigon, going over the treaty item by item, explaining its advantages for South Vietnam and issuing only slightly veiled warnings that a refusal to go along could mean an end to American support. Thieu was not appeased. He was furious he had not been consulted in advance of the negotiations and that he had first learned of the terms through captured NLF documents. Kissinger's crude and arrogant effort to present the South Vietnamese with a fait accompli reinforced their deepening suspicions of the United States.[61]

Thieu made several demands back at US officials, including North Vietnamese withdrawal, the establishment of a demilitarized zone, and the rejection of NLF sovereignty. Kissinger pressured Nixon to press on with the peace process without Thieu. Nixon, according to Herring, was "suspicious of Kissinger's ambitions, he suspected that his aide was rushing a peace agreement to be able to claim credit for the President's reelection, and he feared that a peace settlement on the eve of the election might be dismissed as a political ploy."[62] The US president felt it was imprudent to isolate Saigon since long-term political solutions hinged on GVN performance.

Conclusion – The Tet Offensive as Motivation for Local Compliance

Four primary factors contributed to the likelihood of South Vietnamese compliance with US requests in the Vietnam War. The first factor was an interaction effect between US dependency on its local allies to implement a given policy, and whether US and South Vietnamese interests converged or diverged. For example, when the USA could undertake the policy requested unilaterally and interests converged, there were higher levels of noncompliance due to incentives to free ride. If, however, interests diverged and the USA could act independently if necessary, Washington could coerce Saigon into complying by threatening to act independently and effectively exclude the GVN. Second, the weak capacity of the regime in Saigon limited what could be accomplished by the collective USA–GVN counterinsurgency endeavor. When Saigon's institutions were unable to implement the reforms requested, the GVN was unable to comply with American demands.

Third, the consequences of low-intensity warfare made it more difficult for Saigon to comply with US demands, as complications from the war strained local capacity; however, the shock of an acute enemy offensive exemplified by the 1968 Tet Offensive increased GVN compliance with US demands. Lastly, the withdrawal of US troops and the willingness of National Security Adviser Henry Kissinger to threaten to act unilaterally without Saigon motivated local compliance. Toward the end of the war, a shrinking US commitment provided American policymakers with coercive leverage in dealing with Saigon. US forces had been constructing, advising, and funding Saigon, careful not to undermine the very political project of strengthening their local proxy. As long as bolstering Saigon was the priority, US policymakers felt they could not readily threaten Saigon for what Washington deemed was bad behavior. Yet, the diminished US commitment under the Nixon administration and Washington's subsequent threats to Saigon that motivated compliance, also ultimately undermined the regime in Saigon, and the anti-communist mission collapsed entirely in 1975. This dynamic reflects the complexity of counterinsurgency alliances, and cautions that acquiescence from local allies is not inherently a strategic advantage for intervening forces or the collective counterinsurgency mission, as the sagacity of the policies promoted by the patron to suppress the insurgent enemy remains paramount in determining the outcome of the war.

Notes

1 Robert S. McNamara et al., *Argument without End: In Search of Answers to the Vietnam Tragedy*, 1st ed. (New York: PublicAffairs, 1999), 222; Herring, *America's Longest War*, xi.

2 Kurt Vonnegut, *Armageddon in Retrospect* (New York: Putnam, 2008), 44.

3 Arnold R. Isaacs, *Vietnam Shadows: The War, Its Ghosts, and Its Legacy* (Baltimore, MD: Johns Hopkins University Press, 1997), 9.

4 Lyndon B. Johnson, "Our Objective in Vietnam," in *Public Papers of the Presidents of the United States: Lyndon B. Johnson: Containing the Public Messages, Speeches, and Statements of the President, 1963/64–1968/69*, Book 1 (Washington, DC: US Government Printing Office, 1965), 172.

5 See Chapter 2, Hypothesis 4.

6 For more information about FRUS see http://history.state.gov/historicaldocuments/about-frus. Produced by the State Department Office of the Historian, FRUS contains declassified records from all the foreign affairs agencies. Foreign Relations volumes contain documents from Presidential libraries, Departments of State and Defense, National Security Council, Central Intelligence Agency, Agency for International Development, and other foreign affairs agencies as well as the private papers of individuals involved in formulating US foreign policy. In general, the editors choose documentation that illuminates policy formulation and major aspects and repercussions of its execution.

7 For information regarding coding see Chapter 3.

8 US Department of State, US Embassy Saigon, "Telegram from the Embassy in Vietnam to the Department of State," November 9, 1964. Also sent to CIA, Department of Defense, and the White House and repeated to CINCPAC. *Foreign Relations of the United States, US Department of State, 1964–8, Volume I, Vietnam 1964*, Document 408. See also The White House, The National Security Council, "Instructions from the President to the Ambassador to Vietnam (Taylor)," Washington, December 3, 1964, Johnson Library, National Security File, Aides File, McGeorge Bundy, Memos to the President, Top Secret. The instructions were approved by the president on December 3 (see Document 434) and formally transmitted to Rusk, McNamara, and McCone as Tab. 2 to Document 440. *Foreign Relations of the United States, US Department of State, 1964–8, Volume I, Vietnam 1964*, Document 435.

9 US Army, "Operational Report – Lessons Learned for Period 1, November 1966–31 January 1967," RCS CSFOR – 65 (U), Department of the Army, Headquarters, US Army, Vietnam, 81, https://archive.org/stream/DTIC_AD0386164/DTIC_AD0386164_djvu.txt.
   Nonmilitary aid from the United States topped $230.6 million per year by 1964. James M. Carter, *Inventing Vietnam: The United States and State Building, 1954–1968* (New York: Cambridge University Press, 2008), 147.

10 Sheehan, *A Bright Shining Lie*, 624.

11 Carter, *Inventing Vietnam*, 176.

12 Allan E. Goodman and Lawrence M. Franks, "The Dynamics of Migration to Saigon, 1964–1972," *Pacific Affairs* 48, no. 2 (1975): 202.

13 Cooper et al., "The American Experience with Pacification in Vietnam – Volume I – An Overview of Pacification (Unclassified)," 49.

14 Dan C. Cavanaugh, "Some Observations on the Current Plague Outbreak in the Republic of Vietnam," *American Journal of Public Health* (April 1968): 742–3.

15 "We will expand our recent efforts to respond to problems of wartime expansion of urban population primarily through self-help projects. We will encourage municipal participation in facilities such as bus transportation and garbage collection," US Department of State, "Ambassador Bunker's Seventy-Second Weekly Message from Saigon Briefing President Johnson on the Present Situation in Vietnam," Cable, Secret, October 30, 1968, Declassified October 27, 1994. For contractors and US participation see US Department of State, "Political Stability and Security in South Vietnam," Samuel P. Huntington, Miscellaneous, Secret, December 1, 1967, Declassified September 17, 1990, 97–9; Martin J. Murray, "The Post-colonial State: Investment and Intervention in Vietnam," *Politics & Society* 3, no. 4 (1973): 456; US Agency for International Development (USAID), Joseph A. Medenhall, "U.S. Government, USAID and U.S. CORDS Objectives and Organization in Vietnam," July 22, 1969, 19. And regarding contractors for sanitation services see US Navy, "Danang U.S. Naval Support Activity/Facility Danang – Command History – 1970," OPNAV Report 5750–1, May 10, 1971, 116.

16 This request was coded as partial compliance, because effort was made by the GVN, such as workers hired back and trucks put on the street.

17 In 1966 USAID and the GVN commissioner initially authorized 160–270 classrooms for refugees. A few months later this was deemed inadequate as USAID and GVN officials discussed an additional 780 classrooms to meet increasing demand. US Department of State, "Status of Activities Social, Political, and Economic Pertaining to the Honolulu Conference," Memorandum, Secret, April 7, 1966, Declassified February 25, 1980; US Department of State, "Status Report on Political and Economic Reform in Vietnam," Report, Secret, May 26, 1966, Declassified February 25, 1980; US Department of State, "Response to DOS Telegram 19056 on Refugees," Secret, August 17, 1966, Declassified July 7, 1993.

18 USAID, "United States Economic Assistance to South Vietnam 1954–75," Terminal Report, (Washington, DC: Asia Bureau, Office of Residual Indochina Affairs), December 31, 1975, Volume II, 7. A US Congressional investigation summarized CIP's multiple stages:

AID issues procurement authorizations permitting the purchase of certain approved commodities and sets up dollar credits against these authorizations in U.S. banks. In Vietnam, importers registered with the GVN invite bids ... The United States makes payment to the suppliers in these countries through letters of credit ... The importer obtains a letter of credit from his bank and an import license from the GVN. He then forwards his order to the supplier. The U.S. bank where AID has set up the dollar credits makes payment ... The

Vietnamese importer pays his bank the piaster (local currency) equivalent of the dollar cost ... and then sells it on the local market. The Vietnamese bank pays the piasters received from the importers into a GVN-owned counterpart fund which theoretically is jointly controlled by the United States and the GVN.

Congressional Record, "An Investigation of the U.S. Economic and Military Assistance Programs in Vietnam," 42nd Report by the Committee on Government Operations, House Report 2257, 89th Congress, 2nd Session, October 12, 1966 (Washington, DC: US Government Printing Office), 10–11.

19 Congressional Record, "An Investigation of the U.S. Economic and Military Assistance Programs in Vietnam," 39.

20 US Department of State, "For the President from Bunker," 411.

21 Congressional Record, "An Investigation of the U.S. Economic and Military Assistance Programs in Vietnam," 39.

22 Spencer C. Tucker (ed.), *Encyclopedia of Insurgency and Counterinsurgency: A New Era of Modern Warfare* (Santa Barbara, CA: ABC-CLIO, 2013), 407.

23 The Pentagon Papers, "[Part IV. C. 8.] Evolution of the War. Re-emphasis on Pacification: 1965–1967, 1," ARC Identifier 5890510, Series: Report of the Office of the Secretary of Defense Vietnam Task Force, 6/1967–1/1969, Record Group 330: Records of the Office of the Secretary of Defense, 1921–2008, US National Archives, https://catalog.archives.gov/id/5890510.

24 Ibid., 3. See also p. 5.

25 US Department of State, "Ambassador Bunker's Weekly Report to President Johnson Regarding the Situation in Vietnam," January 24, 1968, Declassified June 12, 1997; US Department of State, "Project TAKEOFF – Action Program," Undated, Central File, Pol27 Viet S., Douglas Pike (ed.), *The Bunker Papers, Reports to the President from Vietnam, 1967–1973*, Vol. I, 52. Also cited in William Conrad Gibbons, *The U.S. Government and the Vietnam War: Executive and Legislative Roles and Relationships, Part IV* (Princeton University Press, 1995), 716.

26 Richard A. Hunt, *Pacification: The American Struggle for Vietnam's Hearts and Minds* (Boulder, CO: Westview Press, 1998), 100.

27 Ibid., 236. Defenders of the Phoenix program point to the inherent brutality of revolutionary warfare and the strategic effectiveness of the program. Yet many consider the program unethical. The Phoenix program received widespread criticism as an example of US human rights abuses in the execution of the Vietnam War.

28 For example in May 1970 the prime minister moved the Phung Hoang central office to the Directorate General of the National Police in order to better centralize efforts and coordinate interagency intelligence collection after US complaints that GVN agencies did not cooperate sufficiently and were not giving the program sufficient prioritization. Hunt, *Pacification*, 237.

29 Lewis Sorley, *A Better War: The Unexamined Victories and Final Tragedy of America's Last Years in Vietnam* (New York: Mariner Books, 2007), 147; Hunt, *Pacification*, 234, citing Stanley Karnow, *Vietnam, A History*, revised ed. (New York: Viking, 1991), 602.

30 Hunt, *Pacification*, 2.

31 US Department of State, US Embassy Saigon, "Telegram from the Embassy in Vietnam to the Department of State," November 9, 1964. And US Department of Defense, "Memorandum from the Director, Far East Region, Office of the Assistant Secretary of Defense for International Security Affairs (Blouin) to the Assistant Secretary of Defense for International Security Affairs (McNaughton)," Washington, March 30, 1964 in *Foreign Relations of the United States, 1964–1968, Volume 1, Vietnam, 1964,* Document 101, http://history.state.gov/historicaldocuments/frus1964–68v01.

32 The Pentagon Papers, "[Part IV. C. 1.] Evolution of the War. U.S. Programs in South Vietnam, November 1963–April 1965," NASM 273 – NSAM 288, ARC Identifier 5890498, Series: Report of the Office of the Secretary of Defense Vietnam Task Force, 6/1967–1/1969, Record Group 330: Records of the Office of the Secretary of Defense, 1921–2008, US National Archives, 60, https://catalog.archives.gov/id/5890498. There had been a steady decline in the strength of RVNAF since October 1963, notably including a decrease of 4,000 in March alone; and the current strength was almost 20,000 below the authorized figure agreed necessary by both governments.

33 Ibid., 104.

34 Robert K. Brigham, "Dreaming Different Dreams: The United States and the Army of the Republic of Vietnam," in *A Companion to the Vietnam War,* ed. Marilyn B. Young and Robert Buzzanco (Malden, MA: Wiley Blackwell, 2002), 148.

35 Herring, *America's Longest War,* 190.

36 Approximately 33 million US dollars using the official piaster exchange rate 118 to $1. However, note that due to inflation, the amount may be considered significantly less. Request for increases in taxes on oil referenced in US Department of State, "Ambassador Bunker's Seventy-Second Weekly Message from Saigon Briefing President Johnson on the Present Situation in Vietnam," 9.

37 David Brown, "The Development of Vietnam's Petroleum Resources," *Asian Survey* 16, no. 6 (June 1976): 553–70.

38 Douglas C. Dacy, *Study S-337, The Fiscal System of Wartime Vietnam,* (Arlington, VA: Institute for Defense Analyses Program Analysis Division, February 1969), 71, www.dtic.mil/docs/citations/AD0692611.

39 Prados, "The Shape of the Table."

40 Asselin, *A Bitter Peace,* 14–15.

41 US Department of State, "Ambassador Bunker's Seventy-Second Weekly Message from Saigon Briefing President Johnson on the Present Situation in Vietnam," 9.

42 Anh Tuân Nguyên, *South Vietnam, Trial and Experience: A Challenge for Development,* Ohio University, Monographs in International Studies, Southeast Asia Series, no. 80 (Athens: Ohio University Press, 1986), 203–5.

43 A lieutenant commander from Medical Service Corps assigned to III MAF Headquarters (Marine Amphibious Force) was charged with organizing medical training efforts and personnel for the Vietnamese. US Department of Defense, "Report for the President," June 28, 1966, Secret, Special Handing Required, Declassified June 28, 1978, Johnson Library, White House Central

File, Confidential File, Subject Reports, Department of Defense, June 1966, Declassified Documents Reference System; *GALE Declassified Documents Database*, titled "[Summary of aircraft sorties flown in Vietnam; conduct of Market Time on South Vietnamese coast; review of selected ground operations in Vietnam; supply of new rifles for troops in Vietnam; testing of missiles and warheads; other subjects.] Report for the President. June 28, 1966. 74 p. SECRET to UNCLASSIFIED. SPECIAL HANDLING REQUIRED. SANITIZED copy. Released June 28, 1978. Johnson Library, White House Central File, Confidential File, Subject Reports, DOD, June 1966," Document Number CK2349410934, http://tinyurl.galegroup.com/tinyurl/6rVNo7.

44 The Pentagon Papers, "[Part IV. C. 9. a.] Evolution of the War. U.S. GVN Relations. Volume 1: December 1963– June 1965," ARC Identifier 5890511, Series: Report of the Office of the Secretary of Defense Vietnam Task Force, 6/1967–1/1969, Record Group 330: Records of the Office of the Secretary of Defense, 1921–2008, US National Archives, 17, https://catalog.archives.gov/id/5890511.

45 US Department of State, "Transmittal Memorandum, Leonard Unger, Dep. Asst. Secy of State for Far Eastern Affairs and Chairman, Vietnam Coordinating Committee, to Chet [Chester Cooper]," Secret, March 20, 1965, Declassified March 17, 1980, Sanitized, Johnson Library, NSF, Countries, Vietnam, Vol. 31, 6. "In order to make maximum use of available manpower resources and deny rural youth to VC recruiters, reduce the draft age from present 20 years to 18 or 17." The USA was still pressing for similar reforms in late 1967: "Improvement of the armed forces. This would include fairer and more effective dependent, survivor and disabled veterans' benefits, lower draft age, improved merit promotion system, and more effective punitive measures for soldiers who mistreat the civilian populace." US Department of State, US Embassy Saigon, "Post-election Priorities in South Vietnam Detailed," Cable, Secret, September 2, 1967, Declassified December 15, 1994, Unsanitized, Complete, 2. See also Pentagon Papers,

By the end of December 1967, MACV was recommending a further increase of 366 advisors for the FY 1969 program, primarily for district level intelligence slots. Meanwhile, on September 28, the JCS had forwarded with their endorsement the MACV-CINCPAC recommendation on RVNAF force increases, of which the RF/PF component was the largest. Requested was an increase in FY 68 RVNAF authorized strength from 622,153 to 685,739, a net of 63,586. Of this number, 47,839 were RF/PF spaces, and only 15,747 were for the regular forces (of which ARVN's share was 14,966). To achieve these higher levels, MACV proposed the reduction of the draft age from 20 to 18 and the extension of tours of duty for active RVNAF personnel.

"[Part IV. C. 8.] Evolution of the War. Re-emphasis on Pacification: 1965–1967."

46 Thomas C. Thayer (ed.), "A Systems Analysis View of the Vietnam War: 1965–1972. Volume 2. Forces and Manpower" (Washington, DC: Office of

the Assistant Secretary of Defense Southeast Asia Intelligence Division), February 18, 1975, https://apps.dtic.mil/dtic/tr/fulltext/u2/a051609.pdf, 58.

47 Michael A. Hennessy, *Strategy in Vietnam: The Marines and Revolutionary Warfare in I Corps, 1965–1972* (Westport, CT: Praeger, 1997), 153.

48 William C. Westmoreland, "Chapter IV – The Year of Decision (1968)," in "Report on Operations in South Vietnam, January 1964–June 1968," in *Vietnam War: After Action Reports* (Beverly Hills, CA: BACM Research, 2009), 170.

49 Keesing's, *South Vietnam: A Political History, 1954–1970* (New York: Scribner, 1970), 139–40.

50 William C. Westmoreland, "Appendix B – Republic of Vietnam Armed Forces," in "Report on Operations in South Vietnam, January 1964–June 1968," in *Vietnam War: After Action Reports* (Beverly Hills, CA: BACM Research, 2009), 217.

51 Komer, "Bureaucracy Does Its Thing," 28–9.

52 Regarding rural development the document notes a two-tiered system that centralizes policy planning and decentralizes resources. "Centralized administrative authority at the Saigon level is necessary to accomplish these objectives in the countryside," while under this primary policy control the GVN should work toward "decentralizing decision making and resources to that district and provincial chiefs can be quickly responsive to the desires of the people in the hamlets and villages." US Department of State, "Blueprint for Vietnam," National Archives Records Administration, RG 59, S/S–S Files: Lot 70 D 48, Misc. VN Rpts. & Briefing Books. Chapter IV – National Development, 10–11. Referenced in "Editorial Note," *Foreign Relations of the United States, 1964–1968, Volume V, Vietnam, 1967*, Document 296. In September 1967 Ambassador Bunker clarified the restructuring priority for rural development and focused US pressure on centralization:

I believe we should use our influence immediately after the election to have [Thieu and Ky] do the following ... 4. Centralizing all rural development efforts in non-Revolutionary Development hamlets (education, agriculture, public works, public health, etc.) under one coordinated control in the same manner in which they are now centralized under the Ministry of Revolutionary Development in Revolutionary Development hamlets; funding resources at the provincial level for non-Revolutionary Development hamlets in the same manner in which they are presently funded for Revolutionary Development hamlets, so that coordinated programs can be established in each village with flexibility and with resources quickly available.

US Department of State, US Embassy Saigon, "Post-election Priorities in South Vietnam Detailed."

53 US Department of Defense, "South Vietnam's Internal Security Capabilities," National Security Study, Secret, May 1, 1969, Declassified March 17, 2004, 20.

54 It was reported that

province and district officials have had problems controlling ARVN and directing the PRU [Provincial Reconnaissance Units], Special Police Branch and National Police Field Forces. All of these attitudinal factors and other difficulties from Saigon to village and hamlet level combine to make it difficult to fully integrate and combine political/military operations which are required by the US/GVN methodology to provide adequate security ... In the past, there has been a constant balancing off of political and military forces by the various incumbent political powers in attempts to ensure the continuance in power. Previous American attempts to convince the GVN to centralize control over various internal security forces in accordance with our organizational principles often have gone astray because they ran counter to the realities of SVN political life. US Department of Defense, "South Vietnam's Internal Security Capabilities," D-11

Several problems soon developed when efforts were made to introduce new organizational structures. "There either was no follow-on government presence to continue the RD work, or there was an existing administrative structure in the hamlet village which resented RD Cadre interference and, further, which was often ignored by the cadre thus exacerbating the differences between the cadre and the existing structure." US Department of Defense, "South Vietnam's Internal Security Capabilities," B-17.

55 Paul Pierson, *Politics in Time: History, Institutions, and Social Analysis* (Princeton University Press, 2004), 43–4.
56 Berman, *No Peace, No Honor*, 54; Lloyd C. Gardner and Ted Gittinger, *The Search for Peace in Vietnam, 1964–1968* (College Station: Texas A&M University Press, 2004), 361.
57 Ronald H. Spector, *After Tet: The Bloodiest Year in Vietnam* (New York: Vintage, 1994), 305. Quoting Clarke Clifford, "Annals of Government: Serving the President, the Vietnam Years," *The New Yorker*, May 13, 1991, 87.
58 Prados, "The Shape of the Table," 360–1.
59 Berman, *No Peace, No Honor*, 54. See also The White House, "Henry A. Kissinger Cable to Ambassador Bunker on President Thieu's Reply to President Nixon Regarding Peace Negotiations with Hanoi and the Withdrawal of North Vietnamese Troops from South Vietnam Meeting with Alexander Haig, Äù," Cable, Top Secret, November 12, 1972, Gerald R. Ford Presidential Library, Declassified May 6, 1997.
60 US Department of State, US Embassy Saigon, "Telegram 10019 from Saigon," June 24, 1971, National Archives, RG 59, Central Files 1970–73, POL 14 VIET S, Telegram 6169 from Saigon; The White House, "Henry A. Kissinger Cable to Ambassador Bunker; US Department of State, US Embassy Saigon, "For the President's Files – Lord, Vietnam Negotiations," Sensitive, Camp David, Cables, 10/69–12/31/71; US Department of State, US Embassy Saigon, "Telegram from the Embassy in Vietnam to the Department of State," Saigon, August 20, 1971. See Documents 250 and 225 – Editorial Note, *Foreign Relations of the United States, 1969–1976, Vietnam, Volume VII, 1971.*
61 Herring, *America's Longest War*, 277.
62 Ibid., 277–8.

7   India in Sri Lanka

India is engaged in threatening the future of Sri Lanka and perhaps the region. Unless India acts positively, the future of Sri Lanka will witness hard times but so will the future of India. You cannot play with fire without worrying about the spread of it to beyond what you can control.   —Sri Lankan President Jayewardene

Labeled "India's Vietnam,"[1] New Delhi's bungled intervention in Sri Lanka lasted from July 1987 to March 1990. Yet, despite its short duration compared to other large-scale counterinsurgency interventions, the war cost India more than $1.25 billion, and left 1,155 Indian soldiers, including 49 officers, killed.[2] Despite these high costs, as Ashok K. Mehta, General Officer commanding Indian forces summarized, the intervention is generally considered an utter "political and military failure."[3]

Summary Findings

Sri Lankan rates of compliance with Indian policy requests during intervention were similar to that of other local allies in large-scale counterinsurgency (COIN) interventions: 38.0 percent full compliance, 24.1 percent partial compliance, and 38.0 percent noncompliance. Similar to the US wars in Vietnam, Iraq, and Afghanistan there was an interaction effect between the convergence or divergence of allies' interests, and the dependency of intervening forces on local actors for implementing a given policy. Furthermore, due to the relative high capacity of the Sri Lankan regime compared to other local counterinsurgency allies, noncompliance due to insufficient capacity was uncommon. However, unlike South Vietnam's increased compliance with US demands in response to the Tet Offensive, an increase in enemy activity made Colombo less likely to comply with New Delhi's requests. Instead, insurgency offenses motivated Colombo to ignore its Indian partners and seek independent security solutions to the insurgent threat, demonstrating

Table 7.1 *Timeline of the Sri Lankan–Indian alliance during the Indian intervention*

July 1983	The Tamil secessionist group, led by Velupillai Prabhakaran, Liberation Tigers of Tamil Eelam (LTTE) attack Sinhalese soldiers in an army post in Jaffna leading to anti-Tamil riots that culminate in an insurgency marking the onset of the First Tamil War.
1987	India intervenes with 65,000 troops, the Indian Peace Keeping Force (IPKF). Its goal is to implement a cease-fire and enforce an agreement between the Tamils and Sinhalese. The cease-fire breaks down in less than two months leading to a violent struggle between India and its one-time ally, the LTTE.
May 1989	Elected in January of 1989, President Ranasinghe Premadasa brokers a cease-fire agreement with the LTTE and India promises to withdraw from Sri Lanka.
1990	In March, India completes its withdrawal from Sri Lanka with 1,155 Indian soldiers dead. In June the thirteen-month-old cease-fire breaks down and violence resumes.
May 2009	With Prabhakaran and the top leadership of the LTTE killed, the Sri Lankan government announces the end of the war and victory over the LTTE.

variation in how local allies can cope with the pressing threat of an insurgency. In order to analyze these findings in alliance politics in the Sri Lankan case, it is first important to summarize the politics behind the Indian counterinsurgency effort from 1987 to 1990 that set the tone for this tense partnership.

Sri Lanka, India, and the Tamil Insurgency (LTTE) – A Briefing

India's shifting alliances between both insurgent and counterinsurgent created complex political dynamics in the Indian military intervention in Sri Lanka. From 1983 to 1990 New Delhi switched teams, initially arming, funding, and training Tamil independence groups, to later sending troops into Tamil regions of Sri Lanka to fight those very Tamil separatists. Hundreds of Indian soldiers were killed by Tamil militants who were trained and armed by Indian intelligence agents just years earlier. New Delhi simultaneously sought to coerce the Sri Lankan government into compromising with Tamil leaders by funding the Tamil Tiger insurgents (LTTE), while also coercing the LTTE by sending troops to uphold a primarily Indian-designed agreement signed by

Colombo and New Delhi seeking to disarm the LTTE. An overesti-
mation of its ability to coercively influence both the LTTE as a proxy
faction and the government in Colombo as a proxy ally backfired, costing
New Delhi both politically and militarily. After almost three years of
fighting, India withdrew when the Tamil insurgents and the Sri Lankan
government counterinsurgents grew tired of India's meddling and covertly worked together to expel Indian forces from Sri Lanka. The Sri
Lankans battled each other for another twenty years until the government
finally triumphed in a notoriously bloody campaign.[4]

Understanding the dynamics of the Indian–Sri Lankan alliance
requires a summary of the dynamics of the insurgency underpinning
the war. Following its 1948 independence from British rule, Sri Lankan
Sinhalese nationalists promoted Sinhalese social, cultural, and political
dominance of the multiethnic Sri Lankan state on behalf of the Sri
Lankan Sinhalese majority. Measures such as making Sinhala the official
language and Buddhism the official religion of the state, as well as
restricting the number of Tamils admitted to universities and disenfran-
chising Tamil tea plantations, marginalized Sri Lanka's Muslim and
Tamil Hindu minorities.[5] The extent of systemic exclusion of Tamils
from government benefits is evident in the requests made by New Delhi
to Colombo as part of the Indian counterinsurgency strategy. In 1989,
for example, New Delhi requested that Colombo recruit more Tamils
into the security forces. Indian military officials were alarmed that
Tamils, who make up approximately 20 percent of the population and
in the 1950s made up 40 percent of the Sri Lankan state military,[6]
constituted less than 1 percent of the total sanctioned strength of
72,665 security personnel.[7] India insisted that, "immediate steps ... be
taken to ensure that the Army, Navy and Air Force reflect the ethnic
composition of the country both among the Regulars and the Volunteers."[8] Sri Lankan officials agreed, but the percentage of Tamils
included in government forces did not notably change.[9]

In 1983 activity by the preeminent Tamil separatist group, the LTTE,
prompted anti-Tamil riots in Colombo. The violence precipitated an
exodus of almost 300,000 Tamil refugees to India. A 60 million-strong
Indian Tamil population was sympathetic to the Tamil cause in Sri
Lanka and petitioned New Delhi to intervene, marking the start of
India's expanding involvement in the Sri Lankan Tamil problem.[10] By
March 1988 more than 100,000 Indian troops were deployed in North-
ern and Eastern Sri Lanka.[11]

Following the 1983 riots, New Delhi started covertly aiding Tamil
separatists in order to pressure Colombo to embrace Tamil rights. New
Delhi hoped that such support would effectively push Colombo toward a

political solution acceptable to the Indian Tamils and inspire Colombo to abandon its attempts to achieve a military victory over the LTTE. Furthermore, India was also concerned that Colombo was growing increasingly friendly to Pakistan, Israel, China, and the USA. According to the strategic logic of Prime Minister Indira Gandhi, strengthening the LTTE would draw Sri Lanka away from Pakistani and Israeli counter-insurgency advisers if they collectively failed to stifle the Tamil independence movement. A stronger LTTE would force Sinhalese leaders in Colombo to compromise with the Tamil insurgents and New Delhi. Since the LTTE drew support from its safe haven in India's southeastern state, Tamil Nadu, New Delhi was key to addressing the insurgency in Sri Lanka. Colombo could not win decisively without addressing the Indian LTTE safe haven. Simultaneously, the Gandhi political dynasty would be better positioned to win elections with Indian Tamil support, a reward for its efforts to promote the rights of the Sri Lankan Tamils.

Although never officially acknowledged by the Indian government, it is widely accepted that by 1986, Indian intelligence was arming as many as 20,000 Tamil militants.[12] Led by India's Research and Analysis Wing (RAW), these insurgent forces were reportedly given offices, weapons, and training in Tamil Nadu in as many as thirty-two different camps. During this period these groups were reportedly better armed and potentially better trained than the Sri Lankan government troops stationed in northern Sri Lanka tasked with confronting the Tamil uprising.[13] The insurgents received assistance from the population and local leaders, but according to unconfirmed sources, also had direct Indian state ties through RAW agents and other prominent Indian officials.[14] Reportedly, "some militant leaders were in fact living in the quarters provided by the State Government to Tamil Nadu politicians."[15] This sanctuary for Tamil militants in India meant that Sri Lankan forces could never militarily defeat the insurgents without Indian assistance. This in turn provided a significant source of leverage for New Delhi. The USA and the USSR would discover something similar regarding Pakistan during their interventions in Afghanistan: fighting Afghan insurgents supported by Pakistan gave Islamabad tremendous influence in Afghanistan's conflict and leverage over two superpowers.

Due to India's support for the Tamil insurgents prior to 1987, there was substantial hostility between Colombo and New Delhi, which came to a head in the spring of 1987. Sri Lankan forces launched an aggressive military campaign to clear the LTTE from the northern city of Jaffna. Indian Prime Minister Rajiv Gandhi condemned the Sri Lankan response. New Delhi reportedly sent Sri Lankan President J. R.

Jayewardene a blunt message: "India will not allow Jaffna to be taken ... India will arm the LTTE with Surface to Air Missiles (SAM-7 and 8)."[16] In part due to this pressure, Sri Lankan forces stopped short of taking the city, but the Jaffna crisis was not over.

On June 1, 1987 Indian High Commissioner J. N. Dixit announced that due to suffering in Jaffna caused by recent Sri Lankan government offensives and Colombo's five-month blockade of the port, India was sending aid to Jaffna.[17] In a public announcement Dixit asked Colombo to allow Indian Red Cross ships through the Sri Lankan naval blockade. Sri Lankan officials were irate. Colombo issued a statement calling the allegations of suffering

unfounded ... While pointing out that the tragic situation in Sri Lanka would not have become acute as at present but for the patronage of separatist terrorism by the State of Tamil Nadu, a constituent of the Republic of India, the Government of Sri Lanka wishes to point out that neither has the Government of Sri Lanka solicited any humanitarian aid nor does the situation obtaining in the North require any assistance from any outside sources as the Government of Sri Lanka is in a position to meet all the requirements.[18]

On June 3, 1987, nineteen Indian boats arrived in Sri Lanka with humanitarian supplies. After a lengthy negotiation between navies, the Indian fleet was denied entry. Indian supplies would have eased suffering in Tamil regions, likely aiding the LTTE, and portrayed Indian forces as rescuers. New Delhi was embarrassed by the denial. Despite its size and regional power, it failed to coerce Colombo to let its ships through. Nonetheless, Indian embarrassment was short-lived. The next day Indian Air Force pilots entered Sri Lankan air space without permission and dropped 25 metric tons of food and supplies over Jaffna, sending a clear message to Colombo. India could do what it wished in Sri Lankan territory, regardless of the resistance put forward by Colombo.[19]

In light of this hostile interaction, as well as India's size, regional domination, military superiority, and a Tamil population three times the size of total population in Sri Lanka, it is not surprising that Colombo considered India its primary security threat.[20] What is surprising then is how New Delhi and Colombo became military and political allies working to suppress the Tamil insurgency less than two months after the Jaffna airdrop.

After several years of providing covert support for Sri Lankan militants, by 1987 New Delhi was losing its ability to influence the insurgents. Prime Minister Gandhi was reportedly told by RAW officials, "India was losing control over the militants. The militants in general and the LTTE in particular, were not responsive to India and were charting their own plan of action … a settlement must be reached quickly."[21] Furthermore, LTTE presence was destabilizing Tamil Nadu, a sizable state that has been labeled "the Kashmir of the South," for its potential for political turmoil. Tamil Nadu's strong Tamil identity, sympathy for the LTTE, and ethno-political fault lines made it a potential "problem for the unity of India," according to historian Rohan Gunaratna.[22] Prime Minister Rajiv Gandhi sought to diffuse the Tamil conflict in Sri Lanka in order to avoid further destabilizing the Tamil-dominated parts of India. If successful, the LTTE's goal of Sri Lankan Tamil independence could potentially pose a threat to India, which perennially struggled to balance powerful regional groups. Additionally, caring for the 150,000 Sri Lankan Tamil refugees in Tamil Nadu was expensive, and a political settlement would enable repatriation.[23] For India, a political solution in Sri Lanka could also put an end to Colombo's military relations with Pakistani and Israeli contractors. New Delhi sought a unified Sri Lanka that provided sufficient concessions to its Tamil minorities to pacify the insurgency and its supporters in Tamil Nadu.

The Indian and Sri Lankan governments held bilateral negotiations on the Tamil issues in the summer of 1987. India "represented" the Tamil groups, who were excluded from the drafting process. On July 29, 1987 the two states signed the Indo-Sri Lanka Accord. Sri Lankan President Jayewardene argued that the July 1987 accord, as well as the 10,000 Indian troops initially deployed to Sri Lanka to impose the ceasefire and LTTE weapons surrender, posed no threat to Sri Lankan sovereignty.[24] Nevertheless, there was significant trepidation regarding India's intrusion into the Sri Lankan civil war. As a prominent congressional member articulated,

These matters have caused a grave sense of fear that Sri Lanka has lost her independence and sovereignty – whatever the Government might say to the contrary … Our people are tired of being lectured to about how independent we are and that our sovereignty has not been violated, when the events that are unfolding daily show proof to the contrary.[25]

The ambitious agreement intended to put an end to Sri Lanka's insurgency. New Delhi believed it could deliver the Tamil militants in a peace negotiation. Sri Lankan President Jayewardene conceded a great deal in the accord, including agreeing to temporarily unify the Eastern and Northern provinces of Sri Lanka that effectively united Tamil areas, granting amnesty to militants, adding Tamil as an official language, and withdrawing all Sri Lankan military forces from the Northern and Eastern regions. It was far more than Colombo had previously agreed to, in no small part because of pressure put on President Jayewardene from New Delhi. As Sri Lankan National Security Minister R. Wijeratne lamented, "If we did not sign [the] Accord we would have had to fight with India. India would not have come forward [to help the Sri Lankan government], but would have given arms [to the Tamil separatists] to shoot down helicopters."[26] In exchange for the concessions given to the Tamils, India promised to enforce a cease-fire, oversee a surrender of arms by Tamil militants, and effectively ensure the unity of Sri Lanka by delivering the militants without giving into Tamil demands for independence.

Like so much in counterinsurgency, the Indo-Sri Lanka Accord did not go according to plan. All sides initially greeted Indian troops landing in Jaffna warmly, including LTTE sympathizers, Tamil civilians, and Sri Lankan government forces.[27] But Tamil goodwill toward the IPKF was short-lived. By the end of August 1987, difficulties arose, and the LTTE began to turn against the IPKF. Among other issues, a mass suicide of LTTE commanders captured by the Sri Lankan Navy inspired remaining LTTE officials to distrust Indian mediation, and to reject IPKF authority to enforce the Indo-Sri Lanka Accord that they never explicitly approved. The IPKF became a direct LTTE target and the Indian mission quickly turned violent.

The ensuing Indian-led counterinsurgency effort to disarm the LTTE continued until the March 1990 withdrawal of Indian forces at the request of newly elected Sri Lankan President Ranasinghe Premadasa. Elected in 1989, Premadasa had worked secretly with the LTTE to come to an agreement to drive Indian forces out of Sri Lanka.[28] They succeeded, but the LTTE did not cooperate with any government for long. Three years later Premadasa and former Sri Lankan National Security Minister Lalith Athulathmudali were both killed in attacks orchestrated by the LTTE. LTTE operatives also assassinated Indian Prime Minister Rajiv Gandhi with a suicide attack in Tamil Nadu, India on May 21, 1991. In 1992 India declared the LTTE a terrorist organization.

In order to investigate trends in Sri Lankan compliance with requests from India during the volatile Indian-Sri Lankan partnerships from 1987 to 1990, I analyze seventy-nine Indian policy requests identified by searching through 500+ government documents from India and Sri Lanka (900+ pages). The documents are available in English in a five-volume document collection, "India–Sri Lanka: Relations and Sri

Lanka's Ethnic Conflict Documents 1947–2000," edited by Avtar Singh Bhasin, the former director of the Historical Division at the Indian Ministry of Foreign Affairs. The collection contains a variety of documents including press releases, transcripts from parliamentary debates, formerly classified memos, and correspondence between the Sri Lankan president and Indian prime minister initially classified as "Top Secret." The seventy-nine policy requests identified date from December 19, 1986 to November 3, 1989, while the intervention lasted from July 1987 to March 1990. As observed in other counterinsurgency interventions fewer demands are made from intervening forces as they prepare to leave. In general, there are fewer policy requests to analyze compared to the wars detailed in previous chapters due to the relatively short duration of the Indian intervention when compared to the US interventions in Iraq, Afghanistan, or Vietnam, which were significantly longer engagements, allowing for additional bargaining encounters between local and intervening forces.

In order to determine the outcome of each request from New Delhi I relied on documents in the previously referenced collection, collaborated with other sources such as Ministry of External Affairs publications as well as media reports, the accounts of retired Sri Lankan and Indian military officials, reports by international organizations, and historical sources. Lastly, I conducted interviews with former Indian officials and Indian counterinsurgency experts in New Delhi in February 2013, including Lieutenant General Amarjeet Singh Kalkat, former commander of the IPKF.

What Kinds of Policies Did India Ask for in Sri Lanka?

Indian forces asked Colombo to institute a variety of reforms, including addressing refugees, accepting Indian proposals for humanitarian assistance, holding elections to provincial councils, recognizing Tamil parties, extending amnesty, and including Indian officials in negotiations. Table 7.2 provides a summary of Indian requests to Sri Lankan partners identified in archival records.

To analyze requests across certain topics, six general categories of subjects were created to classify requests. While helpful for gaining an idea of what New Delhi asked for, the general subject of requests was not found to be significant in determining the likelihood of compliance with or not as other structural factors had a greater influence on compliance outcomes than the general issue addressed.

Table 7.2 *Summary of Indian requests to Sri Lanka*

	Year	Indian request to Colombo
1.	1986	Redefine Eastern Province
2.	1986	Establish provincial council for new Eastern Province
3.	1987	Trincomalee oil tank farm
4.	1987	Rehabilitate militant youths
5.	1987	Return of displaced persons
6.	1987	Transfer paramilitary forces into regular security forces
7.	1987	Sri Lankan forces to stay confined to barracks
8.	1987	Trincomalee or other ports not to be available for military use by any country against India
9.	1987	Foreign military and intelligence personnel – discussions regarding their role in Sri Lanka
10.	1987	Withdraw paramilitaries and homeguards
11.	1987	Lift embargo on Jaffna
12.	1987	Monitor cessation of hostilities
13.	1987	Accept Indian proposal for humanitarian assistance for Jaffna
14.	1987	Stop military activities in the north
15.	1987	North and East Provinces united (temporarily)
16.	1987	Foreign broadcasting not for military or intelligence purposes
17.	1987	Announcement of interim Administration in the NE Province before elections (on or around August 5)
18.	1987	Amend 1956 Official Languages Act to make Tamil an official language
19.	1987	Referendum in the Eastern Province regarding unity with North – on or before December 31, 1988
20.	1987	Monitor referendum by committee headed by Chief Justice
21.	1987	Devolution – increased provincial autonomy
22.	1987	Amend legislation regarding provincial councils
23.	1987	Announce northern and eastern areas united
24.	1987	Lift emergency in NE by August 15, 1987
25.	1987	Law enforcement in NE same as the rest of the country
26.	1987	Agree to LTTE suggestions regarding the composition of the Interim Administrative Council
27.	1987	Affirm proposals presented from May to December, 1986
28.	1987	Hold elections to provincial councils before December 31, 1987
29.	1987	Promote full participation of all voters from NE
30.	1987	Recognize the "multiethnic" character of Sri Lanka
31.	1987	Amnesty to militants who surrender arms
32.	1988	Refugees to be sent to Trincomalee
33.	1988	Get the finance commission functioning
34.	1988	*Pradeshiya Sabhas* [village councils] and district councils – do not dilute role of provincial council
35.	1988	New legislation allowing displaced Tamil citizens a postal ballot
36.	1988	Issue directives to the election commissioner
37.	1988	Initiate the process of elections in North-Eastern Province
38.	1988	Election orders and notices should reference the combined North-Eastern Province
39.	1988	Appoint a governor for North-Eastern Province

Table 7.2 (*cont.*)

	Year	Indian request to Colombo
40.	1988	Recognition of Tamil political parties
41.	1988	Elections to provincial councils around May 1, 1988
42.	1988	Announce provincial council elections and appoint governors
43.	1988	Reiterate amnesty and election eligibility to those abiding by Indo-Sri Lanka Accord
44.	1988	Extend amnesty to Tamil militants abiding to the Indo-Sri Lanka Accord
45.	1988	Determine a timeframe for elections, concluding by the last week in November
46.	1988	Announce all detainees to be released if all groups negotiate
47.	1988	Assure Tamil groups of improvements, amnesty, and strengthened devolution
48.	1988	Announce use of 1982 register for electoral purposes, one NE council, and the release of detainees
49.	1988	Early order postponing nominations
50.	1988	Appoint one high court for North-Eastern Province
51.	1988	Issue notification in May calling for election to a single council – not two separate councils
52.	1988	Change government agent in Vavuniya and Jaffna
53.	1988	Representatives of the election commission of India and the Tamil United Liberation Front (TULF) to inspect voters' list
54.	1988	Exclude names of individuals settled after 1983 in the NE region from voting – base registration on 1982 registers
55.	1988	Provide India with copies of electoral rolls from 1982 to 1986
56.	1988	Postpone deadline for election nominations
57.	1988	Tell India about the outcome of discussions with Tamil groups regarding the election
58.	1988	Notification for nominations to provincial council elections in the NE to be issued in April
59.	1988	Prominent Tamil leaders to participate in elections
60.	1988	Handle logistical and administrative arrangements for elections
61.	1988	Don't blame the IPKF for postponing election nominations
62.	1988	Order the release of all non-LTTE detainees prior to the poll announcement
63.	1988	Announce openings for Tamils in government
64.	1988	No bilateral talks between Sri Lanka and the Tamils
65.	1988	Act as an intermediary sending funds from India to the LTTE
66.	1989	Citizen Volunteer Force (CVF) to be inducted into the reserve and regular police
67.	1989	Armed forces of Sri Lanka to reflect ethnic makeup of society
68.	1989	Increase strength of CVF
69.	1989	Arm police and CVF personnel
70.	1989	Special task force out of Amparai
71.	1989	Maintain law through chief minister and CVF
72.	1989	Sri Lanka Army to vacate and relocate
73.	1989	CVF should take action to evict the LTTE and Sri Lanka Army to assist the CVF
74.	1989	Deputy inspector general responsible for law and order

Table 7.2 (*cont.*)

	Year	Indian request to Colombo
75.	1989	Sri Lanka Army limited to operations approved by the president and vice president
76.	1989	Police should reflect the ethnic makeup of society
77.	1989	Bill for national provincial police commissions
78.	1989	Adviser to the president to liaise with officials of North-Eastern Province
79.	1989	Convene peace committee meeting

Issue Areas of Indian Requests to Sri Lanka

1. *Development* – projects or activities intended to support economic growth and provide social services. Includes provision of services for refugees and youth programs. 4/79 (5.1 percent) of Indian requests.
2. *Economic reform* – actions intended to change economic policies. Includes the establishment of a financial council. 1/79 (1.3 percent) of Indian requests.
3. *Military reform* – actions intended to change military policies and institutions. Includes organizing paramilitary groups and dissolving military institutions. 5/79 (6.3 percent) of Indian requests.
4. *Military strategy* – actions intended to guide military forces in the execution of the war effort. Includes withholding force and the position of Sri Lankan forces. 14/79 (17.7 percent) of Indian requests.
5. *Political reform* – actions intended to change government policies and institutions. Includes electoral, law enforcement, governance, and amnesty protocols. 45/79 (57.0 percent) of Indian requests.
6. *Political-military counterinsurgency strategy* – actions intended to implement counterinsurgency strategy. Includes COIN projects, announcements recognizing Tamil grievances, and pacification activities. Note, there may be certain overlaps with political reform. 10/79 (12.7 percent) of Indian requests.

Development and economic reform. The likelihood of compliance based on the issues addressed was different in Sri Lanka compared to other interventions. For example, there were relatively few Indian requests for economic reform or development compared to the USA in Vietnam or USSR in Afghanistan. India's few development-related requests addressed Tamil refugees. This relative lack of Indian interest in development and economic reform may be attributable to the high capacity of the Sri Lankan state when compared to other local allies. India was unlikely to have felt the need to pressure Colombo to implement economic reforms or development projects, because policies were already in place. Even more critically, the relatively tenuous commitment of New Delhi to its local COIN partner meant, unlike other intervening allies, the Indians were not overly concerned with the long-term development of the Sri Lankan state. New Delhi, therefore, did not have substantial plans regarding financial restructuring. The few requests that were made had a mixed record of compliance, and ultimately were correlated with lower levels of compliance.

Military strategy and military reform. Available diplomatic documents outline five requests from New Delhi to Colombo regarding military reform, including a request to make the armed forces of Sri Lanka, dominated by the Sinhalese, reflect the diverse ethnic makeup of Sri Lankan society, transferring paramilitary forces into regular security forces, inducting the Citizen Volunteer Force (CVF) into the reserve and regular police force, increasing arms available to the CVF, and implementing policies that increased the strength of the CVF. Colombo implemented none of these reforms. Given the history of Indian support for insurgents to weaken Colombo, Sri Lankan officials were distrustful of Indian advice regarding military organization and ignored it entirely.

Colombo had a similar, but not quite as severe response to Indian requests pertaining to military strategy. In the US intervention in Vietnam military strategy was the most likely category to produce compliance by Saigon, yet this was clearly not the case in Sri Lanka. The South Vietnamese were much less likely to view requests for particular military approaches as a threat to their regime, since military actions targeted the enemy, and usually did not require internal reform. However, dynamics were different in Sri Lanka due to the historical Indian alliance with the Tamils. Colombo was inclined to suspect that Indian requests regarding military strategy could be problematic for Sri Lankan state security. Even though such requests targeted the Tamil threat instead of the Sri Lankan regime, Colombo remained wary since a strategic move favoring the insurgents could set back their campaign against the LTTE. Requests included vacating and relocating the Sri Lankan Army following the arrival of the IPKF, the cessation of military activities in the north, ordering the CVF to take action to evict the LTTE and to withdraw paramilitaries and "homeguard" troops.

Political reform and counterinsurgency projects. Similar to other interventions, the majority of requests from intervening forces related to political reforms New Delhi sought from Colombo. These included issues such as amending the 1956 Official Languages Act to make Tamil an official language, increasing provincial autonomy in Tamil regions, making law enforcement protocol in the northeast the same as in the rest of Sri Lanka, passing legislation allowing displaced Tamil citizens a postal ballot, recognizing Tamil political parties, and postponing election nomination deadlines. Colombo fully complied with 20/45 (44.4 percent) of requests and partially complied with 11/45 (24.4 percent).

Other issues such as military reform, military strategy, counterinsurgency strategy, and development were all associated with lower rates of compliance when compared to requests pertaining to political reform. This high rate of compliance with political requests was motivated by Colombo's willingness to cede certain political concessions to the Indians in order to get New Delhi's cooperation in solving the Tamil issue, especially since India had previously been working against Sri Lankan interests. This dynamic boosted India's leverage with Colombo, since India's commitment to the regime in Sri Lanka was tenuous. This enabled New Delhi to push Colombo toward compromise with its demands and even with LTTE insurgents, as evidenced by India's September 1987 demand that Colombo agree to LTTE recommendations regarding the composition of the Interim Administrative Council, a new administrative body designed to provide policy guidance to the newly formed North-Eastern Province.[29] The Sri Lankan government agreed to the LTTE's suggestions regarding the Council.[30] However, the Council did not have any significant impact and as the intervention wore on Colombo became increasingly hesitant to provide such concessions, as it seemed "every concession led to more demands by the LTTE."[31]

Indian requests for Colombo to act as an intermediary to send Indian funds to the LTTE, announce openings for Tamils in government, and refrain from blaming the IPKF for postponing election nominations, were coded as requests related to counterinsurgency strategy. Although several of these requests can also pertain to political or military reforms, this subgrouping is intended to capture requests specifically relating to a counterinsurgency protocol in order to track compliance trends within and across particular counterinsurgency strategies as articulated by political and military thinkers. Ultimately, there was a mixed reaction to these requests in Sri Lanka, as Colombo complied with COIN-related demands less frequently than with other political reforms, but more frequently than with requests for military action. Figure 7.1 compares rates of Sri Lankan compliance across categories of issues addressed in Indian requests.

Explaining the Compliance and Defiance of Local Allies in Sri Lanka

As detailed in Chapter 2, I propose four variables to help explain the likelihood of Sri Lankan compliance with requests from New Delhi.

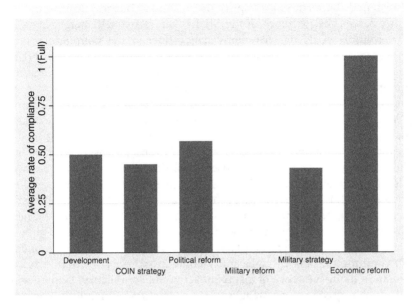

Figure 7.1 Rates of compliance with Indian requests, by subject area, 1986–90

Namely, local capacity to implement the request, the interests of Colombo and New Delhi converging or diverging over the request, the dependency of Indian forces on local Sri Lankan partners to implement the request, and a significant enemy offensive. Similar to other interventions, I find support for the proposed theory that there is an interaction effect between interests and dependencies producing a particular pattern of compliance outcomes that is key to understanding local free-riding behaviors and to specify when intervening forces are likely to have coercive leverage over local clients. Interestingly, while the Indian intervention in Sri Lanka did not experience a significant, coordinated acute enemy offensive similar to the Tet Offensive in Vietnam, an uptick in insurgent violence toward the end of the intervention drove Sri Lankan officials to initiate bilateral talks with the LTTE, excluding Indian representatives. This contradicts the logic proposed by Hypothesis 4, contending that significant enemy threats will draw allies together, and instead provides caution that the relationship between violence and local compliance requires nuanced and contextualized analysis. There are several reasons why enemy operations had the opposite effect on the behavior of local allies in Sri Lanka and Vietnam. In Sri Lanka, the Tamils threatened the territorial integrity of Sri Lanka, but the insurgency did not threaten the existence of the Sri Lankan state itself, suggesting that the political goals of the enemy can influence the counterinsurgent response to insurgency violence. The National Liberation Front (NLF) and North Vietnamese in Vietnam, on the other hand, were not pressing for limited secession or autonomy, but had more existential goals vis-à-vis the regime in Saigon, vying to destroy the US-backed regime and control Vietnam. The potentially devastatingly high cost of such a defeat may have drawn Vietnamese allies closer to US forces since losing meant losing legal sovereignty, not just territorial sovereignty over certain areas. Additionally, the type of threat posed by the increased enemy activity was also substantially different. In Sri Lanka there was a somewhat gradual increase in LTTE activity, while the military action of the Tet Offensive created an unprecedented, coordinated shock in areas of South Vietnam that had previously been mostly subdued.

Furthermore, differences between India and the USA in their respective commitments in Sri Lanka and Vietnam may have also contributed to variation in the influence of insurgent action on compliance. Coping with a long history of Indian bureaucracies undermining Colombo's interests, Sri Lankan officials did not trust that requests from New Delhi would necessarily serve Sri Lanka's interests. LTTE offensives further discredited Indian claims that they could disarm the LTTE. The Sri Lankan government became less willing to provide concessions as Tamil violence expanded, an indication of New Delhi's strategic failure. In Vietnam, in contrast, due to the high level of US commitment, Saigon may have been more willing to go with US requests, as Washington had persistently battled the communist North. Therefore, given the exceptional history in the Sri Lankan case of Indian intervening counterinsurgents having switched sides, Sri Lankan local forces were wary of relying on Indian forces. Increased insurgent violence only confirmed for Colombo that New Delhi could not deliver an acceptable arrangement with the LTTE, and therefore Colombo would be best served by pursing an independent protocol. In this dynamic increased violence drew counterinsurgency allies apart, not together.

Furthermore, the dynamics between India and Sri Lanka were unique in the degree of bargaining that transpired between allies. Colombo was less willing to readily agree to a request from its intervening ally than other local counterinsurgency partners. One example is the January 1988 request that President Jayewardene publicly announce the "Sri Lankan Government's decision to hold Provincial Council elections." The Indian request further specified, the Sri Lankan "President may also appoint Governors for each of the Provinces of Sri Lanka especially one Governor for the North-

Eastern Province who should be a prominent Tamil figure acceptable to a wide cross section of Tamil opinion."[32] President Jayewardene replied that the request was "Approved, except the names of the individuals."[33] There was domestic political pressure not to give in to Indian and Tamil demands that Tamils govern the newly unified North-Eastern Province. It is also likely that Jayewardene would need time to compile a list of officials who would be acceptable to this disparate group of Indian, Tamil, and Sinhalese nationalists. This request resulted in partial compliance as Colombo announced Provincial Council elections, but Jayewardene appointed Nalin Seneviratne, a Sinhalese Sri Lankan Army officer, as governor of the North-Eastern Province in November 1988. This fell short of the "Tamil figure" New Delhi requested.[34]

Local Capacity Limitations – Decreased Likelihood of Compliance

Unlike the US intervention in Vietnam and the USSR in Afghanistan, the Indian intervention in Sri Lanka was not motivated by impending failure of the Sri Lankan state. Although faced with a brutal Tamil insurgency campaign, Colombo was not at risk of collapse in 1987. On the contrary, New Delhi's decision to interfere was motivated by a fear in the mid-1980s that the Sri Lankan military would succeed against the Tamils, driving scores of refugees to India and leaving Indian politicians vulnerable to accusations that they did nothing to protect Tamil rights.

In light of the competence of Colombo in comparison to other regimes analyzed in this study such as Saigon or Kabul, it is not surprising that capacity was not nearly as important an issue affecting the likelihood of local ally complying with requests from intervening forces when compared to other interventions. Colombo's military and political infrastructure was expansive enough to handle most of New Delhi's demands. For only 13/79 (16.5 percent) of requests were capacity shortcomings cited as a factor contributing to a lack of Sri Lankan compliance, leaving 83.5 percent of requests unaffected by capacity limitations.

Again, although there were more similarities than differences in the types of requests made by India, the USA, and the USSR, differences in what was being asked for may also have contributed to capacity being less of a factor in the Indian intervention. New Delhi was generally not aiming for broad institutional reforms. Its requests tended to focus on short-range issues such as the official recognition of the Tamil language or a specific election announcement made by a given date. This, in combination with Colombo's relatively high level of competency, meant that capacity was not as significant a factor in influencing Sri Lanka's rate of compliance.

Complications from local politics, such as parliament blocking the implementation of a policy request from New Delhi, as well as complications from the war itself, can complicate the process of local allies complying with requests from outside allies. These factors demonstrate how compliance decisions are made in dynamic environments, which are at times influenced by the enemy, competing political actors, or internal disagreement within the political institutions or local or intervening allies. The decision whether or not to comply with the request of a large ally can be a complicated process involving multiple actors.

All militaries have internal bureaucratic tensions. As a 1972 Institute for Defense Analysis pacification summary report observed, "the American effort to advise and support the Vietnamese in their pacification program was significantly blunted by institutional rivalries and frictions among MACV, CIA, AID [US Military Assistance Command, Vietnam; Central Intelligence Agency; Agency for International Development], and the embassy itself ... there was little attempt to establish effective overall control, or even coordination, of the various far-flung American programs."[35] Similarly India's force in Sri Lanka suffered internal divisions that affected the alliance with Colombo.

Indian intelligence services (RAW) trained Tamil militants until 1987, and continued to work covertly to negotiate with LTTE leadership until 1990. Nonetheless, Indian military leaders tasked with suppressing the LTTE were often only partially informed about RAW's activities. According to journalist Rohan Gunaratna, the tension between Indian political, military, and intelligence services created several serious problems for the Indian effort. For example, in March 1988, IPKF forces ambushed and killed two LTTE leaders who were in active negotiations with authorities from RAW, and were even reportedly carrying correspondence between LTTE founder and leader Velupillai Prabhakaran and Prime Minister Rajiv Gandhi. When they were killed, "the IPKF was not aware of the RAW plan – even if they knew, it would not have mattered as they had no mutual respect for each other."[36] Indian High Commissioner J. N. Dixit, who worked closely with representatives from Colombo, was also largely excluded from ongoing RAW–LTTE negotiations, despite being the senior-most political officer in the Indian effort.[37] Due to poor interbureaucratic cooperation, Indian efforts would at times work at cross-purposes.

Although important at times, bureaucratic issues within Indian agencies were not frequently cited as impacting Sri Lankan compliance with Indian demands. In fact, according to available documents, only 4/79 (5.1 percent) of policy request discussions cite bureaucratic issues as having influenced Sri Lankan compliance. One example is the request that Sri Lanka maintain law through the CVF.[38] New Delhi asked Colombo to bolster the CVF, only to have Indian intelligence (RAW) take over training and arming, ultimately using the CVF as an instrument to implement Indian interests in Sri Lanka.[39] Colombo initially attempted to follow through with support for the CVF, but when its ranks swelled with over 30,000 Indian-recruited and armed Tamils, Colombo started to distance itself from the CVF.[40]

Moreover, Sri Lankan bureaucratic processes could emerge as a factor affecting compliance. This was the case with the 1988 demand that Colombo prevent individuals settled in northeast Sri Lanka after 1983 from voting in the 1988 election.[41] New Delhi feared "if such persons are allowed to vote in the Provincial Council elections, Tamil opinion across the Board will say that Sinhala colonialization has been legitimized."[42] In other words, if post-1983 registrants were allowed to vote, Tamils that had fled their homes would be excluded, while Sinhalese that had subsequently claimed those areas would have political voice. There was notable Sinhalese pressure not to give in to this demand, as not only would it work against Sinhalese representation, but would also exclude individuals who had turned 18 since 1983, and violate long-standing electoral laws.[43] Colombo did not comply. Yet, Sri Lankan bureaucratic issues were not significant in determining the likelihood of compliance, since bureaucratic hurdles could be overcome and internal division did not hold up the majority of requests.

Since the Sri Lankan government was making decisions regarding complying with Indian demands within the context of a violent insurgency, it is important to note that developments in the war affected Sri Lankan capacity and compliance outcomes. Aspects of the wartime setting, including violence, combat operations, refugee influxes, and overtures or offenses by the enemy were cited as an issue affecting compliance in 34/79 (43.0 percent) of demands. Consider for instance the Indian demand that Colombo "make special efforts to rehabilitate militant youths."[44] This seemingly simple request was complicated by the fact that security concerns were consuming an exorbitant amount of Colombo's federal budget. Sri Lanka is a small state with big security problems. Budget deficits averaged 12.4 percent of GDP from 1985 to 1989, and defense spending escalated from 1.1 percent of GDP in 1982 to 4.8 percent in 1988, despite India taking over responsibility for disarming the LTTE from 1987 to 1990.[45] These expenses left little room for rehabilitation programs such as those requested by New Delhi. The cost of military operations constrained Colombo's ability to implement other policies.

Increased military activity by the LTTE led to a decrease in Sri Lanka's rate of compliance with Indian requests. For example, in

1987 New Delhi requested that Colombo (1) withdraw its paramilitary personnel from the North and East and (2) transfer paramilitaries into regular military services and disband its "Homeguard" forces.[46] The "Homeguards" typically consisted of groups of locals ostensibly organized to protect Muslims and Sinhalese civilians in Tamil areas and border villages.[47] In keeping with the July 1987 Indo-Sri Lanka Accord, Colombo began to disband these groups, but this "was suspended when Muslims and Sinhalese were attacked and retention of Homeguards became essential to protect them."[48] Colombo, furthermore, did not comply with New Delhi's requests that they dissolve the paramilitaries, in large part because increased violence from the war made it politically untenable to do so.

Interests and Dependencies

As was the case in Vietnam, Iraq, and Afghanistan (USA) the potential for short-term benefits for local allies was one of the highest predictors of local compliance.[49] When the requests could potentially serve the interests of the regime or head policymakers in Colombo, it unsurprisingly had a higher likelihood of being adopted than if it couldn't immediately help Colombo. Furthermore, the interests of each ally combine with their dependencies to create patterns in the likelihood of cooperation or defiance as hypothesized in Chapter 2. Indian dependency on Sri Lanka to implement a policy request from New Delhi interacts with the convergence or divergence of interests between allies on the requested policy to either motivate or discourage compliance by Colombo. Therefore, as expected, local ally compliance was lower when the interests of allies diverged and India lacked a unilateral capacity to implement the request.

Interests Diverge and Indian Forces Are Dependent on Local Allies for Policy Implementation: Predicted Sri Lankan Noncompliance

As hypothesized, when New Delhi was dependent on Colombo to implement a policy, a pattern of interaction with interests affected the likelihood of Sri Lankan compliance. For example, consider how the Sri Lankans disregarded the Indian demand that bilateral talks between Colombo and the LTTE cease.[50] Colombo had strong diverging interests from New Delhi and the Indian delegation had little to coerce the Sri Lankan allies with since New Delhi could not implement the request without the participation of the Sri Lankans.

Interests Diverge and Indian Forces Are Not Dependent on Local
Allies for Policy Implementation: Predicted Sri Lankan Compliance

In July 1987, New Delhi demanded that Sri Lankan ports not "be made available for military use by any country in a manner prejudicial to India's interests."[51] Despite its diverging interests, Colombo complied with the request because India, with its regional dominance, sizable navy, military superiority, and record of using force to protect its interests in Sri Lanka, had the physical ability to implement the policy unilaterally if needed. The Indian Navy could block foreign vessels, or even go so far as unilaterally taking over Sri Lankan ports. Colombo was well aware of Indian capabilities, and complied with the Indian demand in order to avoid confrontation. This finding supports Hypothesis 2B, which predicts that if the interests of allies diverge over a policy, yet the intervening force can implement the policy unilaterally, compliance by local allies is likely, since they would seek to avoid being excluded or undermined by unilateral action that could potentially impose costs on them.

Interests Converge and Indian Forces Are Dependent on
Local Allies for Policy Implementation: Predicted
Sri Lankan Compliance

One example of a request from New Delhi where the interests of local and intervening forces converged, and India found itself dependent on Colombo, was the 1988 request that President Jayewardene issue directives to the Sri Lankan Election Commissioner "to set in motion preparations for elections to Provincial Councils, polling booths, list of officers, etc ...."[52] India pushed the President to include various election directives but also left several aspects of election protocol up to Colombo. This openness in the request provided an opportunity for Jayewardene to make choices that would serve his short-term interests, including decisions regarding the election that could appease various domestic interests or consolidate his authority. The request provided an opportunity for Colombo's short-term interests, and could only be fulfilled by Colombo. Unsurprisingly, this resulted in compliance. In September 1988 the president issued specific orders to the Election Commissioner for mid-November elections.[53] This finding supports Hypothesis 3A, which proposes that when the interests of allies converge, and the local ally must participate in order for the request to be fulfilled, local compliance is likely.

Interests Converge and Indian Forces Are Not Dependent on
Local Allies for Policy Implementation: Predicted
Sri Lankan Noncompliance

One example of free riding was the August 1988 request from Prime Minister Gandhi that Sri Lankan officials take responsibility for the "logistical and administrative arrangements for the elections" to provincial councils.[54] Although Colombo complied in part, the IPKF also made independent security arrangements to limit the opportunities for the LTTE to disrupt the process.[55] Notice that despite the security provided, the elections were not a wholesale success. Voter turnout was extraordinarily low as numerous political parties boycotted the council elections because the councils had been demanded by New Delhi.[56] It is also important to note that due to the relatively short duration of the intervention, and the distrust between Indian and Sri Lankan allies, free riding by Colombo on Indian efforts was less frequent than observed by local allies in other wars including the USA in Vietnam or Iraq.

This interaction between Indian dependencies and the convergence or divergence of interests between India and Sri Lanka is indicated in Table 7.3. If New Delhi could act unilaterally and Sri Lankan and Indian interests converged, Colombo was more likely *not* to comply due to incentives to free ride. If New Delhi could act unilaterally, and allied interests diverged, there was a higher probability of compliance as Colombo was more willing to agree in order to protect its interests since the action may be implemented without them anyway. The reverse was true if unilateral action was not possible. If independent action was not possible, and interests converged, there was a higher likelihood of compliance, as Sri Lanka undertook the action for the sake of its own

Table 7.3 *Ordered probit – local Sri Lankan compliance with Indian requests, 1986–90, interaction effect between Indian dependency on Sri Lanka for policy and allied interests*

Local Sri Lankan capacity	Indian unilateral potential	Allied interest convergence/ divergence[a]	Interaction with unilateral ability and allied interests
-0.911^{**}	0.901^{*}	1.912^{**}	-1.568^{**}
(0.356)	(0.376)	(0.376)	(0.569)

Notes: $^{*}p < 0.05$; $^{**}p < 0.01$.
[a] As discussed in Chapters 4–6, the convergence/divergence of interests is produced by combining two variables related to interest, "Private Benefit for Local Allies," and "Threat to Private Benefits for Local Allies."

interests. If alliance interests diverged and India could not implement the reform without Sri Lankan assistance, noncompliance was likely. Each of the seventy-nine identified requests from India to its partners in Colombo were coded to yield a picture of broad trends related to the proposed variables and the likelihood of Iraqi compliance. An ordered probit statistical model is provided (Table 7.3), accounting for capacity limitations and the proposed interaction effect between interests and dependencies in Sri Lanka. Additional quantitative information is provided in Appendix C (Table C.4) that specifies additional checks conducted, including fixed effects by year, and considers interests alone without dependencies as an alternative explanation. Additionally, the data was reanalyzed using robust standard errors clustered by the issue area of request, which provided a statistically and substantively comparable result to the results presented in Table 7.3.

Sri Lankan Compliance with Indian Requests over Time

Figure 7.2 charts the number of requests made by New Delhi by year juxtaposed with the Sri Lankan rate of compliance with those policy requests.[57] As illustrated, the number of requests made and the number

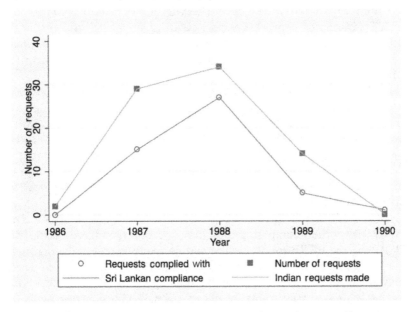

Figure 7.2 Rates of Indian requests and Sri Lankan compliance, 1986–90

of requests complied with follow the same trajectory: an increase in requests corresponds with an increase in the number of requests complied with. While this pattern appears straightforward, it is actually distinctive when compared to the alliance dynamics of other counterinsurgency partnerships including the Soviet intervention in Afghanistan and the USA in Vietnam, Iraq, and Afghanistan. In other case studies compliance fluctuated, not necessarily with the number of requests made by intervening allies, but with developments in the conflict itself. The symmetry between the rate of Indian requests and the rate of Sri Lankan compliance can be attributed to several factors.

First, the US and Soviet interventions in Vietnam and Afghanistan each lasted almost a decade. Indian military involvement in Sri Lanka was less than three years, from July 1987 to March 1990, which is less than one-third of the duration of Vietnam or Afghanistan for both the USA and USSR. Part of the reason the intervention was relatively short is due to Colombo making the decision in 1990 to work with the Tamil insurgent enemy to expel intervening Indian forces. Interestingly all three intervening states, India, the United States, and the USSR, believed their military deployment would be substantially shorter than what actually developed. When committing troops, Moscow and Washington never believed that they would be fighting for a decade. Based on estimates provided by Indian intelligence (RAW), Rajiv Gandhi thought the LTTE disarmament process would begin in July 1987 and would be completed by the end of 1987 or early 1988.[58] Counterinsurgency interventions tend to be significantly longer and bloodier interventions than anticipated.

Second, the objective of the Soviet and US missions in Vietnam and Afghanistan differed from the Indian mission in Sri Lanka. Washington and Moscow intended to rapidly reinforce a local regime that could defend itself and the interests of their US or Soviet sponsors. India, however, was not necessarily interested in bolstering the government of Sri Lanka. In fact, India's years of sponsoring the LTTE insurgency prior to changing missions in support of Colombo's counterinsurgency was an attempt to weaken Colombo and coerce Sinhalese leadership to compromise with Tamil interests. Although there is more similarity than dissimilarity when comparing the types of requests made by India, the USA, and the USSR to local allies, certain requests reveal divergence in the foreign missions of these intervening allies. For example, New Delhi only made one appeal to Colombo relating to Sri Lankan economic reform – a request for a finance commission.[59] This contrasts with 12/105 (11.4 percent) of US requests to South Vietnam regarding state economic policies, 26/106 (24.5 percent) to Baghdad, and 25/148

(16.9 percent) from the USA to Kabul. Similarly, policy requests from intervening allies regarding development projects also revealed Sri Lanka to be a notable outlier. Indian demands regarding development made up just 4/79 (5.0 percent) of requests, whereas the US requests in Vietnam regarding such projects made up 27/105 (25.7 percent) of requests in Vietnam, 14/106 (13.2 percent) in Iraq, and 27/148 (18.2 percent) in Afghanistan.

Conclusion – A Complex Set of Allies and Enemies

New Delhi had both intersecting and divergent interests with both insurgents and counterinsurgents in Sri Lanka. India used its common interests and resources to attempt to leverage concessions from all sides of the Sri Lankan conflict, but was unsuccessful getting either side to fully promote the Indian vision for Sri Lanka. New Delhi's position on the transfer of power from the centralized state apparatus in Colombo to localized government institutions in Tamil areas, often referred to as devolution, illustrates how India's attempts to force compromise failed. After pushing Colombo to transfer power to more local levels, in 1988, New Delhi asked President Jayewardene to ensure "that *Pradeshiya Sabhas* [Village Councils] and District Councils do not dilute role of Provincial Council and Provincial Government."[60] However, this request for a middle-range concentration of authority in Tamil areas failed to satisfy the Tamils, who sought localized control, while also failing to please Sri Lankan officials, who sought centralized authority. The focus on the provincial level was largely based on India's own governance model. This solution effectively took power from Colombo without significantly strengthening the Tamil position, and it left both sides distrustful of New Delhi's proposed solutions.

Furthermore, despite India's expected leverage over Colombo and the LTTE, Colombo's rate of compliance with Indian demands was consistent with rates of local ally compliance observed in other conflicts (see Chapter 10, Table 10.1). The LTTE was also similarly effective at evading Indian petitions. New Delhi was nevertheless frequently able to coerce Colombo when it could credibly negotiate with the LTTE insurgents. This is particularly evident in India's initial demands that Colombo provide the Tamils with certain rights, such as recognition of the Tamil language, and acknowledge the North and East areas as traditional Tamil homelands. Meaningful dialogue with the enemy was similarly useful for the USA at the end of the US war in Vietnam. Henry Kissinger coerced South Vietnamese President Thieu to agree to uncomfortable aspects of the Paris Peace Accords because Kissinger was in direct contact with Hanoi and could potentially cut Saigon out of negotiations completely. This is similar to what officials in New Delhi were able to threaten in 1987 before the LTTE made it clear that they were not interested in cooperating with the IPKF, and Colombo no longer found New Delhi's threats to work with the insurgents credible. Relating back to the relationship between commitment to an ally and leverage for intervening forces, an intervening military that can credibly negotiate with an insurgent enemy can use that connection to insurgents to leverage concessions from local counterinsurgents. However, as cautioned by the Sri Lankan case, such a complex and varying web of allies and enemies can create political volatility that can easily contribute to, instead of mitigate, the likelihood of violence.

Notes

1 Neil DeVotta, "Control Democracy, Institutional Decay, and the Quest for Eelam: Explaining Ethnic Conflict in Sri Lanka," *Pacific Affairs* 73, no. 1 (2000): 71.

2 Asoka Bandarage, *The Separatist Conflict in Sri Lanka: Terrorism, Ethnicity, Political Economy* (New York: Routledge, 2009), 153. The cost of this intervention has also been cited as 1,000 Crore or more than ten billion Indian Rupees.

3 Ashok K. Mehta, "India's Counterinsurgency in Sri Lanka," in *India and Counterinsurgency: Lessons Learned*, ed. Sumit Ganguly and David P. Fidler (New York: Routledge, 2009), 155.

4 Ahmed Hashim, *When Counterinsurgency Wins: Sri Lanka's Defeat of the Tamil Tigers* (Philadelphia: University of Pennsylvania Press, 2013).

5 A. M. Navaratna-Bandara, "Ethnic Relations and State Crafting in Post-independence Sri Lanka," in *Sri Lanka: Current Issues and Historical Background*, ed. Walter Nubin (New York: Nova Publishers, 2002), 63; Mehta, "India's Counterinsurgency in Sri Lanka," 156; Erik K. Jenne, "Sri Lanka: A Fragmented State," in *State Failure and State Weakness in a Time of Terror*, ed. Robert I. Rotberg (Washington, DC: Brookings Institution Press, 2003), 227. Most Sri Lankan Tamils are Hindu.

6 Chris Smith, "South Asia's Enduring War," in *Creating Peace in Sri Lanka: Civil War and Reconciliation*, ed. Robert I. Rotberg (Washington, DC: Brookings Institution Press, 1999), 35.

7 "Decisions of the Security Co-ordination Group Regarding Security of the Tamils," 2414–15.

8 Ibid., 2417–18.

9 Ibid. Indian officials concluded that this reform was "not implemented." See also Smith, "South Asia's Enduring War," 35.

10 Mehta, "India's Counterinsurgency in Sri Lanka," 157.

11 P. A. Ghosh, *Ethnic Conflict in Sri Lanka and Role of Indian Peace Keeping Force (IPKF)* (New Delhi: APH Publishing, 2000), 108; Gunaratna, *Indian Intervention in Sri Lanka* , 269.

12 Gunaratna, *Indian Intervention in Sri Lanka*, vii.

13 Bandarage, *The Separatist Conflict in Sri Lanka*, 114.

14 Gunaratna, *Indian Intervention in Sri Lanka*, 2.

15 Ibid.

16 Ibid., 178.

17 "Message of the Indian Government to the Sri Lankan Government on the Urgent Need to Send Relief Supplies to Jaffna through the Indian Red Cross," New Delhi, June 1, 1987, in *India–Sri Lanka: Relations and Sri Lanka's Ethnic Conflict Documents – 1947–2000*, ed. Avtar Singh Bhasin, Vol. III, Document 680 (New Delhi: Indian Research Press, 2001), 1902.

18 "Reply Message of the Sri Lankan Government to the Indian Government," Colombo, June 1, 1987, in *India–Sri Lanka: Relations and Sri Lanka's Ethnic Conflict Documents – 1947–2000*, ed. Avtar Singh Bhasin, Vol. III, Document 681 (New Delhi: Indian Research Press, 2001), 1902.

19 Ghosh, *Ethnic Conflict in Sri Lanka and Role of Indian Peace Keeping Force (IPKF)*, 89; Gunaratna, *Indian Intervention in Sri Lanka*, 183.

20 Ghosh, *Ethnic Conflict in Sri Lanka and Role of Indian Peace Keeping Force (IPKF)*, 48.

21 Gunaratna, *Indian Intervention in Sri Lanka*, 164–5.

22 Ibid., 89.

23 Bandarage, *The Separatist Conflict in Sri Lanka*, 112.

24 The 54 Air Assault Division made up the core of the IPKF. By August all three of the Division's Brigades were in Sri Lanka, but all at less than half strength. Two battalions of the 91st Infantry Brigade were near Jaffna, with an additional battalion inside Jaffna city, 47th Infantry Brigade was in Vavuniya, south of Jaffna, and the 76th Infantry Brigade at Batticaloa in the east. The 340 Independent Brigade was in Trincomalee, in eastern Sri Lanka. Smith, "South Asia's Enduring War," 21; Kenneth Conboy and Paul Hannon, *Elite Forces of India and Pakistan* (Oxford: Osprey Publishing, 1992), 15.

25 "Statement of the Leader of the Opposition in the Sri Lanka Parliament Anura Bandaranaike Regarding Statements Made by the Commander of the IPKF," Colombo, December 8, 1987, in *India–Sri Lanka: Relations and Sri Lanka's Ethnic Conflict Documents – 1947–2000*, ed. Avtar Singh Bhasin, Vol. IV, Document 789 (New Delhi: Indian Research Press, 2001), 2178.

26 Roshani M. Gunewardene, "Indo-Sri Lanka Accord: Intervention by Invitation or Forced Intervention," *North Carolina Journal of International Law and Commercial Regulation* 16, no. 2 (1991): 224.

27 Gunaratna, *Indian Intervention in Sri Lanka*, 194; Harkirat Singh, *Indian Intervention in Sri Lanka: The IPKF Experience Retold* (Colombo: Vijitha Yapa Publications, 2006), 36.

28 Mehta, "India's Counterinsurgency in Sri Lanka," 167.

29 "Letter of the Indian High Commissioner in Colombo to the Sri Lankan President J. R. Jayewardene Conveying the Agreement Arrived at with the

LTTE on the Interim Administrative Council in the Northern and Eastern Province," Colombo, September 28, 1987, in *India–Sri Lanka: Relations and Sri Lanka's Ethnic Conflict Documents – 1947–2000*, ed. Avtar Singh Bhasin, Vol. IV, Document 753 (New Delhi: Indian Research Press, 2001), 2083.

30 Sabha, India, Parliament, Lok, *Lok Sabha Debates* (Lok Sabha Secretariat, 1987), 457.

31 K. M. Silva, *Regional Powers and Small State Security: India and Sri Lanka, 1977–1990* (Washington, DC: Woodrow Wilson Center Press, 1995), 258.

32 "Points of Verbal Message of the Indian Prime Minister Rajiv Gandhi," 2184–5.

33 Ibid.

34 Dagmar Hellmann-Rajanayagam, *The Tamil Tigers: Armed Struggle for Identity* (Stuttgart: F. Steiner, 1994), 17; Michael Roberts, "Language and National Identity: The Sinhalese and Others over the Centuries," *Nationalism and Ethnic Politics* 9, no. 2 (2003): 75.

35 Cooper et al., "The American Experience with Pacification in Vietnam – Volume I – An Overview of Pacification (Unclassified)," 21–2.

36 Gunaratna, *Indian Intervention in Sri Lanka*, 266–7. Citing S. C. Sardesh-pande, *Assignment Jaffna* (New Delhi: Lancer Publishers, 1992).

37 Gunaratna, *Indian Intervention in Sri Lanka*, 266.

38 "Decisions of the Security Co-ordination Group Regarding Security of the Tamils," 2414.

39 Gunaratna, *Indian Intervention in Sri Lanka*, 355–6.

40 Ibid. See also "Extract from the Statement of the Sri Lankan Minister of Foreign Affairs," 2356–8.

41 "Points of Verbal Message of the Indian Prime Minister Rajiv Gandhi," 2184–5.

42 Ibid.

43 Gunaratna, *Indian Intervention in Sri Lanka*, 349. See also "TOP SECRET Note of the RAW Agent to the Chief of the Sri Lankan Intelligence and Security RE," 2263.

44 "Question in the Sri Lankan Parliament," 2386.

45 Saman Kelegama, "Economic Costs of Conflict in Sri Lanka," in *Creating Peace in Sri Lanka: Civil War and Reconciliation*, ed. Robert I. Rotberg (Washington, DC: Brookings Institution Press, 1999), 73.

46 "Question in the Sri Lankan Parliament," 2386–7.

47 S. H. Hasbullah and Barrie M. Morrison, *Sri Lankan Society in an Era of Globalization: Struggling to Create a New Social Order* (New Delhi: SAGE, 2004), 260.

48 "Question in the Sri Lankan Parliament," 2387.

49 See Appendix C.

50 "Press Briefing of the Indian High Commissioner in Colombo J. N. Dixit on the Peace Efforts to Bring about a Negotiated Settlement between the LTTE and the Sri Lankan Government," Colombo, February 27, 1988, in *India–Sri Lanka: Relations and Sri Lanka's Ethnic Conflict Documents – 1947–2000*, ed. Avtar Singh Bhasin, Vol. IV, Document 806 (New Delhi: Indian Research Press, 2001), 2206–7.

51 "Exchange of Letters between the Prime Minister of India and the President of Sri Lanka, Prime Minister of India," New Delhi, July 29 1987, in *India–Sri Lanka: Relations and Sri Lanka's Ethnic Conflict Documents – 1947–2000*, ed. Avtar Singh Bhasin, Vol. IV, Document 723 (New Delhi: Indian Research Press, 2001), 1949.

52 "Points of Verbal Message of the Indian Prime Minister Rajiv," 2185.

53 "Communiqué of the Sri Lankan Presidential Secretariat Underlining the Importance of Holding Elections to the North-East Provincial Council," Colombo, September 19, 1988, in *India–Sri Lanka: Relations and Sri Lanka's Ethnic Conflict Documents – 1947–2000*, ed. Avtar Singh Bhasin, Vol. IV, Document 857 (New Delhi: Indian Research Press, 2001), 2312.

54 "Message of the Indian Prime Minister Rajiv Gandhi to the Sri Lankan President J. R. Jayewardene Delivered through the Indian High Commissioner in Colombo J.N. Dixit," New Delhi, August 28, 1988, in *India–Sri Lanka: Relations and Sri Lanka's Ethnic Conflict Documents – 1947–2000*, ed. Avtar Singh Bhasin, Vol. IV, Document 851 (New Delhi: Indian Research Press, 2001), 2292.

55 "Statement of the Indian Minister for External Affairs P.V. Narasimha Rao in the Lok Sabha: 'North-Eastern Provincial Council Elections in Sri Lanka'," New Delhi, November 22, 1988, in *India–Sri Lanka: Relations and Sri Lanka's Ethnic Conflict Documents – 1947–2000*, ed. Avtar Singh Bhasin, Vol. IV, Document 870 (New Delhi: Indian Research Press, 2001), 2335.

56 Sankaran Krishna, *Postcolonial Insecurities: India, Sri Lanka, and the Question of Nationhood*, new ed. (Minneapolis: University of Minnesota Press, 1999), 198–9.

57 Compliance includes partial compliance and full compliance.

58 Silva, *Regional Powers and Small State Security*, 253.

59 "Aide Memoire Containing Points Conveyed by the Indian High Commissioner J.N. Dixit," 2257–9.

60 Ibid.

8   The USSR in Afghanistan

> Soviet officials reportedly occupy the senior positions in every Afghan Ministry except the Foreign Ministry; where Afghans – because of the Ministry's visibility and its dealings with foreigners – occupy Deputy Director positions ... All decisions are Soviet, and most Afghan civil servants simply sit at their desks and collect their paychecks.
>
> —US Department of State, *Soviet Dilemmas in Afghanistan*

The 1978–89 Soviet intervention in Afghanistan was unsuccessful, and ultimately so costly for the Kremlin that it has been cited as a catalyst for the collapse of the Soviet Empire.[1] This chapter analyzes the political dynamics of the partnership between Soviet intervening forces and the Afghan regime they supported.

Summary Findings

The Afghans allied with Soviet forces agreed to all Soviet requests, in private as well as in public. A reporter asked Afghan President Babrak Karmal in 1982 if he "'disagree[d] with the Soviet Union on any matter of policy or principle?' 'Not at all,' he dutifully replied. And in 1987 President Najibullah was asked 'Are there any differences between the U.S.S.R. and Afghanistan on how to resolve the Afghan problem?' 'No, none whatsoever,' he said. 'There is complete unanimity of views between us.'"[2] Yet, despite this stated harmony between local and intervening allies, there was no marked increase in the rate of substantive compliance from Afghans when compared to other local counterinsurgents responding to requests from intervening allies. Partial and full substantive compliance taken together in the USSR-Afghan war was 14/22 (63.6 percent), which is strikingly similar to the other interventions including Iraq 67/105 (63.2 percent), Vietnam 71/105 (67.6 percent), Sri Lanka 49/79 (62.0 percent), and Afghanistan during the US intervention 92/148 (62.2 percent).

Table 8.1 *Timeline of the Soviet–Afghan alliance during the Soviet intervention*

April 1978	Nur Mohammad Taraki overthrows Mohammed Daoud Khan and establishes a communist regime, the Democratic Republic of Afghanistan (DRA), run by the People's Democratic Party of Afghanistan (PDPA). Taraki reaches out to the Soviet Union, asking for military and financial assistance.
September 1979	President Taraki is assassinated by agents working for his deputy, Hafizullah Amin. Both Taraki and Amin belong to the Khalq faction of the PDPA.
December 1979	The USSR invades Afghanistan, Soviet KGB officials assassinate Amin and install Babrak Karmal as head of the PDPA. Karmal is a leader of the non-Khalq faction of the PDPA, Parcham.
May 1986	Mohammad Najibullah, head of Afghan Intelligence (KhAD), replaces Karmal as secretary general of the PDPA. Najibullah is also in the Parcham faction of the PDPA.
April 1988	Geneva Peace Accords signal the end of Soviet involvement in the war.
February 1989	Soviet troops withdraw from Afghanistan.

The USSR–Afghan Alliance – Local Sovereignty Denied

Data for the Soviet–Afghan war is limited to 1978–80, despite the war continuing for another nine years. After 1980, the Soviet Politburo stopped asking its Afghan partners for new reforms, choosing instead to use internal Soviet agents embedded in Kabul to carry out requests.

The US intervention in Afghanistan was rife with bureaucratic contradictions and a lack of consistency in its overall strategic mission. The Soviet war in Afghanistan, which predates the American war by twenty years, was marred by arguably even deeper strategic and organizational inconsistencies. The very style with which Soviet troops initiated the war in December 1979 was indicative of the violence and bureaucratic blundering that would undermine the future counterinsurgency effort. Just prior to invasion, KGB agents infiltrated the Afghan presidential palace and poisoned President Hafizullah Amin.[3] Soviet diplomats, including Soviet Ambassador Fikriat Tabeev, were uninformed of the plot and Soviet doctors rushed to revive Amin with intravenous liquids.[4] When the KGB learned that Amin was on the path to recovery, Soviet troops stormed the Afghan presidential palace to assassinate Amin, who still had a pack of Soviet-manufactured intravenous fluids tucked into his underwear to save him from the KGB poisoning. As Soviet forces fired their weapons into the presidential palace, one of the Soviet doctors who had attended Amin was shot and later died, along with Amin and his children.[5]

Following Amin's death, Babrak Karmal was installed as president and USSR officials embedded themselves within the Afghan state. As historian Anthony Arnold summarized, "by the close of 1979 the PDPA no longer ruled Afghanistan; the CPSU [Communist Party of the Soviet Union] did."[6] The number of Soviet counselors increased, and by early 1984 reportedly 10,000+ advisers were at work in various positions in Afghan military and civilian organizations.[7] By 1987, 9,000 Soviets worked in the Afghan civilian bureaucracy alone.[8] According to M. Hassan Kakar, who was imprisoned by the Karmal regime in 1982, thousands of Soviets "worked not only as advisers but also as executives in all the military and civilian departments to which they were assigned."[9] KGB documents confirmed Russian involvement in the Afghan government. "The KGB used one adviser as an example. N.K. Grechin said that he wanted 'to be a shadow minister,' to run the economic and financial side together with the minister of planning, to be responsible for fulfilling the plans as well as drawing them up, in fact to be jointly responsible for everything."[10]

Due to the degree of involvement in Afghan affairs, the Soviet takeover of the Afghan government was not secret for long. A US intelligence official writing in 1983 described how

Afghan bureaucrats, including those of ministerial rank, found that even the most routine of orders had to be approved and countersigned by the ubiquitous Soviets. In fact the roles of adviser and advisee had been reversed; in 1980 it was the Afghans who advised (if tolerated to do even that) and the Soviets who decide. Typical was the new role of Vassiliy Safronchuk, now assisted by eight subordinates: no cable could be sent from the Foreign Ministry without being approved (if not actually written) by one of this group. The situation was essentially the same in other ministries.[11]

Soviet involvement extended into the top echelons of Afghan government. President Karmal was surrounded by a staff appointed by the Kremlin and was under constant KGB surveillance. Having witnessed Moscow's disposal of President Amin, Karmal was careful to agree to Soviet proposals and, according to Henry Bradsher, a historian on USSR-Afghan relations, the Afghan president felt subjugated. Late in his career Karmal lamented how "Soviet 'advisers were everywhere ... I was not a leader of a sovereign state. It was an occupied state where in fact [Moscow] rules."[12] Bradsher also displays the extent of Soviet control by analyzing the proceedings of a November 13, 1986 Moscow Politburo meeting. The Politburo specified several minor issues for the

Afghan leadership to decide on directly, implicitly indicating that Afghans were not expected to weigh in on any topics aside from those minor assignments.[13]

The KGB Factor

The KGB was the most influential Soviet political organization in Afghanistan. The bulk of Soviet advisers in Afghanistan were KGB operatives, accustomed to the KGB's extraordinary powers that enabled it to dominate other Soviet institutions. Vasiliy Mitrokhin explained how KGB personnel fell into one of two groups. The "Residency" worked from the Soviet embassy while "Representatives" consisted of "KGB officers sent to assist the Afghan government in various functions – generally, but not exclusively, connected with security, covert operations, sabotage, intelligence, and prisons. In addition came hundreds of KGB intelligence operatives, both Afghans and Soviets, whose identities were not known to the Communist Afghan government."[14] In effect, KGB officials were either working in the embassy, embedded in the Afghan government, or spying on everyone else who did. Over the years the KGB expanded its control within the Afghan government and worked to ensure Afghans loyal to the KGB moved up in the ranks.[15]

Moscow's takeover of the Afghan state had notable consequences on Afghan government institutions and the political conditions of the war. Shortly after the invasion and KGB infiltration of Kabul there was a massive desertion of Afghan civilian and military personnel. They were often defecting to the insurgency, providing intelligence and strengthening the bureaucratic and political capacity of the emerging mujahadeen opposition. Furthermore, many Afghans that stayed at their posts grew weary of the Soviet system. As Ed Giradet reported in 1985, "many party members … have become disillusioned and deeply resent the way they are being treated by the Soviets."[16] This contributed to an acute shortage of Afghans in government, a direct result of "the exodus of qualified personnel and strife from within the communist party … By the end of 1983, well over four-fifths of the country's career diplomats had quit their posts, been forced to retire or transferred to other ministry."[17] This exodus deepened Kabul's dependency on Soviet staff to fill in the gaps throughout the government. The Afghans that remained were often promoted for their loyalties and many eventually gave up on their duties. One Russian official bitterly complained that the Afghan "leadership thinks that the USSR will solve all the economic and military problems. All they can think about is motorcars, positions and amusements."[18] Yet it is unclear what substantive affect motivated Afghans could have on the system if Soviet personnel dominated all the principal policy positions.

To compound the problem of overreliance on the Soviets, the Afghan military suffered remarkably high losses in manpower. Entire divisions deserted and combat casualties remained staggeringly high. The Afghan military shrunk "from an estimated 90,000 troops in 1978 to 30,000 in 1981."[19] Soviet forces grew frustrated with their local proxies. The KGB reported that Afghan

officers openly disobeyed orders, co-operated with the Dushmen [mujahadeen] and went over to their side. Over 17,000 men had deserted from the army by 30 April 1980. There were also vast numbers of deserters in subsequent years. In 1981 30,000 deserted from the army, and in 1982 from 2,500 to 3,000 each month. The number of deserters was six times greater than the number killed which was also very high.[20]

What remained were hollow agencies that required increasing levels of Soviet support.

The Soviet Bureaucracy in Afghanistan

The KGB falsified reports being sent up the chain of command and failed to implement requests from the Soviet Politburo for reforms in Afghanistan. Under these circumstances, Afghan "noncompliance" would actually be better understood as Soviet institutional failings. One example pertains to requests from Moscow that laws regarding arrest, trial, and detention be followed in Afghanistan. Moscow feared kidnappings and extrajudicial executions were becoming harmful to public support for the faltering regime. Mitrokhin accused KGB Director Yuri Andropov of outright fraudulence in Andropov's presentations to other Politburo members. According to Mitrokhin, the "KGB compiled reports on the events in the DRA and the new leadership. A memorandum, No.2519-A, dated 31 December, 'On the Events in Afghanistan on 27 and 28 December 1979' for the CPSU Central Committee was signed by Andropov, Ustinov, Gromyko and Ponomarev."[21] Furthermore, according to Mitrokhin,

The memorandum was written according to the rules of disinformation. Facts were distorted and rearranged, and a false interpretation of the situation was given. Andropov was the only signatory who knew the whole truth about the events. He had prepared and influenced them and had stage-managed what had happened. The rest knew only part of the truth and their role had been subsidiary as a form of insurance.[22]

This was one of several dysfunctional trends in the Soviet bureaucracy impacting implementation of Politburo requests in Afghanistan.

Corruption within Soviet institutions was also on the rise throughout the 1980s, and rivalries emerged between the Soviet embassy, military, KGB, and Communist Party officials, causing agencies to "work at cross-purposes" in Afghanistan and elsewhere.[23] By 1980 the Soviet Army had adopted policies that alarmed the CPSU and Afghan officials,

D. Panjshiri, a member of the Politburo and chairman of the Party Control [Commission] of the PDPA CC [Central Committee], was very upset that the Soviet army was beginning to fight against the Afghan people, was displaying unheard of cruelty and ruthlessness and was acting on the principle of "the worse it is, the better." The soldiers and officers had no aversion to marauding and speculating with military property and fuel. They disregarded the traffic rules.[24]

There are important parallels and differences regarding bureaucratic issues generated by the Soviets and those created by the Americans in their respective interventions in Afghanistan. US forces have not been accused of plundering the country, and the counterinsurgency strategy adopted by the USA did not attempt to alter Afghan behavior by inducing collective suffering, therefore was not as strategically ruthless when compared to Soviet coercive methods. However, similar to the Soviets, there was a notable lack of interagency coordination between US organizations that impacted the US alliance and the greater counterinsurgency effort. Historian Conor Keane describes an American approach to Afghanistan "that was chaotic and disordered, which was driven by both inadequate leadership and bureaucratic factions with different interests, perceptions, cultures and power."[25] Steve Coll similarly documented the contradictory US approach, detailing how American policy was "laced with contradictions ... the C.I.A. put warlords and [US-supported Afghan President Hamid] Karzai aids on its payroll for information, security and stability. Simultaneously, [US General David] Petraeus's command, Justice prosecutors, and Treasury investigators tried to put some of the same men in jail."[26] Therefore, as the Soviet case is examined, it is fascinating to note similarities and differences with the US experience, including a parallel regarding disjointed bureaucratic approaches to intervention.

Interestingly, because the Afghan state was staffed by Soviets, Afghan institutions soon mirrored their Soviet institutional counterparts – regardless of whether or not the structure or operating procedures of the Soviet organization in charge were appropriate or strategically sound for Soviet counterinsurgency goals. In 1982 the US Central Intelligence Agency (CIA) reported, "the Soviets are helping the Afghan Communists set up the same kind of party and government institutions that the USSR uses to control its own population."[27] In addition, the notorious

Afghan intelligence and security agency Khadamat-e Etela'at-e Dawlati (KhAD) came to directly resemble the KGB, adopting intimidation campaigns and extrajudicial killings, the very policies that Soviet leaders were advocating against in the Afghan context, in a push for greater rule of law and due process to boost the popularity and legitimacy of the regime in Kabul. In effect, the institutions being established in Afghanistan by Soviet officials were fundamentally mismatched with the vision of the Soviet Politburo as outlined in released Politburo documents. These Afghan organizations were replicas of Soviet institutions and often failed to comply with the Politburo's demands.

Mohammad Najibullah, the Afghan front of KhAD, would take over as president from Babrak Karmal in 1987. Najibullah headed KhAD, but according to one historian, KhAD was "controlled by the KGB. A Soviet journal said KhAD's KGB advisors 'used tools of Stalin's great terror – secret denunciations, anonymous spies, "confessions" extracted by torture, secret trials for tens of thousands and public show trials for a few, unannounced executions and long prison terms'."[28] KhAD quickly became a formidable political force. One of its victims later reported that KhAD "had the power and the means to torture men and women to the point of death with impunity. Although by law the execution of a prisoner after his trial in court was the prerogative of the head of state, KhAD determined the case one way or another."[29]

Kidnappings and brutality by KhAD were common and interrogation personnel were rewarded for confessions, regardless of their validity. According to Mohammed Kakar,

the establishment of the truth, which was likely to lead to the acquittal of the detainee, would deprive the interrogator of the rewards (promotion, cash, trips to the Soviet Union) that he was granted when he made the detainee confess to the crime of which he or she was accused. It was in his interest to make the detainee guilty.[30]

Counterinsurgency theory, however, notes the importance of reliable intelligence to enable government forces to pinpoint insurgents without alienating the rest of the population.[31] False intelligence that led to false arrests and imprisonment of innocent Afghans likely harmed the war effort.

Finding a Scapegoat for Counterinsurgency Failure

The Soviet Politburo sought unity between Afghan communists, yet KGB and KhAD policies reinforced divisions between communist Afghan groups. In their introduction to Mitrokhin's KGB files,

Cold War historians Christian Ostermann and Odd Arne Westad comment that

what is most striking (and most useful) about Mitrokhin's text is the pervasive sense it gives of the distrust that the KGB fomented and spread among Afghan and Soviets alike. While it is clear that Moscow's interest in the critical year 1979 lay in finding ways for the two main PDPA factions to cooperate against their increasingly efficient Islamist enemies, the KGB's operations achieved exactly the opposite – by concocting rumors and slander, the KGB contributed significantly to the destruction of the PDPA (complete in most senses before the Soviet December invasion) and to the dysfunctionality of Soviet Afghan policies.[32]

Afghan leaders made the same speeches regarding reform and unity throughout the decade of Soviet intervention, but little, if anything was ever done differently and divisions persisted.[33] The Soviet leadership had unwittingly deepened the rift between Afghan communists by overthrowing Amin, a Khalqi, and replacing him with Babrak Karmal, a Parchami. This did not unite the PDPA under Parcham domination, as the USSR intended. Instead, because Afghan military leadership contained a sizable presence of Khalqis, a consequence of Taraki and Amin's anti-Parcham military purges, there was perpetual internecine conflict among communists. Furthermore, Karmal was quick to undermine Khalq forces in an effort to promote Parcham agendas and to neutralize Taraki and Amin's policies. The anti-Amin Soviet move intended to minimize division ultimately reinforced it. Regardless of the policy approach they adopted, the Soviets simply could not force the Khalq and Parcham factions of the PDPA to unite in a meaningful fashion. Factionalism continued until the communist Afghan regime collapsed in 1992.

A pattern emerged in 1978–80 between Soviet and local Afghan communist allies. The Soviet Politburo would make a policy request, Afghan proxies would agree to it, and the KGB in Kabul would either implement the policy or ignore it depending largely on how the policy fit with KGB interests and standard operating procedures. By 1980 the Afghans were largely cut out of decisions regarding policy implementation altogether and so cannot be credited or blamed for outcomes since they did not exercise any meaningful legal sovereignty. The Soviet assumption of responsibilities largely deprived their Afghan allies of the ability to make decisions about the state, but it also allowed the Afghans to deny responsibility for the political or military state of affairs, focusing instead on aspects of policy that they could potentially influence. As Mitrokhin notes,

No resolutions were passed unless they had been prepared beforehand by the Soviets. The Afghans paid an unreasonable amount of attention to internal party intrigues. They presumed that other matters would be settled by the Soviets. They behaved like dependants but, at the same time, were noticeably insincere with their Soviet comrades. Babrak held the view that an increase of Soviet influence and intervention in Afghanistan would increase his prestige and importance but not allow the Soviets to control him as they might wish.[34]

The Soviets similarly found it convenient to blame their Afghan counterparts. This was particularly useful in the later stages of the war once it became evident that the intervention was failing. According to historian Mohammed Kakar, once

> Karmal's inability to consolidate his government had become obvious, Mikhail Gorbachev, then general secretary of the Soviet Communist Party, said, "The main reason that there has been no national consolidation so far is that Comrade Karmal is hoping to continue sitting in Kabul with our help." Colonel Nikolai Invanov, a Soviet military writer, however wrote that "he [Karmal] was a nobody." Both statements reflect a failure of Soviet foreign policy. It was because of this policy that Karmal was unable to achieve "national consolidation," that he had become "a nobody."[35]

Karmal was initially chosen by the Soviets precisely because he wasn't "a nobody." He was a founding member of the PDPA, was a high-profile member of parliament, and a well-known moderate figure in Afghan political life in the 1960s and 1970s. His ineffectiveness and irrelevance was at least partially due to Soviet decisions to marginalize Afghans in the Afghan state, a dynamic that speaks to the risks of denying local regimes legal sovereignty in counterinsurgency interventions. Nevertheless, the use of the Afghan political leadership as scapegoats for the failure of the Soviet agenda in Afghanistan only intensified after Mohammad Najibullah took over as president in 1987.[36]

What were the consequences for Moscow's heavy-handed political approach toward its Afghan allies? The evidence suggests the Soviet takeover of the decision-making process harmed the USSR's military and political agendas in Afghanistan. The annexation of Kabul caused the PDPA to become increasingly reliant on Russian advisers, assistance, decisions, money, and military operations. Kabul increasingly lost credibility among Afghans looking for a national political force. Continued Soviet intervention in the internal processes of the Afghan state made it increasingly difficult for Moscow to withdraw, and for the Afghans to take leadership roles. The more the Afghan communists were shut out, the more indifferent they became. According to Bradsher, Najibullah described Afghan cabinet meetings as dominated by the KGB. Najibullah recalled that each Afghan "'minister had a Soviet adviser. As the conference goes on the debate gets higher and the advisers move closer to the table, while the Afghans move away, and finally the Soviets are left to quarrel among themselves.' He added the 'large number of advisers ... caused ... a spirit of stagnation, laziness, irresponsibility, and corruption inside the [PDPA] party'."[37]

The Soviets promoted Afghans that were loyal to the Soviets. Although this policy likely contributed to greater Afghan-Soviet cooperation, it raises complex questions regarding interalliance cooperation and the question of tension and local sovereignty on counterinsurgency efforts. Can the Afghan-Soviet dynamic even be labeled cooperation if agreement was accompanied by the confiscation of sovereignty and a coercive surveillance of partners? What does it mean if local allies are loyal to intervening forces perhaps even above other political and national associations? KGB archivist Mitrokhin touched on this issue while discussing one of the KGB's favorite local officials, Abdul Kadyr. Mitrokhin noted that Kadyr

was completely trusted by the Soviet organs and military. The Residency had a high opinion of him and considered him to be a man of principle and devoted to the USSR ... Babrak was aware of his strong personality and secretly disliked and mistrusted him ... A. Kadyr is loyal to the Soviet Union and will not make any important military or political moves without orders from the Soviets or their agreement.[38]

The Soviets assumed that Kadyr's loyalty to the KGB above Karmal and the Afghan ruling leadership benefited them and the war effort. But having glimpsed into the potential effects of the Soviet takeover of the Afghan state on the political dynamics of the counterinsurgency, it is questionable whether local figures being steadfastly loyal to intervening forces helps the greater war effort.

Additionally, the Soviet–Afghan alliance illustrates that plans offered by intervening forces may not necessarily be strategically superior to those offered by local actors and that compliance with requests from intervening forces may or may not lead to greater rates of success in war, depending on the strategies proposed. There were substantial reports that Soviet approaches to defeating the insurgency were poorly designed. In 1982 the CIA wrote:

In our view, Babrak's policies have generally failed because they have the conflicting goals of winning popular support and turning Afghanistan into a socialist state on the Soviet model. The land reform program illustrates most of the problems government efforts have encountered. The Taraki government intended the program to win peasant support and destroy the power of the "feudal" landowning elite through redistribution of land, to increase production, and to lay the basis for organizing Afghan agriculture on the Soviet model. Like many other government programs, it reflected doctrinaire Marxist misconceptions about Afghan society. Most peasants had little reason to support the program.[39]

The Afghan and Soviet counterinsurgency coalition experienced multiple policy failures, including important measures to negotiation with
the mujahedeen. As Kakar explained:

On 15 January 1987, while inaugurating the policy of "national reconciliation,"
Najibullah invited political groups for a dialogue about the formation of a
coalition government. He also invited leaders of the Islamic groups, but in
reply they reiterated their view: "the continuation of armed jehad until the
unconditional withdrawal of Soviet troops, the overthrow of the atheistic
regime, and the establishment of an independent, free and Islamic
Afghanistan." The former king Mohammad Zahir also rejected the call. Even
within the PDPA opposition was felt. The followers of Karmal, who numbered
more than the followers of Najibullah, set up a separate faction, the SNMA
[Organization for the National Liberation of Afghanistan].[40]

National reconciliation failed and the insurgency against the Soviets and
their local partners continued to intensify. Soviet efforts to negotiate with
the Afghan resistance were unsuccessful. Negotiations with enemy forces
can be a way for intervening forces to coerce concessions from local allies,
as observed in Vietnam. Yet, the Soviets couldn't find a partner to negotiate with among mujahedeen insurgents. For example, Ahmad Shah
Masoud, the commander who later led the anti-Taliban Northern Alliance
until his assassination by al-Qaeda on September 9, 2001, had been con-
tacted by Soviets seeking to negotiate. Masoud sent the following response:

Mister Adviser!
    I already wanted to go to the place to meet the Soviet representatives when
I received your latest letter. I should say for the sake of clarity: we have endured
war and your presence of 10 years. God willing, we will endure it a few more
days. But if you begin combat operations then we will give you a fitting rebuff.
That's all. From this day we will assign our detachments and groups the mission
of being in full combat readiness.
    With respect, Ahmad Shah Masoud
    26 December 1988[41]

The failure of large-scale counterinsurgency intervention is of course
not unique to the Soviets. The Soviet–Afghan case, however, had some
unique characteristics, namely the harsh military-focused Soviet
approach combined with Soviet dominance of the Afghan state and
denial of local legal sovereignty that deprived the Afghans of a sense of
investment and responsibility, which only deepened dependencies on
Soviet leadership, organization and financing, and made independence
impossible.[42]

US strategists in Vietnam, Iraq, and Afghanistan were reluctant to
dominate local political functions for an extended period of time due to
fear that doing so would undermine the mission to create an independent

local government capable of resisting the insurgency without American aid. As the US Department of Defense summarized for Vietnam in a 1972 lessons-learned study, "If we could turn back history, the process of 'Vietnamization' would have been started in 1962, not 1969."[43] In contrast, the Soviet intervention in Afghanistan was guided by a very different philosophy. According to KGB records by the end of the invasion, Soviet advisers "delved into all the crevices of the [Afghan] ministries."[44] These institutionalized dependencies on Soviet personnel made it increasingly difficult to establish a long-term self-sustaining pro-Soviet political solution for Afghanistan.

High-level primary source documents were analyzed to identify specific policy requests from the USSR to local Afghan allies. The fall of the Soviet Union in December 1991 led to an unexpected opening of archives containing files from the Soviet Politburo. This was granted by President Yeltsin to disgrace the Communist Party after the Soviet collapse and allowed scholars in the 1990s to collect and translate critical decision documents on the Afghan war.[45] Over 200 high-level Soviet documents have been organized and made publicly available through the Cold War International History Project (CWIHP) at the Woodrow Wilson International Center for Scholars as well as through electronic briefing books by the Russia project at the National Security Archive at the George Washington University.[46] These documents were used in this study to compile a list of twenty-two demands placed on the Afghan communist regime, the PDPA by the Soviet Politburo.[47]

Analyzing the outcome of Soviet requests similarly relied on primary source documents as well as secondary sources, including statements from former KBG officials, former Afghan political prisoners, declassified US government documents analyzing the war, and historical accounts of the Afghan–USSR alliance. Documents from the Afghan regime were not included because they were largely destroyed when the regime collapsed in 1994. Even if PDPA documents survived and were available, there would be good reason to doubt their validity. Former KGB archivist Vasiliy Mitrokhin who defected to the UK with six cases of transcribed KGB documents commented,

For the sake of personal interests the [PDPA] history of the period before and after the coup was re-written and falsified. The role and place of the party and its leaders in the life of the country were deliberately distorted. Documents, articles and letters were rewritten and altered. Approval of a person was given for personal and subjective reasons rather than on a realistic basis.[48]

The Soviets, on the other hand, being accustomed to unwavering state control and absolute secrecy, could afford to be more candid in documents. After all, Soviet officials writing from 1979 to 1989 would have had difficulty imagining a scenario where internal policy documents would become public. This is reflected in released documents that include highly sensitive details including specific covert operational orders to intelligence agencies such as the 1979 order to send "a parachute battalion disguised in the uniform (overalls) of an aviation-technical maintenance team. For the defense of the Soviet Embassy, send to Kabul a special detachment of the KGB U.S.S.R. (125–150 men), disguised as Embassy service personnel."[49]

Twenty-two policy requests from Moscow to Kabul were identified by examining hundreds of documents from Soviet Politburo meetings between 1978 and 1980.[50] After 1980, Soviet Politburo members were working directly with Soviet handlers in Afghanistan or not discussing specific policy demands during meetings with Afghans. Because not all documents from Soviet Politburo meetings throughout the war are available, it is possible that Soviet requests to the Afghans from 1981 to 1989 are not yet available. However, there are several reasons aside from access that are likely underpinning the dearth of post-1980 Soviet Politburo policy requests. First, Soviet implantation in the Afghan state meant that high-level Soviet officials did not have to negotiate with Afghan leaders on most issues as Soviet subordinates in Kabul had the ability to bypass Afghans and directly implement any sought-after reforms. Second, the record shows discussion throughout the war regarding reforms the Soviets first demanded from 1978 to 1980. The Soviets acting on behalf of the Afghans typically could not conduct these reforms without Afghan assistance. This dynamic indicates that records on Soviet requests to local allies are available, but the USSR was not asking for new reforms after 1980 because it controlled the Afghan state through an entrenched network of advisers in decision-making positions. A summary of the requests issued by Moscow prior to their takeover of the state is provided in Table 8.2.

Explaining the Compliance and Defiance of Local Allies in Afghanistan (USSR)

From the Politburo documents twenty-two requests from Moscow to Kabul were identified including requests to strengthen the border patrol, implement land reform, and negotiate with tribal leaders. For 11/22 (50.0 percent) of Soviet demands the Afghans were fully compliant, for 3/22 (13.6 percent) they were partially compliant, and for 8/22 (36.4 percent) they were noncompliant. This is not as high of a rate of compliance as one might expect given Soviet intimidation techniques and the

Table 8.2 *Summary of Soviet requests to Afghan allies*

	Year	Soviet request to Afghans
1.	1978	Stop repressive techniques against other groups
2.	1979	Communist political programs for the Afghan Army
3.	1979	Strengthen border patrol
4.	1979	Implement land reform
5.	1979	Rely on rule of law
6.	1979	Stop purging the military of political enemies
7.	1979	Organize poverty committees – win over peasants
8.	1979	Publicize Soviet economic aid
9.	1979	Close the border with Pakistan
10.	1979	Do not harm Taraki supporters
11.	1979	Do not harm Taraki
12.	1979	Unite the Afghan Communist Party
13.	1979	Fractionalize Mullahs – divide and conquer
14.	1980	Enact a constitution
15.	1980	Negotiate with tribal leaders
16.	1980	Negotiate with religious leaders
17.	1980	Organize youth programs – especially for students
18.	1980	Economic focus – raise standard of living
19.	1980	Talks with neighboring nations – Pakistan and Iran (bilateral)
20.	1980	Expand local government structures
21.	1980	Offer political settlement – national reconciliation

infiltration of Afghan government even in the early period of the war, from which these requests date. However, these numbers are largely consistent with rates of compliance across the other wars investigated in this study including Vietnam, Iraq, Afghanistan, and Sri Lanka. This section of the chapter discusses the factors that influenced these compliance outcomes in the Soviet–Afghan alliance.

There is no evidence of instances where the PDPA said "no" to requests from the Soviet CPSU.[51] Soviet domination over Afghan bureaucracies and pervasive KGB intimidation techniques gave Afghan officials incentives to agree to Soviet demands, regardless of their intention to follow through or not. Agreement with Soviet policies came from the very top of the Soviet and Afghan leadership as the KGB reported that President "Babrak [Karmal] asked the KGB representatives to assure [KGB head] Andropov that he would unswervingly carry out all Andropov's suggestions and advice."[52] There were also ritual acts of agreement from lower levels of the Afghan bureaucracy toward their Soviet advisers.

Niyaz Muhamad, the head of the economic department of the PDPA Central Committee, who was in Moscow for medical treatment, said, in confidence, in

December 1980 that the Afghans had been instructed to tell Soviet officials they met that there was unity in the party, the safety of the population of the whole country had been secured and good conditions for economic activity had been established.[53]

Moscow was demanding unity in the PDPA, Afghanistan's communist party, which nevertheless remained perpetually divided between two groups, the Parcham and Khalq.[54]

Capacity, Interests, and Dependencies

Insufficient capacity of the Afghan state to carry out Soviet-requested reforms was cited for 6/22 (27.3 percent) of requests. Some of the requests that cite a lack of capacity illustrate the early naiveté of Soviet planners. In March 1979, for example, Soviet officials asked the Afghans to seal the Afghan border with Pakistan.[55] Following the invasion, Soviet troops took over this duty and found it fundamentally impossible.

An assessment of the interaction effect between Soviet unilateral ability to implement a policy and the interests of the Soviets and their Afghan allies to compare this war to other counterinsurgency interventions was not possible due to the relatively small number of Soviet requests. Nevertheless, more general observations are evident. The PDPA was more likely to undertake the action requested by the Soviets if there was a clear short-term benefit from the request or use the action to better solidify their hold on power, in support of principal–agent models advocating for the alignment of interests and incentives to promote compliance. The Soviet request that Kabul implement communist political programs within the Afghan Army provides an example. The request provided Kabul with an opportunity to expand its political base of support and inculcate the military.[56]

For 12/22 (54.5 percent) of policy requests the Soviets had the ability to unilaterally accomplish the task on their own without the Afghans; the rest of the time (45.5 percent), they were unable to do so. Usually this incapacity to act without the Afghans was due to political considerations and local legal sovereignty – while it lasted. Despite their deep involvement in the Afghan state, after 1979, there were several activities the Soviets did not want to carry out unilaterally since doing so risked admitting their annexation of the Afghan state and denying local legal sovereignty. The USSR wanted to be careful not to support US or mujahedeen propaganda chastising Moscow as an aggressive occupier. The Soviets decided they needed the Afghans to cease repressive techniques against other Afghan communist or politically sympathetic groups in Afghanistan, to follow rule-of-law procedures instead of extrajudicial proceedings, and to enact a constitution. Theoretically, for example, the Soviets could have drafted the constitution for the Afghans, although this was deemed unnecessary since the version of the constitution passed in April 1980 was based on a draft written under Amin in 1979.[57] Regardless, prior to 1980 the Soviets were still dependent on the Afghans to go through the motions of enacting it, passing it into law, including presenting, voting on, and signing the constitutional proposal. This they could not do without Afghan participation unless they wanted to publically admit they had commandeered the Afghan state denying local sovereignty immediately.

Because it was strategically important to uphold the appearance of the PDPA government as a local communist revolutionary movement governing Afghanistan for as long as possible, Moscow could not threaten to destroy the Afghan communist political party, or its flawed leaders. Furthermore, as early as 1980 the Russians were looking for a way out of the conflict. Eliminating the Afghan communist leadership was seen as problematic because it would only embed them deeper in Kabul. This effectively meant that although Afghan communist leaders could agree to Moscow's demands that they meaningfully unite as a single party and refrain from purging potential rivals from the military, Afghans in power were frequently able to avoid complying with such demands. To the Soviets' frustration, there was nothing they could do to change this set of circumstances, as long as they wanted the PDPA to continue to function. After only a year since the Soviet intervention, in 1980, the KGB was reporting that Karmal was

adopting the tactics of positive inaction. He listened attentively to advice, but did little to put it into action. He often complained: "It was not my idea that I should sit at the same table as the Khalqists. What unity is this?" ... [Karmal's] only concern was how to compromise the leaders of the Khalq faction. He was not concerned about consolidating the situation in the country as he considered that the Soviets must do this.[58]

In sum, unsurprisingly, the diplomatic record indicates that the Afghans were less likely to comply with Soviet requests if they did not have the capacity to implement the given request. However, Soviet allies in Kabul were more likely to adopt the given demand if they had the potential to profit, or if the USSR could undertake the given demand without their help. Overall, there was more compliance than noncompliance with Soviet requests, but nevertheless over one-third of requests were not fulfilled, a finding that aligns with the alliance dynamics in other large-scale foreign interventions.

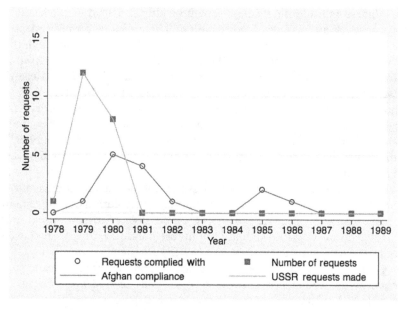

Figure 8.1 Rates of USSR requests to local allies, and rates of Afghan compliance, 1978–89

Afghan Compliance with Soviet Requests over Time

Figure 8.1 graphs the number of requests made by the Soviet Politburo in Moscow to its Afghan allies in the war. Again there are no new requests made from high-level Soviet leadership to the Afghans from 1981 to 1989 because Soviet forces were directly entrenched in the Afghan bureaucracy, denying Afghanistan legal sovereignty. Certain requests, including promoting a comprehensive program of national reconciliation to insurgent forces, publicizing political settlement opportunities, or pushing propaganda in the Afghan Army were requested by Soviet leaders early in the conflict (1979–80), but nevertheless not complied with until later in the war.

Conclusion – Less Interalliance Negotiation, More Domination

The Soviet experience in Afghanistan illustrates an important dynamic impacting compliance with requests from intervening forces in counter-insurgency wars. The greater the intrusion of foreign forces into local governance institutions, and the more forceful the means used by the foreign power in coercing its local partners, the more agreement intervening forces is likely to receive. There are no records of officials in Kabul ever telling Moscow they wouldn't do what Moscow asked. Yet 36.4 percent of requests ended in noncompliance, a figure comparable to the percentages found in Afghanistan under US occupation, Vietnam, Iraq, and Sri Lanka.

As James C. Scott stated, "the more menacing the power, the thicker the mask."[59] The more heavy-handed the intervening state is in its dealings with local proxies, the more "yeses" it will hear regarding reforms, but the less those positive affirmations will indicate about action. The likelihood of noncompliance with requests from foreign patrons is likely to hover around the same figure (32–38 percent) due to inherent capacity failures and divergent interests between local and foreign allies that time and again push compliance with the requests of intervening allies below 50 percent.

Notes

1 Paul A. Winters, *The Collapse of the Soviet Union* (San Diego, CA: Greenhaven Press, 1999), 122; Fred Coleman, *The Decline and Fall of the Soviet Empire: Forty Years That Shook the World, from Stalin to Yeltsin* (New York: Macmillan, 1997), 209; Anthony Arnold, *The Fateful Pebble: Afghanistan's Role in the Fall of the Soviet Empire* (Novato, CA: Presidio, 1993); Robert Strayer, *Why Did the Soviet Union Collapse? Understanding Historical Change* (New York: M. E. Sharpe, 1998), 80–1, 129.

2 Henry S. Bradsher, *Afghan Communism and Soviet Intervention* (Oxford University Press), 124.

3 Mohammed Kakar, *Afghanistan: The Soviet Invasion and the Afghan Response, 1979–1982* (Berkeley: University of California Press, 1997), 23–4; Panagiotis Dimitrakis, *The Secret War in Afghanistan: The Soviet Union, China and Anglo-American Intelligence in the Afghan War*, 1st ed. (London: I. B. Tauris, 2013), 58–60; Gregory Feifer, *The Great Gamble: The Soviet War in Afghanistan*, 1st ed. (New York: Harper Collins, 2009), 60–2.

4 Feifer, *The Great Gamble*, 70.

5 Dimitrakis, *The Secret War in Afghanistan*, 62.

6 Anthony Arnold, *Afghanistan's Two-Party Communism: Parcham and Khalq* (Stanford, CA: Hoover Press, 1983), 99.

7 Girardet, *Afghanistan*, 136; also cited in Kakar, *Afghanistan*, 70.

8 Bradsher, *Afghan Communism and Soviet Intervention*, 122.

9 Furthermore, "Bureaucrats of the regime found that even routine orders had to be approved and countersigned by the Soviets," Kakar, *Afghanistan*, 70.

10 Mitrokhin, "The KGB in Afghanistan, English Edition," 43.

11 Arnold, *Afghanistan's Two-Party Communism*, 99–100.

12 Bradsher, *Afghan Communism and Soviet Intervention*, 122.

13 Ibid., 123.

14 Mitrokhin, "The KGB in Afghanistan, English Edition," 3.
15 Ibid., 101.
16 Girardet, *Afghanistan*, 136.
17 Ibid., 138.
18 Mitrokhin, "The KGB in Afghanistan, English Edition," 126.
19 Douglas Borer, *Superpowers Defeated: Vietnam and Afghanistan Compared* (London: Frank Cass, 1999), 175.
20 Mitrokhin, "The KGB in Afghanistan, English Edition," 122.
21 Ibid., 101.
22 Ibid., 104.
23 Christian Ostermann and Odd Arne Westad in Mitrokhin, "The KGB in Afghanistan, English Edition," 3; Bradsher, *Afghan Communism and Soviet Intervention*, 130–1, 203; Feifer, *The Great Gamble*, 171–8; see also Steve Coll, *Directorate S: The C.I.A. and America's Secret Wars in Afghanistan and Pakistan* (New York: Penguin, 2019), 666 for an interesting parallel discussion in the US intervention in Afghanistan as there were "stovedpiped, semi-independent campaigns waged simultaneously by different agencies of American government."
24 Mitrokhin, "The KGB in Afghanistan, English Edition," 120.
25 Conor Keane, *US Nation-Building in Afghanistan* (London: Routledge, 2016), 55; see also James L. Creighton, "How Bureaucracy Impedes Victory in Afghanistan," *World Policy Blog*, World Policy Institute, April 16, 2012, www.worldpolicy.org/blog/2012/04/16/how-bureaucracy-impedes-victory-afghanistan.
26 Coll, *Directorate S*, 496.
27 Central Intelligence Agency, "Afghanistan: The Revolution after Four Years," July 1982, NSEA 82-10341, iii.
28 Bradsher, *Afghan Communism and Soviet Intervention*, 137.
29 Kakar, *Afghanistan*, 156.
30 Ibid., 161.
31 US Army and Marine Corps, *Counterinsurgency, FM 3-24*, chapter 3; Tse-Tung, *On Guerrilla Warfare*.
32 Mitrokhin, "The KGB in Afghanistan, English Edition," 3–4.
33 Bradsher, *Afghan Communism and Soviet Intervention*, 131.
34 Mitrokhin, "The KGB in Afghanistan, English Edition," 128.
35 Kakar, *Afghanistan*, 74.
36 Bradsher, *Afghan Communism and Soviet Intervention*, 132–3.
37 Ibid., 123.
38 Mitrokhin, "The KGB in Afghanistan, English Edition," 133.
39 Central Intelligence Agency, "Afghanistan: The Revolution after Four Years," 2.
40 Kakar, *Afghanistan*, 261.
41 "Letter from Ahmad Shah Masoud to the Soviet Chief Military Adviser," December 26, 1998, Cold War International History Project (CWIHP), Record ID 112471, http://digitalarchive.wilsoncenter.org/document/112471.
42 Grau, *The Bear Went over the Mountain*.
43 Cooper et al., "The American Experience with Pacification in Vietnam – Volume I – An Overview of Pacification (Unclassified)," 33.

44 Mitrokhin, "The KGB in Afghanistan, English Edition," 43.

45 Serge Schmemann, "Soviet Archives: Half-Open, Dirty Window on Past," *The New York Times*, April 26, 1995, www.nytimes.com/1995/04/26/world/ soviet-archives-half-open-dirty-window-on-past.html; George and Bennett, *Case Studies and Theory Development in the Social Sciences*, 100.

46 Over 150 Soviet documents on Afghanistan are available at www .wilsoncenter.org produced under the guidance of Christian Ostermann. From the National Security Archive, see "Afghanistan Déjà vu? Lessons from the Soviet Experience," "Afghanistan and the Soviet Withdrawal 1989: 20 Years Later: Tribute to Alexander Lyakhovsky Includes Previously Secret Soviet Documents," and "The Diary of Anatoly Chernyaev: Former Top Soviet Adviser's Journal Chronicles Final Years of the Cold War."

47 Most documents in the CWIHP's Afghanistan collection are translated into English. However, any remaining nontranslated documents related to this period are not addressed here.

48 Mitrokhin, "The KGB in Afghanistan, English Edition," 31.

49 "Gromyko-Andropov-Ustinov-Ponomarev Report to CPSU CC on the Situation in Afghanistan," June 28, 1979, CWIHP, Collection: Cold War in the Middle East, Soviet Invasion of Afghanistan, Document identifier: 5034D714–96B6–175C-9A356BA28915CAE5, www.wilsoncenter.org.

50 See the CWIHP declassified and translated Soviet documents on Afghanistan.

51 Two requests did not have an available translated record of whether or not there was correspondence from the Afghans agreeing or disagreeing to the policy requested.

52 Mitrokhin, "The KGB in Afghanistan, English Edition," 98.

53 Ibid., 125.

54 "Special File Record of Conversation of L.I. Brezhnev with N.M. Taraki"; "CPSU CC Politburo Decisions on Afghanistan (Excerpts) Calling on Puzanov to Meet with Taraki and Try to Mend the Rift between Taraki and Amin," September 13, 1979, CWIHP, The Woodrow Wilson International Center for Scholars, Cold War in the Middle East, Soviet Invasion of Afghanistan, Document identifier: 111561, https://digitalarchive.wilsoncenter.org/document/ 111561.

55 "Special File Record of Conversation of L.I. Brezhnev with N.M. Taraki."

56 Ibid.

57 Kakar, *Afghanistan*, 73.

58 Mitrokhin, "The KGB in Afghanistan, English Edition," 127, 132.

59 James C. Scott, *Domination and the Arts of Resistance: Hidden Transcripts* (New Haven, CT: Yale University Press, 1990), 3.

9   When Small States Intervene

> Cubans have given their lives for Angola ... For us, a small, distant
> country that faces its own dire threats, these are heavy sacrifices. Have
> they been entirely useful? Has the [Angolan] MPLA [People's
> Movement for the Liberation of Angola] used our internationalist aid
> correctly?   —Fidel Castro to Jose Eduardo dos Santos

Costly, large-scale military interventions are usually considered foreign
policy options limited to large powers. Yet, Vietnam, Egypt, Syria, and
Cuba each engaged in long and costly counterinsurgency entanglements
supporting local allies in large-scale military interventions. Emerging
from their own histories contending with colonialism, these four smaller
interveners offer a different perspective to foreign counterinsurgency
interventions, as they draw from their experiences of occupation, revolution, and insurgency. Furthermore, these wars reveal how alliance
dynamics between local and foreign counterinsurgents shift when the
asymmetries in capabilities between allies are less significant than in the
interventions examined in previous chapters. Smaller interveners are less
apt to rely on technological solutions and more likely to maintain modest
agendas for development, which can make fulfilling counterinsurgency
policies and maintaining institutions easier for local partners with limited
capacity.

Despite the secrecy policies of Vietnam, Egypt, Syria, and Cuba significantly limiting the availability of primary source materials on alliance
dynamics during their interventions, several findings emerge. Namely,
there is significant variety in the approaches taken to managing local
allies when comparing these diverse wars. Similar to the USSR in
Afghanistan, Vietnam in Cambodia and Egypt in Yemen, for example,
embedded themselves deeply into the local regimes they were aiding,
thus ignoring the norm of promoting the legal sovereignty of local
regimes. Syria in Lebanon, however, was less focused in its intervention,
aiding multiple groups in order to assert its interests, as opposed to
commandeering the government in Beirut, in part due to Israel's efforts
to constrain Damascus. Similarly, Cuban forces did not occupy the

Angolan state, partly due to the influence of the Soviet Union, which financed the intervention, and partly because Fidel Castro's anti-imperialist stance made the Cubans wary to appear as an occupying force, and finally due to the Angolan government's other sources of support through oil revenues and its ties to sympathetic European governments such as Sweden.

Vietnam in Cambodia, 1978–1991

On December 25, 1978, a year before the Soviet invasion of Afghanistan, 150,000 Vietnamese troops crossed into Cambodia to overthrow Pol Pot and the Khmer Rouge. On January 10, the People's Republic of Kampuchea (PRK) – the new pro-Vietnamese regime – was formed, tasked with rebuilding Cambodia after a genocide. In just four years Pol Pot had killed nearly 2 million people, including 80 percent of schoolteachers and 95 percent of doctors.[1] For the Vietnamese, solidifying the post-Khmer Rouge PRK regime and fighting a counterinsurgency war against remaining Khmer insurgents would take more than a decade. Hanoi withdrew from Cambodia in September 1989, having lost 15,000 soldiers, but generally achieving its strategic goals of solidifying a regime in Cambodia friendly to Vietnamese interests. By the end of the Cold War, as historian Thu-Huong Nguyen-Vo noted, Vietnam had essentially turned both Laos and Cambodia "into good neighbors. Indochina was tuned into a Vietnamese bloc, the next best thing to a Vietnamese empire."[2]

Following a similar model to that of the Soviet Union's occupation of Afghanistan, the Vietnamese occupation authorities dominated decision-making processes in the PRK largely denying Cambodian legal sovereignty. As stated by Nguyen-Vo, "To ensure the survival of both the infant regime in Kampuchea and Vietnamese dominance in Indochina, Hanoi felt it necessary to hold Phnom Penh's hand through all decisions big and small."[3] For the Vietnamese this effectively avoided managing an "alliance" between two political organizations and the complex process of interstate negotiations as conceptualized in Chapter 2 and examined in Chapters 4–7. Similar to the Soviets in Afghanistan after 1981, and the USA for the first fourteen months of US involvement in Iraq, the Vietnamese dictated domestic policy in Cambodia by specifying high-level policies for Cambodian bureaucracies. Based on his analysis of high-level PRK documents while serving as a legal liaison in Cambodia, Evan Gottesman noted that "long after Vietnamese advisors had set up the PRK, its Party structure, and its state bureaucracy, they continued to wield tight control over the regime. Deference to the advisors was a matter of official policy."[4] Vietnamese advisers, as their

Table 9.1 *Timeline of the Vietnamese–Cambodian alliance during the Vietnamese intervention*

1970	In Cambodia on October 9, General Lon Nol overthrows Prince Norodom Sihanouk. Nol cracks down on suspected members of Vietnam's communist insurgent group National Liberation Front (NLF) in Cambodia.
	The Khmer Rouge, a Cambodian communist guerilla organization, escalates violence in opposition to Nol. Eventually, with support from the North Vietnamese (NVA) and NLF allies, the Khmer Rouge seize much of rural Cambodia.
	On April 29, the US-ARVN (Army of the Republic of Vietnam) attack NVA–NLF sanctuaries in Cambodia. By May 6, about 30,000 US troops and 48,000 ARVN soldiers are operating in Cambodia. American ground forces are withdrawn in June, but US bombing continues.
1973	The Khmer Rouge blockade the capital Phnom Penh in the summer while US bombing intensifies.
1974	The Khmer Rouge capture the provincial capital of Oudong on March 18. After several attempts, Nol's forces finally recapture the capital on July 9.
1975	On New Year's Day, the Khmer Rouge launch an offensive cutting off the Mekong River and highways from Phnom Penh. Fighting intensifies, and on April 1 General Nol flees. On April 17, the Khmer Rouge seize Phnom Penh, bringing Pol Pot into power. Pol Pot conducts mass killings in an alleged attempt to create a classless Cambodia.
1978	Moderate Khmer Rouge leaders flee toward the border with Vietnam where about 100,000 are killed along with other pro-Vietnamese Cambodians.
1979	Vietnam invades Cambodia to overthrow Pol Pot and establish the PRK. The Khmer Rouge and two noncommunist factions wage a guerilla war against the PRK and Vietnamese forces in Cambodia.
1989	The insurgency weakens and in September Vietnamese troops withdraw from Cambodia.
1991	The PRK reaches an agreement with warring parties to end the war.

KGB counterparts in Afghanistan, infiltrated all levels of the local bureaucracy during the occupation.

In particular, three organizations were charged with implementing Vietnamese policy in Cambodia. Foremost was A-40. Led by Le Duc Tho, A-40 consisted of high-level political experts from Hanoi's Central Committee tasked with advising key PRK officials. B-68, the second committee, consisted of a mid-level group tasked with overseeing everyday government functioning in the PRK. And lastly A-50, a group promoting Vietnam-friendly policies locally throughout the provinces.[5] As foreign policy analyst Stephen Morris concluded, there was "direct political control of the Phnom Penh administration – the People's Republic of Kampuchea (PRK), later renamed the State of Cambodia (SOC) in 1989 – by the Vietnamese. According to the accounts of defectors from the regime, the PRK operated under the tutelage of Vietnamese advisers at all levels."[6]

Vietnamese advisers drafted the PRK constitution and selected higher-ranking PRK officials, but turned to Cambodian nationals to staff fledgling bureaucracies.[7] This practice led to a good deal of nepotism as positions were determined by loyalty and based on connections.[8] The Vietnamese attempted to combat this trend by instituting protocols for personnel authorization and promoting party figures they favored. According to Nguyen-Vo, the "maze of Vietnamese advisers making most decisions in Kampuchean public life was further augmented by Vietnamese selection and training of Kampuchean party and government personnel. Vietnamese advisers screened potential party members. Those selected would be sent to Vietnam for training."[9]

The Vietnamese approach to counterinsurgency in Cambodia is particularly interesting, given that Hanoi had spent decades fighting an insurgency against the French, American, and South Vietnamese forces. As a result, policymakers in Hanoi had an insider perspective, which influenced their counterinsurgency policies in Cambodia. Hanoi invited reformed Khmer Rouge elements into political life to defuse the political basis of the insurgency, adopted an ostensibly inclusive brand of communism embedded in symbols of Cambodian nationalism, and was wary of PRK corruption and heavy-handed approaches that could foster sympathy for antigovernment forces.[10]

Unfortunately, specifics on the Vietnamese–Cambodian alliance are difficult to locate since Vietnamese documents are unavailable and the only Cambodian files released to date are from mid- and low-level PRK records.[11] Available materials suggest that the Vietnamese determined that denying both legal and territorial sovereignty to Cambodians was preferable to risking the possibility of local Cambodian allies implementing policies that ran counter to Vietnamese interests, and were less concerned that dominating Cambodian affairs would delegitimize their local proxy regime. For Hanoi, controlling the allied Cambodian state minimized the chances that insurgents could exploit Phnom Penh's failures. This policy was notably more successful for the Vietnamese in Cambodia than the Soviets in Afghanistan, despite the Soviet ability to sink significantly more resources into its project supporting a proxy regime in Afghanistan.

Egypt in Yemen, 1962–1970

Egypt's intervention in Yemen adopted a similarly intrusive approach comparable to the Vietnamese in Cambodia. Cairo dominated its

Yemeni allies, which produced mixed results in its counterinsurgency campaign. The Egyptian military dominated the political decision-making process in Yemen in the early years of the Yemen Arab Republic (YAR). With Cairo's blessing, on September 26, 1962, a coup led by Abdullah al-Sallal overthrew Iman Muhammad al-Badr. Al-Badr had inherited North Yemen a week earlier from his deceased father, Imam Ahmed, a leader described in the US media as "the bulging-eyed tyrant of Yemen."[12] The newly established Republican regime led by al-Sallal was shortly thereafter propped up by an Egyptian military occupation. Those loyal to the overthrown Imam regrouped in the highlands to mount an insurgency funded by Saudi Arabia. Saudi leaders were threatened by perceived Egyptian adventurism, Cairo's geostrategic interests in oil, and announcements that "soon after the deposition of the imam that they [Egypt] were also interested in creating a Republic of the Arabian Peninsula."[13] Disturbed by the antimonarchist revolution orchestrated on its border by the Egyptians, the Saudis provided cash and arms to the royalist insurgents.

Due to the weakness of the al-Sallal regime in Sana'a, Egyptians quickly dominated the Yemeni Republican government. As one historian summarized, "the Egyptians bore the major responsibility for the republic (from 1962 to 1968), all the functions of a modern state bureaucracy and administration were carried out essentially by (and even for) the

Table 9.2 *Timeline of the Yemeni–Egyptian alliance during the Egyptian intervention*

1962	On September 25, Imam Mohammed al-Badr is overthrown in an Egyptian-backed coup led by Colonel Abdullah al-Sallal. Al-Badr flees to the mountains, forming a Royalist insurgency to challenge the new Republican government. Leading the United Arab Republic (UAR), Egypt intervenes on behalf of the fledgling Republic.
1964	By 1964, 40,000 Egyptian troops are stationed in Yemen. Saudi Arabia and Jordan send arms, supplies, and advisers to the insurgent Yemeni Royalists.
1967	Colonel al-Sallal is replaced by the more moderate Abdul-Rahman al-Iryani. UAR expeditionary forces withdraw from Yemen in the aftermath of Egypt's defeat in the Six-Day War with Israel. The Royalists gain control of three-fourths of the nation and bring the capital San'a under siege.
1968	In February, the siege of San'a is broken with support from Syrian air support and veterans of South Yemen's National Liberation Front.
1970	Upon Saudi Arabia's urging the Royalist army agree to end its resistance and peace is brokered on April 14, 1970.

Egyptians, and then only in those areas that were under the government's control."[14] The Egyptians fielded as many as 70,000 troops in Yemen at a time, losing upwards of 15,000 soldiers over six years of fighting. Cairo considered Yemen "to be a backward and medieval country," and was concerned about the survivability of the regime.[15] Dana Adams Schmidt, a reporter with *The New York Times* who documented the Yemeni experience under the Egyptian occupation, colorfully described the Egyptian presence:

At all times the Egyptians were at once the [Yemen Arab] republic's mainstay and its greatest handicap. The republic needed Egyptian help and protection, but the condescension of Egyptian teachers, experts, and officials infuriated the quick, intelligent Yemenis; the Egyptian soldiers, though well behaved and docile, were increasingly resented. In Sana the Egyptians entirely dominated the scene; on the outskirts stood acres and acres of tents surrounded by barbed wire; in the streets their fleshy bodies, clad in yellow-brown uniforms, forever bent over counters, broad buttocks facing the streets endlessly choosing rolls of cloth to send home to their wives ...[16]

As is typical for foreign counterinsurgency interventions, the task of securing local allies became increasingly protracted and complex for the Egyptians.[17] According to an analysis drawing from Arab accounts, "with Egyptians holding key positions in every ministry and throughout the bureaucracy, and the YAR's economy increasingly tied to Egypt's, the military task force became an external prop for what was essentially an Egyptian-sponsored experiment in republican politics."[18] Eventually, a combination of factors convinced Cairo to withdraw, including escalating costs, the British withdrawal from South Yemen, growing public distrust of censored reports in Egypt, and the humiliating military defeat at the hands of the Israelis in the 1967 Six-Day War. Simultaneously, the Saudis backed negotiations known as the Compromise of 1970, which enabled the consolidation of a republican and pro-Egyptian state, while at the same time reserving positions for pro-Saudi royalists.[19] Similar to the Vietnamese and the Soviets, the Egyptians dominated their local counterinsurgency allies. During occupation, Egypt was institutionally embedded in the decision-making processes of Yemen's domestic political organizations. In the long run, however, this did not result in lasting Egyptian control or even thwarting insurgent activity. The negotiated solution involving both Egypt and Saudi Arabia did not fit the pattern of outcomes in either Cambodia or Afghanistan. North and South Yemen were subsequently united in 1990, but instability in Yemen, largely unrelated to the period of Egyptian occupation, is ongoing into the present.

Syria in Lebanon, 1976–2005

The Syrian intervention in Lebanon lasted almost twenty-nine years.[20] Counterbalanced by Israel, unlike the Vietnamese in Cambodia or Egyptians in Yemen, Damascus did not deny Lebanese legal sovereignty entirely, instead systematically strong-arming key Lebanese political organizations, creating patterns of substantial Syrian influence in Lebanese affairs. Lebanese representatives regularly traveled to Damascus before making policy announcements and Lebanese President Ilyas al-Hirawi reportedly quipped, "we now disagree on the appointment of a doorman and go to Damascus to submit the problem to the brothers [there]."[21] Though Syria also, allegedly, directed Lebanese foreign policy decisions,[22] and influenced electoral processes by picking candidates and building domestic Lebanese political coalitions.[23] Yet, despite these influential roles, Syria was restrained, seeking to maximize its influence over fractured Lebanese politics while carefully avoiding direct conflict with the Israelis.

Most accounts of Syrian influence in Lebanon describe Damascus' strategy as a balancing act between leveraging fractured Lebanese political actors against each other to strengthen Syria's influence, while trying to avoid provoking Israel, which was also deeply enmeshed in Lebanon. Syria maintained between 20,000 and 40,000 troops in Lebanon between 1976 and 2005.[24] Through military and political pressure Damascus effectively made itself "the godfather of Lebanese politics," rigging elections and subverting Lebanese foreign policy.[25] Yet, due to the influence of Israel as well as Iranian interests promoted through Hezbollah, Syria was not the only foreign force with influence on the ongoing civil war in Lebanon. Its strategy was to maneuver within the splintered political landscape and to maintain clients across Lebanon that would enable Damascus to bypass the Lebanese state whenever national politics became unfavorable to Syrian interests.[26] These processes differ from other counterinsurgency interventions that more typically support one side and work to strengthen a regime against an insurgency. To assert its influence in Lebanon, Damascus, as one historian put it, would exploit "the rivalries of the various Lebanese factions while using the anarchy they created to make its peacekeeping indispensable and deter the involvement of rival external powers."[27]

Syria's policies of historically supporting insurgent forces, such as factions within the PLO, then instigating a large-scale military intervention in alliance with counterinsurgent forces to broker an agreement between groups that would be favorable to their interests has parallels to the Indian intervention in Sri Lanka. Initially, New Delhi covertly backed the Tamil insurgents, and then intervened unsuccessfully to

Table 9.3 *Timeline of the Syrian–Lebanese alliance during the Syrian intervention*

1976	Conflict erupts between the allied Palestinian Liberation Organization (PLO) and Lebanese Muslims against the Lebanese Christians. Fearful of a PLO-Leftist victory, Syrian President Hafez al-Assad sends 12,000 troops and 300 tanks into Lebanon. Eventually, Syria commits 25,000 troops to fight alongside 25,000 Christians. In opposition are 25,000 PLO guerillas, 3,000 leftist Lebanese fighters, and 2,000 Iraqi volunteers.
	Syrian forces launch an offensive, forcing the PLO-Leftists to agree to a cease-fire in November. Violence lessens, but does not stop entirely.
1978	In July, violence breaks out between Christians in East Beirut and their former Syrian allies. Syrians retaliate against the Christians for their cooperation with Israel as well as what Damascus claims is a Phalangist attempt to divide Lebanon into Christian and Muslim states.
	Violence escalates between the Syrians and the Maronite Christians in September, as the Syrians subject Beirut to the worst shelling in its history. A cease-fire is brokered in October.
1989	General Michael Aoun, a Maronite Christian commanding the Lebanese Army, declares a "war of liberation" against Syria. Muslim officials in Lebanon refuse to support Aoun, creating conflicting administrations in Lebanon.
1990	Syria launches air and ground offensives and Aoun surrenders.
1991	Syria continues to occupy Lebanon, and a treaty is signed authorizing Syria's military presence.
2000	Syrian President Hafiz al-Assad dies and is replaced by his son Bashir al-Assad.
2005	Former Lebanese Prime Minister Rafik Hariri is assassinated in February. Syrian operatives are allegedly involved in the killing.
	Hariri's assassination inspires a series of protests known as the Cedar Revolution calling for the withdrawal of Syrian troops from Lebanon.
	Due to the pressure of the protests, Syrian troops withdraw in April 2005.

contain those insurgents. As the historian Itamar Rabinovich summarized regarding Syria's intervention in Lebanon, "An invasion calculated to put a swift end to an embarrassing political situation thus served in fact to aggravate it."[28] New Delhi and Damascus shared a similar logic, first funding insurgents in neighboring countries, then intervening on the side of counterinsurgent forces to advance their own influence. Arguably, this strategy based on playing insurgent and counterinsurgent forces against each other led to costly and only marginally effective interventions by both India and Syria.

Cuba in Angola, 1975–1991

Smaller intervening states usually do not travel far from home. Fielding an army at great distances is enormously costly. Syria borders Lebanon,

Vietnam borders Cambodia, and Egypt has access to Yemen via the Red Sea. Cuba's intervention in Angola, however, was unprecedented among small states because it traversed an ocean, a feat usually reserved for global powers.[29] The logistics were facilitated by Soviet support for Cuba and its wars in Africa. But Cuba was never a Soviet proxy in Angola. As Piero Gleijeses, a leading historian on Cuban-African relations in this period summarized, Cuba was "the engine ... It was the Cubans who pushed the Soviets to help Angola. It was they who stood guard in Angola for many long years, thousands of miles from home."[30] Soviet support sustained Havana's forays into Africa, but the Kremlin was not the primary decision maker in the intervention in Angola. The US National Security Council, for example, noted in 1976 that "Cuba is not involved in Africa solely or even primarily because of its relationship with the Soviet Union. Rather, Havana's African policy reflects its activist revolutionary ethos and its determination to expand its own political influence in the Third World at the expense of the West."[31] In fact Havana defied the USSR by intervening in Angola, and through subsequent decisions to reinforce this intervention. As Gleijeses notes, "Castro defied the Soviet Union. He sent his troops on Cuban planes and Cuban ships without consulting Brezhnev, hoping that Moscow would come around, but without any assurance that it would. He was no client."[32] It would take Moscow two months to assist Havana in transporting Cuban troops to Angola to pursue the mission Castro envisioned.[33]

Unlike the Soviet, Egyptian, or Vietnamese interventions, the Cubans did not overtly dominate the local allied regime during their intervention in Angola. Doing so would have contradicted Havana's anti-imperialist rhetoric and its contention that its interventions in Africa were not comparable to colonialist or capitalist meddling in the developing world. Castro and other Cuban leaders called Cuba an "Afro-Latin American Nation," declaring that Cuban support for the Angolan counterinsurgency was based on Cuba's "internationalist duty" to oppose imperialism.[34] According to Fidel Castro,

even when the imperialist threats against our own country were enormous, we didn't hesitate in sending much of our most modern and best military equipment to the Southern Front of the People's Republic of Angola ... Wherever Cuban internationalists have gone, they have set examples of respect for the dignity and sovereignty of those countries ... our exemplary selflessness and concern for others is remembered everywhere.[35]

According to Gleijeses, this emphasis on cooperation went beyond Castro's rhetoric. Looking at Cuban documents, Gleijeses comments,

Table 9.4 *Timeline of the Angolan–Cuban alliance during the Cuban intervention*

1975	Three groups, the Popular Movement for the Liberation of Angola (MPLA) led by Agostinho Neto, the National Union for the Total Independence of Angola (UNITA) led by Jonas Savimbi, along with the National Front for the Liberation of Angola (FNLA) led by Holden Roberto battle for control of Angola.
	The UNITA–FNLA alliance is supported by South African and Zairian troops along with personnel from Britain, the United States, Western Europe, and Rhodesia.
	The MPLA is aided by Cuban troops as well as exiles from the rebellious Zairian province of Katanga, troops from Mozambique, as well as East German personnel and Soviet advisors.
1976	Cuban forces devastate the FNLA in the north and UNITA in the south, but nonetheless, UNITA is able to regroup to perpetuate a guerilla war in the south with support from the USA and South Africa.
1977	Supporting UNITA, South African forces engage in combat with Cuban forces supporting MPLA.
1989	Cuba begins to withdraw forces.
1991–2002	Cuba withdraws its last forces in May 1991, ending a sixteen-year military intervention. UNITA and the MPLA negotiate a power-sharing arrangement to end the civil war.
	However, UNITA refuses to accept defeat to the MPLA in national elections, leading to a resurgence of violence. Fighting continues, with another cease-fire brokered in 1994, which in 1998 again collapses.
	UNITA finally agrees to a cease-fire in 2002 after Jonas Savimbi is killed in a government ambush.

"what is striking is the respect with which the Cubans treated the Angolan government."[36]

In order to avoid a public relations nightmare for Castro if Havana was accused of exploitation, Cuba opted decidedly against taking over the Angolan state. Moreover, it is unlikely that Havana would have been able to sustain such a takeover even if Castro had set out to do so. By maintaining strong ties with multiple outside powers, including the Soviet Union, Sweden, and western oil companies such as American-owned Gulf Oil and Chevron, Cuba's partners in Angola were in a unique position for a small, local ally to have alternative allies.[37] In light of this array of diplomatic and commercial relationships, it is not surprising that Angolan President Agostinho Neto was treated well by Cuban diplomats. As Gleijeses summarized, "Neto was at no moment a Cuban client – and at no moment did the Cubans try to treat him as one. On the contrary, the Cubans bent over backwards to treat Neto with respect and deference."[38]

A former Portuguese colony with oil resources, Angola gained independence on November 11, 1975. During the period of Portuguese exodus, three Angolan nationalist groups vied for control of the new state, the MPLA, backed by Cuba and the USSR, the FNLA aided by Zaire (and at points the United States and China), and UNITA, supported by apartheid-era South Africa and the United States. In the summer of 1975 the MPLA attempted to broker an alliance with UNITA, but was turned down as UNITA instead opted to join with South African troops and the FNLA in an aggressive push to expel the MPLA from Luanda, the Angolan capital, ahead of the upcoming independence slated for November 1975.[39] In response to the aggressive military offensive the MPLA appealed for assistance from Cuba and the USSR. Cuba sent troops, the Soviet Union invested arms, equipment, and advisers. By the summer of 1976 the MPLA, backed by Havana, had successfully driven the other two groups out of Luanda, and by October the MPLA was holding twelve of sixteen district capitals. Uprooted, the FNLA and UNITA tentatively allied together to contest the MPLA in an insurgency funded by South Africa and the United States.

At independence, Angola was fractured. Competing paramilitary groups were taking aim at each other, as well as coping with their own internal stresses.[40] In response to South African and US intervention in Angola, Cuba, with Soviet support, mounted successive military interventions, first in conventional conflict from 1975 to 1976 and then against the protracted insurgency from 1976 to 1991. Like many counterinsurgency interventions, the conflict dragged on much longer than anticipated. Castro intended to withdraw Cuban troops within three years.[41] Instead the Cubans stayed for fifteen years, staging as many as 55,000 troops at a time, complemented by 2,000 Soviet and Eastern bloc advisers and several thousand Cuban civilian advisers.[42] By the end of the intervention over 400,000 Cuban soldiers had fought in Angola.[43] How many military and civilian casualties the Cubans suffered in Angola is surprisingly difficult to estimate. In fact, tallying Cuban deaths has been called "the single most contentious issue of the Angolan operation."[44] Accounts range from 2,200 to 15,000 Cuban casualties.[45]

According to one source, only approximately 7 percent of Cubans in Angola were reportedly engaged in humanitarian or political projects although Cuban personnel regularly traded off between civilian and military roles.[46] While the Cubans who were involved in civilian projects were often credited for sustaining Angolan government services, they did not implement policies to ensure they dominated the local state like the Soviets, Vietnamese, and Egyptians did in their interventions.[47] In 1977, the Central Intelligence Agency (CIA) claimed, "Cubans are frequently criticized for being arrogant and for ignoring African sensibilities. They are also faulted for their tendency to take command of a given situation instead of acting as advisers."[48] Piero Gleijeses claims, however, that Cuban contemptuousness was uncommon. "There were Cuban advisers who were arrogant and patronizing, but the evidence – documents and interviews – suggests that they were atypical."[49]

Cubans were embedded in a variety of organizations in the Angolan government, advising military and providing assistance in education and medicine.[50] Although Cuba branded itself as an exemplary member of the developing world on equal ground with struggling movements like the MPLA, the capabilities of Cuba's bureaucratic institutions far exceeded Angolan capacity.[51] In order to manage its expansive forces, Havana established independent government bureaucracies in Angola for military, health, education, and civil engineering.[52]

Historian Christine Hatzky, analyzing Angolan documents related to education during the Cuban intervention, referred to these unilateral Cuban institutions as a "state within a state," while noting that these Cuban bureaucracies in Angola were not designed "to establish any kind of neocolonial or imperial position of dominance in Angola, but rather to supervise and manage independently the large number of Cuban civilians working there."[53] Yet, Hatzky also asserts that despite a lack of intent to dominate its partnership with Angola, Cuba's relatively superior institutional strength motivated unilateral policies, which separated allies, and undermined local institutional growth. She notes that there was a "tendency among the Cubans to implement the cooperative program autonomously and independently of the Angolans. This tendency in turn, led to the Angolans losing sovereignty and control."[54] This is a pervasive dynamic in asymmetric counterinsurgency alliances. In the US experiences in Vietnam and Afghanistan, for example, there was persistent debate about US unilateral action to implement counterinsurgency protocol undermining the very local political institutions required to win the war in the long run.

Interestingly, Hatzky's account of the Cuban-Angolan partnership upends the typical model of interalliance bargaining in counterinsurgency interventions. According to Hatzky, the local Angolan regime requested services and the foreign intervening Cuban forces complied or defied those Angolan requests. Usually, the patron–client dynamic in interventions is modeled the other way, as specified in Chapter 2, with foreign allies making requests and local allies complying or defying. Hatzky, however, claims "the Angolan government's reliance on Cuban support did not automatically mean that cooperation constituted an asymmetrical relationship of dependency, with the MPLA at a disadvantage. Cooperative agreements resulted from the demands of the Angolans, to which the Cubans generally responded positively."[55] In part, Hatzky's claim regarding this shift in roles derives from her claim that Havana was paid, at least in part, by the Angolan government for its intervention to support the MPLA.[56]

This claim, however, has been strongly contested by other scholars, including Piero Gleijeses, who argues that Angola's contribution to the cost of the Cuban mission was minimal. According to Gleijeses' analysis of Cuban documents, Angola never compensated Havana for Cuban military salaries, and only paid for a portion of Cuba's civilian assistance programs. Even then, Gleijeses claims Angolan payments for Cuban civilian assistance programs lasted for only six years from January 1978 to October 1983, despite those programs running for the duration of the intervention, 1976–1991.[57] Moreover, looking to local resources, such as local oil revenues, to offset the costs of an intervention is not unique to Cuba in Angola. In fact, US officials in Congress and the Department of Defense often proposed that Iraqi oil profits should be used to offset the costs of the US war and reconstruction effort.[58]

The absence of a complete collection of primary source materials on the Cuban intervention has left significant uncertainties in existing accounts regarding the extent and importance of the Angolan compensation paid to the Cubans.[59] Nevertheless, more general conclusions about the Cuban-Angolan alliance are possible with the given materials. First, as with other intervening forces, Havana balanced its commitment to its allies with other considerations, including the costs and duration of that assistance, seeking to maintain its commitment, while minimizing costs wherever possible. As the CIA noted, like many intervening forces, "Cuban policy is not free of contradictions."[60] Second, Havana sought compensation from its Angolan partners for certain types of assistance, but it is unlikely that reimbursement was a critical component of the intervention or alliance. All available evidence indicates the assistance was limited to a subset of civilian activities such as education, was only provided to Cuba for a certain portion of the war, six years according to Gleijeses and fourteen years according to Hatzky, while both Cuba and Angola received additional support, including weapons and funding from the Soviet Union.[61] The payment Angolans provided to Cuba for services does not appear significant enough as to reverse the patron–client dynamic, as claimed by Hatzky, which would imply the Cubans fielded a mercenary development corps directed by Angola.

Interestingly, Hatzky concluded that the Cubans did not always live up to their promises and Havana conducted autonomous actions that were "regarded as an infringement of [Angolan] sovereignty," and "generally failed to respond to Angolan demands for better communication and coordination, to the point that the Angolans accused the Cubans of 'unilaterialism' during the meetings of the bilateral commissions."[62] Unilateral activity by intervening forces is a persistent issue in counter-insurgency alliances. As hypothesized in Chapter 2, unilateral capability to implement policies is often the most reliable path through which intervening forces can coerce local allies to reform when the interests of allies diverge. Conversely, when local allies are required to participate to implement a counterinsurgency policy, they can leverage their indispensable position to their political advantage.

Furthermore, even though this was Cuba's intervention, the role of the Soviet Union as a second intervening partner aiding the MPLA altered the nature of the Angolan–Cuban alliance. Consumed by Afghanistan, the Soviets were intent on avoiding a full-scale intervention in Angola, yet were prepared to provide advice and weapons for the MPLA and Cuban effort. The Soviets supplied between 1,500 and 1,700 advisors, invested over $4.9 billion, and by the mid-1980s held a small, but important number of command positions in the Angolan Army (FAPLA, or The People's Armed Forces of Liberation of Angola). Soviet personnel included instructors, professors, and advisors, "particularly in the Finance Ministry and the Central Bank and have reportedly replaced Cubans in the Transportation, Fisheries and Trade ministries."[63] Yet, unlike the Cubans, the Soviets in Angola were prohibited from participating in combat.[64]

Due to Havana's financial dependencies on Moscow, the Soviet Union influenced Cuban policy in Angola, at least in part, including how intimately Havana involved itself in the local regime in Luanda and the amount of leverage Cuba had over the Angolans.[65] As Castro acknowledged to Soviet officials, "we could not have survived ... without the aid that we received from the Soviet Union."[66] Consider, for example, historian Edward George's account of Cuban relations with FAPLA:

Castro insisted that the FAPLA needed to be playing a more prominent role in [counterinsurgency operations], but when [Angolan President Agostinho] Neto suggested that the Cubans simply take over command of the FAPLA, Castro resisted, fearful of upsetting his Soviet patrons who were not only supporting the weaponry for the Cuban operation, but who were also involved in strategic planning at the FAPLA's highest levels. As Castro saw it the Soviets would not take kindly to Cuba taking control of the army they were bankrolling, and he did little more than admonish Neto to take the UNITA threat more seriously.[67]

Castro held back on taking a more dominant role over the Angolan military, at least in part, in order to avoid confrontation with Moscow. Furthermore, commanding Angolan units would undermine the anti-imperialist, revolutionary message deeply embedded in Castro's foreign policy, which the intervention in Angola was intended to support, not undermine.

Additional dynamics of the Cuban intervention in Angola have parallels in other counterinsurgency interventions. For example, as specified in Hypothesis 1 discussed in Chapter 2, Havana found combat and the instability produced by conflict often strained local capacity, exacerbating the limitations of the Angolan state.[68] Furthermore, like all allies, the Angolans, Cubans, and Soviets balanced a variety of overlapping and diverging interests, which shaped bargaining encounters and established the terms of cooperation.[69] As offered in Hypotheses 2A–3B, the dependencies of intervening forces on local partners for policy implementation affected bargaining dynamics and the likelihood of local compliance. The Cubans depended on the Soviets for funding, the Angolans relied on both the Soviets and the Cubans for expertise and support, while the Cubans and the Soviets depended on the Angolans as local proxies to implement policies aligning with their interests. Speaking to the Angolan dependency, Hatzky noted, "discord and conflicting interests ... were rarely aired openly because the Angolan government had few alternatives to accepting Cuban support. Even internal records frequently document this state of dependency, and the corresponding comments indicate that, despite the problems, the Angolans were – at least in general terms – grateful for Cuban support."[70] There were diverse interests among all three counterinsurgent allies in Angola, which created interesting bargaining dynamics in implementing counterinsurgency policies.

Conclusion – Challenges and Opportunities for Small Interveners

The alliances between local and intervening partners in counterinsurgency interventions undertaken by small states such as Vietnam in Cambodia, Egypt in Yemen, Syria in Lebanon, or Cuba in Angola, exhibit several distinct characteristics when contrasted with interventions conducted by larger powers such as the USA in Afghanistan or Iraq. Small states often have more modest development expectations for local allies when compared with the development expectations of larger, resource-rich intervening forces.[71]

Interestingly, foreign bureaucracies tasked with expanding the governance capacity of local proxies tend to recreate aspects of their organization, which may or may not best serve the institutional needs of the regime receiving the assistance. This dynamic is exacerbated with larger, wealthier interveners who are prone to promoting bureaucracies similar to those in large, wealthy states, which may not be well suited or sustainable for small local regimes facing large-scale insurgency. Robert Komer, a pivotal figure in US pacification policy in Vietnam, thought these sorts of bureaucratic tendencies were so critical for understanding the dynamics of foreign intervention that he titled his well-known after-action treatise on the US failure in Vietnam, "Bureaucracy Does Its Thing: Institutional Constraints on U.S.–GVN Performance in Vietnam." Komer summarized, "behavior patterns of the organizations involved influenced not only the decisions made but what was actually done in the field."[72] For Komer, US and South Vietnamese governments failed in their counterinsurgency mission, at least in part, because, "in true bureaucratic fashion, each U.S. and GVN agency preferred doing more of what it was already doing, rather than change accepted patterns of organization or operation."[73]

Another example of this process is detailed in Chapter 8 on the Soviet experience in Afghanistan as the KGB fashioned critical Afghan institutions in their image, even if this undermined policy requests emanating from the Kremlin for better-fitting Afghan institutions. A parallel process is also apparent in the US war in Afghanistan. As noted by John Sopko, the Special Inspector General for Afghanistan Reconstruction, certain basic questions should have been asked before American-style development was imposed in Afghanistan. "'Is this program sustainable?' You should answer that question before you build it. Before you build schools that have power generation that they cannot afford. Before you build a dam, keep in mind where you are. You are not in Kansas."[74] As the journalist interviewing Sopko remarked, "We built an Afghanistan they can't afford."[75] These cases suggest that large states tend to look for governance solutions that are familiar, yet may not be a best fit for local economic, social, or political conditions.

Interventions by smaller states such as Cuba, however, may promote more modest development expectations for local allies, which may be more easily sustained. A narrower gap between the resources of local allies compared with foreign intervening allies can also affect the anticipated benefits that local allies expect will result from foreign intervention, as well as the extent of corruption and economic disruption. Consider, for example, how managing popular local expectations for development in Iraq and Afghanistan has been difficult for Washington.[76] Many Afghans and Iraqis expected the American intervention to produce a rapid rise in the local standard of living. When these expectations were not met, there was dissatisfaction with both foreign and local counter-insurgency partners, potentially fueling sympathy for insurgents.[77]

In contrast, using Cuba as an example, after 1978 in Angola, Piero Gleijeses writes, "many Angolan officials preferred [working with the Europeans] to relying on the austere Cubans. Fidel Castro epitomized the ethos of Cuban assistance during his March 1977 visit to Angola. Addressing Cuban advisers in the Angolan Foreign Trade Ministry, Castro said, 'Help them to buy well, to buy cheap ... Eliminate all luxury products.'"[78] For local allies tied to smaller intervening states, there aren't the same expectations for amenities and kickbacks, which limits opportunities for corruption and dashed expectations.

Furthermore, the size of the intervening state can influence funding for the insurgency. Large intervening powers can unintentionally create fundraising opportunities for enemy insurgents, as other states seeking to balance the power of the intervening state can take advantage of an insurgency to weaken the intervening power.[79] Such enhanced outside support for an insurgency has the potential to be less significant in interventions undertaken by smaller states. While it is only one example, this dynamic is evident when comparing the amount of aid Washington provided the mujahedeen insurgents fighting the Soviets in Afghanistan contrasted to Washington's more modest assistance to UNITA insurgents battling Cubans in Angola. For the USA, providing significant aid to mujahedeen to weaken the Soviets, effectively returning the favor of Soviet support for Hanoi during the US war in Vietnam, was considered to be worth the inherent risks, whereas arming insurgents to weaken Cuba, a state already relatively weak, required a much smaller US investment.

Lastly, smaller intervening states, especially those with recent experiences waging an insurgency, such as communists in Vietnam, are less prone to limit themselves to conventional strategies to combat unconventional forces. In contrast, wealthy military forces tend to prefer relying on technology and firepower. One example is the difference in the strategic advice received by the MPLA in Angola partnered with both the USSR and the Cubans. As a senior Angolan officer recalled,

The Soviet military mission wanted us to pursue conventional warfare. The Cuban military advocated counter-insurgency tactics. They knew a lot about this, and it was more relevant for the situation we faced. There were things that the Soviets didn't understand well. We were closer to the Cuban position, but it was the U.S.S.R. that gave us weapons. This was the reality, and we had to find a compromise so we wouldn't quarrel with the Soviets.[80]

This example demonstrates how the policy requests of intervening allies may vary widely depending on the size, and budget, of the intervener.

Furthermore, by virtue of having less asymmetric resource disparities between local and intervening forces, smaller states run the risk of becoming quickly exhausted by the costs of intervention yet may also have certain advantages compared to larger intervening powers. Small states such as Vietnam or Cuba may bring their experiences utilizing insurgent tactics that are strategically useful on the battlefield. In addition, interventions by superpowers tend to foster unrealistic local expectations for development that are difficult to meet, and shock local economies by injecting large sums of cash fostering inflation and corruption. Controlling more moderate resources means that intervening states like Vietnam, Syria, Egypt, or Cuban may be less burdened by those political and economic drawbacks of wealth. Lastly, smaller intervening states are less likely to create significant fundraising opportunities for insurgents from third parties seeking to balance great powers, similar to how the Soviet intervention led to massive US support for the anti-Soviet insurgency in Afghanistan as a way to diminish Soviet power. While we should not romanticize a lack of resources, it is important to acknowledge that wealth produces certain advantages, as well as posing certain risks in intervention. These processes add depth to the larger claims considered in Chapters 1 and 10 noting that power over allies in counterinsurgencies is highly contextualized and is not necessarily an inevitable product of resources and military capabilities.

Notes

1 Joel Brinkley, *Cambodia's Curse: The Modern History of a Troubled Land* (New York: PublicAffairs, 2012), x.
2 Thu-Huong Nguyen-Vo, *Khmer-Viet Relations and the Third Indochina Conflict* (Jefferson, NC: Mcfarland, 1992), 145.
3 Ibid.
4 Evan R. Gottesman, *Cambodia after the Khmer Rouge: Inside the Politics of Nation Building* (New Haven, CT: Yale University Press, 1996), 143.
5 Nguyen-Vo, *Khmer-Viet Relations and the Third Indochina Conflict*, 146; Nayan Chanda, *Brother Enemy: The War after the War* (New York: Free Press, 1988), 373–4; Gottesman, *Cambodia after the Khmer Rouge*, 143–5.
6 Stephen Morris, *Why Vietnam Invaded Cambodia: Political Culture and the Causes of War*, 1st ed. (Stanford University Press, 1999), 224.
7 Gottesman, *Cambodia after the Khmer Rouge*, 4–5, 50, 53.
8 Ibid., 51.
9 Nguyen-Vo, *Khmer-Viet Relations and the Third Indochina Conflict*, 146.
10 Gottesman, *Cambodia after the Khmer Rouge*, 13.
11 Ibid., xii.
12 Dana Adams Schmidt, *Yemen: The Unknown War* (London: Bodley Head, 1968), 20.
13 Manfred W. Wenner, *The Yemen Arab Republic: Development and Change in an Ancient Land* (Boulder, CO: Westview Press, 1922), 133.

14 Ibid., 140; see also John E. Peterson, "The Yemen Arab Republic and the Politics of Balance," *Asian Affairs* 12, no. 3 (November 1981): 257; Robert D. Burrowes, *The Yemen Arab Republic: The Politics of Development, 1962–1986* (Boulder, CO: Westview Press, 1987).

15 Dawisha, "Intervention in the Yemen: An Analysis of Egyptian Perceptions and Policies," 48.

16 Schmidt, *Yemen: The Unknown War*, 80.

17 Dawisha, "Intervention in the Yemen: An Analysis of Egyptian Perceptions and Policies," 48.

18 Jesse Ferris, "Egypt, the Cold War, and the Civil War in Yemen, 1962–1966" (PhD diss., Princeton University, 2008); Wenner, *The Yemen Arab Republic*, 140–1.

19 Wenner, *The Yemen Arab Republic*, 136.

20 James Sturcke, "Syria Confirms Full Troop Withdrawal from Lebanon," *The Guardian*, March 30, 2005, www.guardian.co.uk/world/2005/mar/30/syria .unitednations.

21 Daniel Pipes, *Syria beyond the Peace Process* (Washington, DC: Washington Institute for Near East Policy, 1996), 47.

22 William Harris, "Syria in Lebanon," *MERIP Reports* no. 134 (July 1, 1985): 9; Robert B. Chadwick, "Lebanon: The Uncertain Road to Reconstruction" (Thesis, Naval Postgraduate School, Monterey, CA), available from National Technical Information Service, Springfield, VA, 1997, http://archive.org/ details/lebanonuncertain00chad, 24.

23 Itamar Rabinovich, *The War for Lebanon, 1970–1985* (Ithaca, NY: Cornell University Press, 1985), 55; Middle East Contemporary Survey (The Moshe Dayan Center, 1993), 540; Central Intelligence Agency, "National Intelligence Bulletin, Lebanon: Sarkis' Unanimous Election Is Setback for Jumblatt," May 10, 1976, CREST Database, National Archives and Records Administration; Stephen Talbot, "Syria/Lebanon: The Occupier and the Occupied," *PBS Frontline World*, August 3, 2004, www.pbs.org/frontline world/elections/syria.lebanon/.

24 Harris, "Syria in Lebanon," 9.

25 Talbot, "Syria/Lebanon: The Occupier and the Occupied."

26 Raymond Hinnebusch, "Pax-Syriana? The Origins, Causes and Consequences of Syria's Role in Lebanon," *Mediterranean Politics* 3, no. 1 (1998): 152.

27 Ibid., 145.

28 Rabinovich, *The War for Lebanon, 1970–1985*, 55.

29 Piero Gleijeses, *Visions of Freedom: Havana, Washington, Pretoria and the Struggle for Southern Africa, 1976–1991* (Chapel Hill: University of North Carolina Press, 2013), 9.

30 Ibid., 15.

31 Piero Gleijeses, *Conflicting Missions: Havana, Washington, and Africa, 1959–1976* (Chapel Hill: University of North Carolina Press, 2002), 392.

32 Ibid., 379; see also Gleijeses, *Visions of Freedom*, 13–14; Christine Hatzky, *Cubans in Angola: South-South Cooperation and Transfer of Knowledge, 1976–1991* (Madison: University of Wisconsin Press, 2015), 58.

33 Gleijeses, *Visions of Freedom*, 30.

34 Hatzky, *Cubans in Angola*, 11; Gleijeses, *Visions of Freedom*, 83.

35 Fidel Castro, Raul Castro, and Nelson Mandela, *Cuba and Angola: Fighting for Africa's Freedom and Our Own*, ed. Mary-Alice Waters, 1st ed. (New York: Pathfinder Press, 2013), 52.

36 Gleijeses, *Visions of Freedom*, 105.

37 Tor Sellström, *Sweden and National Liberation in Southern Africa: Solidarity and Assistance 1970–1994* (Uppsala: Nordic Africa Institute, 1999); Gleijeses, *Visions of Freedom*, 159–60, 382.

38 Gleijeses, *Visions of Freedom*, 106.

39 William Minter, *Apartheid's Contras: An Inquiry into the Roots of War in Angola and Mozambique* (Johannesburg: Witwatersrand University Press, 1994), 20.

40 Gerald J. Bender, "Angola, the Cubans, and American Anxieties," *Foreign Policy* No. 31 (Summer, 1978): 8; Minter, *Apartheid's Contras*, 19.

41 Gleijeses, *Visions of Freedom*, 377.

42 Stephen L. Weigert, *Angola: A Modern Military History, 1961–2002*, 1st ed. (New York: Palgrave Macmillan, 2011), 5; Edward George, *The Cuban Intervention in Angola, 1965–1991: From Che Guevara to Cuito Cuanavale*, reprint (London; New York: Routledge, 2012), 282; Gleijeses, *Visions of Freedom*, 9; Hatzky, *Cubans in Angola*, 4; Gleijeses, *Conflicting Missions*, 233–72.

43 Hatzky, *Cubans in Angola*, 4.

44 George, *The Cuban Intervention in Angola, 1965–1991*, 162.

45 W. Martin James, *A Political History of the Civil War in Angola: 1974–1990* (New Brunswick, NJ: Transaction Publishers, 2011), 111–17; Gleijeses, *Visions of Freedom*, 521; The Associated Press, "10,000 Cubans Reported Killed in Angola War," *Los Angeles Times*, June 16, 1987, http://articles .latimes.com/1987-06-16/news/mn-7734_1_del-pino. It is also likely that accidents and diseases such as malaria caused a sizable percentage of Cuban deaths in Angola. See George, *The Cuban Intervention in Angola, 1965–1991*, 162.

46 George, *The Cuban Intervention in Angola, 1965–1991*, 157.

47 Ibid., 157, 159–61; Bender, "Angola, the Cubans and American Anxieties," 9.

48 Central Intelligence Agency, "Memorandum for: Director of Central Intelligence, from: Acting NIO for Latin America, Subject: Cuban Involvement in Angola, NI-1589-77."

49 Gleijeses, *Visions of Freedom*, 105.

50 Ibid., 80.

51 Hatzky, *Cubans in Angola*, 169, 188.

52 Ibid., 179.

53 Ibid., 178.

54 Ibid., 188.

55 Ibid., 151–2.

56 Ibid., 152, 170.

57 Gleijeses, *Visions of Freedom*, 77–84, 521; Gleijeses, "H-Diplo Review Essay 135 on Cubans in Angola," H-Diplo, February 10, 2016, https://networks.h-net.org/system/files/contributed-files/re135.pdf.

58 *Can Iraq Pay for Its Own Reconstruction?: Joint Hearing before the Subcommittee on International Organizations, Human Rights and Oversights and the Subcommittee on the Middle East and South Asia of the Committee on Foreign Affairs, United States House of Representatives*, 110th Congress, 1st Session, March 27, 2007; Sultan Barakat, "Reconstructing Post-Saddam Iraq: An Introduction," *Third World Quarterly* 26, no. 4 (2005): 565–70.

59 The Cuban and Angolan governments have not made concerted efforts to provide comprehensive documents on the intervention. However, some materials exist. Piero Gleijeses, for example, has released around 3,400 pages of Cuban documents relating to Angola available through the Wilson Center at www.wilsoncenter.org/publication/visions-freedom-new-documents-the-closed-cuban-archives. For his book *Visions of Freedom* Gleijeses analyzed roughly 15,000 pages of Cuban documents. Christine Hatzky relied on documents garnered from the Ministry of Education in Luanda and the Department of International Cooperation that was responsible for coordinating joint Angolan–Cuban programs in Angola. On Angolan compensation paid to the Cubans see also Bender, "Angola, the Cubans, and American Anxieties," 9; Owen Ellison Kahn, "Cuba's Impact in Southern Africa," *Journal of Interamerican Studies and World Affairs* 29, no. 3 (Autumn 1987): 48; George, *The Cuban Intervention in Angola, 1965–1991*, 150.

60 Central Intelligence Agency, "The Cuban Foreign Policy," SNIE 85–79, NIC Files, Job 79R01012A, 6; quoted in Gleijeses, *Visions of Freedom*, 524.

61 Gleijeses, *Visions of Freedom*, 521; Gleijeses, "H-Diplo Review Essay 135 on Cubans in Angola"; Hatzky, *Cubans in Angola*, 182.

62 Hatzky, *Cubans in Angola*, 191.

63 Thomas Collelo, *Angola: A Country Study* (United States Marine Corps, SSOC 03000 Operations and Readiness, Washington, DC: Federal Research Division, Library of Congress, 1989), 225; Central Intelligence Agency, "Soviet Military Support to Angola: Intentions and Prospects," October 1985, Special National Intelligence Estimate, Director of Central Intelligence, SNIE 71/11-85, www.foia.cia.gov/sites/default/files/document_conversions/89801/DOC_0000261290.pdf, 7; George, *The Cuban Intervention in Angola, 1965–1991*, 192–203; Central Intelligence Agency, National Foreign Assessment Center, "Angola's Relations with the Soviet Union," November 15, 1978, RP M 78-10429, CREST Database, National Archives and Records Administration, Declassified May 25, 2006, 3; Gleijeses, *Visions of Freedom*, 343.

64 Gleijeses, *Visions of Freedom*, 343.

65 Central Intelligence Agency, "Soviet Military Support to Angola: Intentions and Prospects," 10. The shift in the war described by the CIA in the mid-1980s corresponded with Soviet ascendency into select command positions in the Angolan Army, likely a change in administration in response to military stalemate in the Cuban intervention. "A contingent of Soviet advisors (perhaps 100 officers) maintained a presence in the FAPLA High Command – overseeing the planning of operations and weapons deliveries." However, even in this era of select Soviet command, Moscow made it clear it would not stray from the "blueprint" where the Soviets funneled in cash and

weapons, while the Cubans provided instruction and manpower. See George, *The Cuban Intervention in Angola, 1965–1991*, 122, 192.

66 Gleijeses, *Visions of Freedom*, 514.

67 George, *The Cuban Intervention in Angola, 1965–1991*, 125.

68 Gleijeses, *Visions of Freedom*, 373.

69 Hatzky, *Cubans in Angola*, 58; Gleijeses, *Visions of Freedom*, 514.

70 Hatzky, *Cubans in Angola*, 192.

71 George, *The Cuban Intervention in Angola, 1965–1991*, 159.

72 Komer, "Bureaucracy Does Its Thing," iii.

73 Ibid., ix.

74 Sheila MacVicar, "SIGAR: We Built an Afghanistan They Can't Afford," *Al Jazeera America*, June 17, 2015, http://america.aljazeera.com/watch/shows/compass/articles/2015/6/17/john-sopko-sigar.html.

75 Michelle Parker, "Programming Development Funds to Support a Counterinsurgency: Nangarhar, Afghanistan," National Defense University, DTP-053, September 2008, http://ndupress.ndu.edu/Media/News/News-Article-View/Article/1227862/dtp-053-programming-development-funds-to-support-a-counterinsurgency-nangarhar/; Tom A. Peter, "Afghanistan War: Gap Grows Between U.S. Efforts, Afghan Expectations," *Christian Science Monitor*, January 19, 2010, www.csmonitor.com/World/Asia-South-Central/2010/0119/Afghanistan-war-gap-grows-between-US-efforts-Afghan-expectation.

76 US Senate Foreign Relations, "Evaluating U.S. Foreign Assistance to Afghanistan," 12; Thomas H. Johnson and Barry Scott Zellen, *Culture, Conflict, and Counterinsurgency* (Stanford Security Studies an imprint of Stanford University Press, 2014), 207; Kilcullen, *Counterinsurgency*, 55.

77 US Senate Foreign Relations, "Evaluating U.S. Foreign Assistance to Afghanistan," 12; Johnson and Zellen, *Culture, Conflict, and Counterinsurgency*, 207; Kilcullen, *Counterinsurgency*, 55.

78 Gleijeses, *Visions of Freedom*, 105.

79 Ryan Grauer and Dominic Tierney, "The Arsenal of Insurrection: Explaining Rising Support for Rebels," *Security Studies* 27, no. 2 (April 3, 2018): 263–95.

80 Gleijeses, *Visions of Freedom*, 352.

10   Conclusion

> The officials of the host country are more often than not harassed, underpaid, and bewildered in the face of new problems. If they cannot avoid frequent confrontations with eager, demanding American counterparts, they tend to resort to supine acquiescence (which is rarely translated into action), stone walling, dissembling, or playing one American official off against another. We have learned, or should have learned in Vietnam, the bootlessness of trying to cajole local officials into pressing forward with American-sponsored programs that are not actively supported by their own government.
>
> —Chester Cooper et al., *The American Experience with Pacification in Vietnam*

Power is the crux of politics, but "power" is a remarkably ill-defined concept in political science. The confusion, as David Baldwin noted, stems from the habit of conflating two related yet distinct ideas about power.[1] The first concept focuses on what power *is*, namely resources such as money, soldiers, and weapons, while the second emphasizes what power *does*, namely influence the behavior of others.[2] This conception of power as both a possession of wealthy actors and a relational process of influence between actors strengthens erroneous and misleading ideas that resources will unfailingly produce influence and that political clout can be reliably purchased. Thus, while resources certainly can be persuasive, history also demonstrates that money and capabilities often fail to produce leverage, even in critical security situations where significant resources are expended. Protracted counterinsurgency interventions have often proved to be just such perplexing situations in which wealthy counterinsurgents cannot seem to readily transform their resources into reliable political influence against either insurgent adversaries or even the local partners they are assisting. As one former Central Intelligence Agency (CIA) officer commented, the US war in Vietnam succeeded in making "a mockery of our vast power, advanced technology, and willingness to spend resources lavishly," effectively describing the surprise and humiliation of resource asymmetries failing to produce influence in costly counterinsurgency interventions.[3]

Counterinsurgency partnerships are labeled asymmetric alliances due to the substantial disparity in resources between intervening forces and local allies.[4] Counterinsurgency interventions typically involve large, affluent states coming to the aid of very small, ailing regimes.[5] Consider the United States in South Vietnam or the Soviet Union in Afghanistan: one ally is a global superpower, while the other faces certain collapse without significant external support. But resource asymmetries are not the only forms of imbalances in these partnerships affecting the bargaining leverage of one ally over another. Local partners have several unique advantages, such as understanding local politics and how to approach difficult political issues. As Hamid Karzai told his American counterparts, "I know I have many flaws … But I do know my people."[6] Furthermore, local counterinsurgents have a strong motivation to prevail in interalliance negotiations, considering these are wars battled on their territory that will determine their survival. And lastly, local allies enjoy a key political advantage in alliance politics because their success is critical for intervening forces to win. Intervening allies that plan to withdraw as soon as local proxies can carry on the effort independently have pinned their success on the long-term survivability of these local proxies. As a 2014 study by the Rand Corporation noted, this has been a significant political dynamic limiting the ability of US forces to issue threats to local proxies in modern counterinsurgency:

U.S. credibility and the reputations of key U.S. actors become intertwined with the fate of the partner regime. Withdrawing U.S. support for a partner might lead to the partner's defeat, which would be seen as a blow to the United States' global reputation for supporting its allies and would impose domestic political costs on those decisionmakers who had initially supported the partner regime.[7]

Intervening forces cannot win the counterinsurgency in the long run unless their local partners prosper, which provides tremendous leverage for local allies over their wealthy foreign patrons.

Therefore, while the collective counterinsurgency effort hinges on foreign forces for immediate security and military success, the fate of long-term political victory is determined by local forces. As long as intervening forces recognize that at least in concept, local partners have legal sovereignty, local regimes have considerable political leverage, as they are thus indispensable to long-term political consolidation and pacification. Considering how politics is archetypally considered more important than military effort, in counterinsurgency, this political asymmetry favoring local counterinsurgency (COIN) partners is not trivial. As John Nagl observed, the local knowledge of host partners provides political "advantages [that] make local forces enormously effective counterinsurgents. It is perhaps only a slight exaggeration to suggest that, on their own, foreign forces cannot defeat an insurgency; the best they can hope for is to create the conditions that will enable local forces to win it for them."[8] Likewise, a summary document provided to cadets at the US Marines Corps officer training school concludes with a warning regarding the imperative political position of local allies:

Ideally, the host nation is the primary actor in defeating an insurgency. Even in an insurgency that occurs in a country with a nonfunctioning central government or after a major conflict, the host nation must eventually provide a solution that is culturally acceptable to its society and meets U.S. policy goals. The conclusion of any counterinsurgency effort is primarily dependent on the host nation and the people who reside in that nation.[9]

While local regimes are dependent on intervening forces for their survival in the short term, intervening allies are reliant on local partners for meaningful political victory. Unsurprisingly, both local and intervening allies will attempt to leverage their critical position to extract concessions from the other. Intervening allies will seek political favors from local regimes in return for money and security, whereas local allies are looking to leverage their political positions to extract money and as much military assistance as possible. Local allies that offer political inputs necessary for the counterinsurgency effort can capitalize on their indispensable role to extract concessions and assert influence over intervening wealthy patrons. The differences in the kinds of contributions made by local and intervening allies provide ample opportunities for both partners to assert themselves in interalliance bargaining encounters and, in the process, to grow frustrated with one another over their divergent priorities.

There are interesting parallels between the process of negotiation between asymmetric allies and bargaining between asymmetric enemies. After all, as Thomas Schelling observed, the differences between coercing a friend and coercing a foe are minimal, as both involve issuing threats.[10] In asymmetric war, insurgents typically operate with significantly fewer resources than their adversaries, yet survive in part by emphasizing politics over military confrontation. A similar process can be observed within asymmetric COIN alliances. Local allies can emphasize their political contributions in order to pressure larger partners, leaving the larger partners to contend with military challenges.

Several additional observations on asymmetric warfare are relevant to asymmetric counterinsurgent coalitions. Consider, for example, David Galula's assertion that "in a fight between a fly and a lion, the fly cannot deliver a knockout blow and the lion cannot fly."[11] Flies and lions allied together will offer different assets, and each ally will see its contributions as sources of leverage to pressure the other, as each wants to ensure that alliance-wide policies serve its interests first. Given their dissimilar tools for extracting concessions, partners will have advantages under different circumstances, based on which tools are most salient in the particular bargaining encounter.

In this book I analyzed conditions such as dependencies, interests, capacity, and enemy threats in order to answer the question: When are local counterinsurgency allies likely to comply with requests from intervening partners, and when are they likely to defy wealthy foreign patrons? An analysis of thousands of primary source documents shows that local allies tend to behave according to particular patterns in response to requests from intervening forces. Overall, there is remarkable consistency in the rate of local ally compliance across this subset of wars, with approximately one-third of requests from intervening forces resulting in compliance, one-third in partial compliance, and one-third in noncompliance.[12]

Full compliance with requests ranged from 50.0 percent in the Soviet intervention in Afghanistan to 33.3 percent in the US intervention in Iraq, with an average rate of 39.8 percent full compliance and 63.8 percent combined full and partial compliance. Noncompliance varied from 38.0 percent in the Indian intervention in Sri Lanka to 32.4 percent in the US war in Vietnam, resulting in an average 36.2 percent rate of local noncompliance.

The findings illustrate that local allies are neither puppets nor dictators in counterinsurgency coalitions and that local responses to requests from intervening forces reflect a roughly even mix of compliance, partial compliance, and defiance. This surprisingly consistent and even distribution in compliance outcomes across different wars is likely the consequence of multiple factors, including local regimes facing threats to their tentative hold on power from several sources, including insurgents, political rivals, and intervening allies that are pressing potentially costly reforms. The outcome of roughly equal parts compliance, partial compliance, and noncompliance likely speaks to local proxies balancing pressures to combat insurgents, neutralize local political rivals within and outside their regimes, as well as placating allied patrons. Enfeebled regimes cope with these diverse threats by accepting some, but not all policies proposed by intervening partners in order to appease these assorted factions that are asserting pressure to promote their interests.

These findings illustrate the complexity of counterinsurgency by proxy, and should temper expectations about how much reform can be coerced from local partners.[13] There are often limited opportunities for

Table 10.1 *Comparison of local compliance with requests from intervening allies by war*

War	Rate of full compliance	Rate of partial compliance	Rate of partial and full compliance combined	Rate of noncompliance
South Vietnam	46/105	25/105	71/105	34/105
(US intervention)	(43.8%)	(23.8%)	(67.6%)	(32.4%)
Afghanistan	11/22	3/22	14/22	8/22
(Soviet intervention)	(50.0%)	(13.6%)	(63.6%)	(36.4%)
Sri Lanka (Indian	30/79	19/79	49/79	30/79
intervention)	(38.0%)	(24.1%)	(62.0%)	(38.0%)
Iraq[a]	35/105	32/105	67/105	38/105
(US intervention)	(33.3%)	(30.5%)	(63.8%)	(36.2%)
Afghanistan	50/148	42/148	92/148	56/148
(US intervention)	(33.8%)	(28.4%)	(62.2%)	(37.8%)

Note: [a] In the US intervention in Iraq, 106 US policy requests were identified, but compliance outcomes could only be determined for 105 of them. One request, a proposal for a MANPADS (Man-Portable Air-Defense Systems) Reduction Program remains classified; see US Department of State, US Embassy Baghdad, "Post Proposes Shoulder-Fired Missile Abatement Program for Iraq (MANPADS Reduction)." Public sources do not make clear whether the Iraqi government adopted the program.

intervening forces to impose their political agendas, and to date no intervening military has been able to achieve over 50.0 percent full compliance with its requests without annexing the local regime (see Table 10.1) and denying the legal sovereignty of the local regime, a risky political move that can discredit an intervention at home, internationally, and in the field. As noted in Chapters 8 and 9 on the Soviet, Egyptian, and Vietnamese models of intervention, commandeering a local regime often merely trades one set of diplomatic challenges for another.

While there were important contextual differences between the alliances examined, generally as hypothesized, local allies tended to either comply with or defy the policy requests of intervening allies based on four primary variables: (1) the capacity of the local regime to undertake the requested activity, (2) whether the respective interests of the allies converged or diverged over the policy, (3) the dependency of intervening forces on local partners for policy implementation, and (4) a significant enemy offensive. These variables affect contexts that made local compliance with requests from allies more or less likely, in particular, I argue, the interaction between interests and dependencies.

This book has attempted to theorize upon a complicated set of dynamic and context-specific political and military processes in counterinsurgency alliances. While addressing certain pressing questions, I have neglected additional important factors that merit future examination. Specifically, moving forward it will be important to break apart assumptions that local and intervening allies are rational unitary actors, and account for variation in factors such as coercive diplomatic strategies employed by intervening forces that may affect compliance outcomes, as well as examining different ways intervening forces have embedded themselves into local institutions as a potential source of leverage that can be used to pressure compliance. Additionally, future scholarship can examine not only rates of local compliance, but the successes or failures of those policies in the war effort and better specify the relationship between alliance politics and war outcome. There is also interesting work to do comparing regime types and intervention as well as exploring what factors within local regimes (aside from capacity) make local allies more or less likely to succeed at counterinsurgency as a way to potentially screen for partners that are more or less likely to be able to solidify territorial control with foreign assistance. Lastly, while the model presented addresses the effect of a significant enemy offensive, this is a very limited measurement of a decisively important actor in counterinsurgency, namely the insurgent. Future research can better account for the relationship between the types of enemies counterinsurgency allies are facing and the effect of insurgents on interalliance cooperation and operational success or failure. While this book has aimed to examine one potential set of variables influencing alliance politics in large-scale counterinsurgency wars, it is clear there is significant work to be done to test, contextualize, and build on the work presented.

What Motivates Compliance?

Shared interests between local and intervening allies are important in motivating cooperative behavior, but in the interventions examined, interests interacted with dependencies to produce incentives for local allies to fulfill or defy requests. Specifically, if the interests of local and intervening allies converged over a policy, *and* intervening forces were dependent on local partners to implement the policy, compliance was likely. But the odds of local compliance decreased significantly, even when allies agreed on a proposed policy, if local participation was *not* required to fulfill the policy. I hypothesize that this is due to incentives for local allies to free ride on the efforts of larger partners when their participation is not required, but they support the policy being implemented.

Opportunities existed for intervening forces to pressure or coerce local partners to comply, but these moments were limited to particular contexts. Across the wars examined, there were three pathways that increased the likelihood of local compliance with requests from intervening forces. These opportunities emerged when the intervening ally (1) had the capacity to execute the action unilaterally and threatened to do so, (2) manipulated the short-term interests of the local ally by emphasizing local political benefits for implementing requested policies, or (3) demonstrated a diminished commitment to the war and issued threats to abandon the local ally. However, this last option usually undermined the objectives of the counterinsurgency effort even if it successfully motivated local compliance with a particular policy.

Consider the first set of conditions. When local and intervening allies disagreed over a proposed policy, intervening allies could coerce compliance if they had a unilateral capacity to implement the disputed policy and could potentially exclude the local ally. Local counterinsurgents reliably opted to comply under these conditions, even when they opposed the policy, in order to minimize the potential damage of being excluded by unilateral action by intervening forces. It is significant that these opportunities were limited to policy requests where intervening allies had the unilateral capacity to implement the policy, such as conducting military operations or development programs, that foreign personnel could manage independently. Policies where the participation of local allies was required, often due to norms of local legal sovereignty over policies adopted by the local government, including personnel decisions or bureaucratic reforms within local regimes, were not subject to threats of unilateral action by intervening forces, and thus were less likely to be fulfilled if they did not align with the interests of local proxies.

However, since dependencies and interests interact to promote compliance or defiance, the short-term interests of local allies can also be nudged to affect the likelihood of their compliance with policies that require their participation. This offers a second pathway to cooperation. For example, in Iraq, the USA was able to use initiatives aimed at creating a free press to promote Iraqi anticorruption reforms. In an attempt to project an image as tough on corruption, Iraqi officials surprisingly often complied, at least partly, with certain anticorruption requests from the USA. These included prosecuting officials who broke the law and requiring Government of Iraq (GOI) officials to submit financial disclosure statements.[14] Implementing these requests provided an immediate political benefit for Iraqi politicians, enabling them to advertise their reputations for being tough on corruption. Consider the following 2010 US assessment of Iraqi motivations for implementing a US-promoted anti-corruption reform:

[Iraqi Interior Minister Jawad] Al-Bulani is generally given credit for removing some corrupt police officers and other undesirables from the Ministry's workforce. His revelation about the FPS [Facilities Protection Service's more than 70,000 fictitious] "ghost" employees was doubtlessly designed in part to promote his anti-corruption credentials as he pursues political ambitions in the run-up to the March parliamentary elections.[15]

This example illustrates how the interests of local allies are a potential means to influence their response to policy requests. If foreign patrons rely on local clients to implement particular policies, rearranging incentives to create immediate political benefits can increase the likelihood that local allies will cooperate. The idea of structuring short-term local political payoffs, however, is a different logic from trying to entice local allies to implement reforms by providing cash inducements, a strategy that has not been reliably effective at motivating compliance. [16]

A third pathway for increasing the likelihood of compliance involves minimizing the commitment of intervening forces to the war and to the local ally. For example, the USA was able to compel the South Vietnamese to make painful concessions during the 1973 peace talks because the USA was in the process of withdrawing forces.[17] By decreasing its commitment to both the war and the local ally, Washington was able to threaten to leave Saigon in an even more precarious post-withdrawal position should they defy the remaining US requests.

Similarly, during preparations for drawdown from Iraq, the USA was able to coerce Baghdad into providing outstanding arrest warrants for 1,600 detainees. Withdrawing troops decreased the US commitment, enabling US officials to present Baghdad with a stark ultimatum should it fail to comply with the request to process the detainees. According to documents, the message delivered to the Iraqis:

was tough, but necessary. Time and time again, the U.S. has raised the threat of the release of AQI [al-Qaeda in Iraq]-affiliated and other high-threat detainees from U.S. detention facilities, but to no avail. The GOI appears concerned at this prospect, but either does not have the will or the ability to speed up the transfer paperwork process ... The U.S., however, made it clear that, although it does not wish to release high-threat detainees from its facilities, the GOI had a decision to make – either produce the transfer paperwork in the coming weeks or prepare for an increase in violence upon the release of these high-threat detainees ... the release of high-threat detainees immediately following the national elections will only serve to increase violence against Iraqi civilians and governmental entities.[18]

Such moments of leverage over local allies during withdrawal illustrate that setting clear and painful consequences for noncompliance can be effective. However, issuing threats of retaliation for noncompliance is more difficult when commitment to an intervention remains high and intervening forces risk undermining their own political agendas by threatening to undermine local proxies. Most actions that potentially weaken local allies will inevitably also harm the intervening ally's interest in protecting and stabilizing the local regime. Concessions leveraged from local allies using threats during withdrawal tended to focus on facilitating the extraction of intervening forces, rather than combating the insurgency, and did not necessarily serve the long-term counterinsurgency interests of either intervening or local forces. On the issue of transferring detainees in Iraq, for example, Michael Weiss and Hassan Hassan report that "plenty of incorrigible AQI jihadists were also let out of jail after the end of U.S. oversight of Iraq's wartime penal system ... among them Mohammad Ali Mourad, [AQI and ISIS founder Abu Musab] al-Zarqawi's former driver."[19] Coercing Baghdad to take over the processing of detainees exacerbated the insurgency that reinvented itself after American withdrawal. Therefore, although preparing to remove foreign forces provides opportunities for intervening forces to coerce local allies, requests made by intervening partners during withdrawal have tended to prioritize withdrawal over fighting the insurgency and thus may be politically useful, but of questionable strategic utility in achieving the objectives of the war.

Does Compliance by Local Allies Matter?

There are strategic consequences for the pattern of local allies responding to the body of policy requests of intervening allies with a roughly even percentage of compliance, partial compliance, or noncompliance. These mixed results over time lead to compromised and half-baked policy outcomes, which in turn undermine the strategic design of the counterinsurgency. David Galula asserted that military, police, judicial, and political operations are *all* required to win a counterinsurgency. For Galula, "the expected result – final defeat of the insurgents – is not an addition but a multiplication of these various operations; they all are essential and if one is nil, the product will be zero."[20] Following Galula's logic, excessively military-focused approaches, half-measures, or partial compliance are detrimental to long-term counterinsurgency success. Yet it was precisely this sort of middle-of-the-road approach, in which policies are just as likely to be ignored or partially instituted as they are to be fulfilled, that has emerged as a norm of counterinsurgency through proxies, at least across the counterinsurgency interventions examined here.

Furthermore, for several reasons it is important to note that noncompliance or partial compliance by local allies should not be presumed to signal incompetence or impending counterinsurgency failure. First, the rates of local ally noncompliance with requests from intervening forces are strikingly consistent across counterinsurgency interventions. Yet, there is nevertheless variation in the outcomes of COIN interventions, indicating that local non-compliance is not necessarily an indicator of success or failure (see Table 3.1). Many interventions, such as the US war in Vietnam, ended in defeat; some concluded as mixed successes, such as Syria in Lebanon or Vietnam in Cambodia; while a few interventions managed to succeed in achieving their primary objectives, including Cuba in Angola.

Local noncompliance is often linked to practices and conditions that are detrimental to a successful counterinsurgency, such as endemic corruption and protecting ineffective personnel. However, as illustrated in Chapter 7 on the Indian intervention in Sri Lanka, local noncompliance with requests from intervening partners does not unfailingly harm the counterinsurgency effort. In Sri Lanka, Colombo's defiance was an indication of the local regime's capacity and determination to resist pressure from intervening patrons, which proved to be a political asset. The Sri Lankans defied several Indian demands not because of weakness, free riding, or incompetence, but because Colombo sought independent control over the counterinsurgency campaign and employed a strategic vision that deviated from New Delhi's prescriptions. It took Colombo decades, but eventually Sri Lanka successfully won the counterinsurgency campaign, and did so without the aid of foreign troops.[21]

This dynamic speaks to a paradox within counterinsurgency interventions, where intervening forces are seeking local partners that are simultaneously legitimate and self-sufficient, yet compliant and deferential to foreign interests and the policy prescriptions of intervening partners. Yet, the stronger the local regime, the less likely it is to be submissive and compliant. As David Sedney observed in Afghanistan, the USA wanted Hamid Karzai to be "strong and subservient" at the same time, which is inherently contradictory.[22] A stronger local ally is more likely to combat local insurgents, while also being more likely to resist pressure from intervening allies and fiercely defend its legal sovereignty and autonomy.

Furthermore, the efforts of intervening forces to reinforce local autonomy frequently end up exacerbating the dependencies of local partners. US Department of Defense official Alain Enthoven commented on US strategy in Vietnam:

If we continue to add forces and to Americanize the war, we will only erode whatever incentives the South Vietnamese people may now have to help themselves in this fight. Similarly, it would be a further sign to the South Vietnamese leaders that we will carry any load, regardless of their actions. That will not help us build a strong nation.[23]

Efforts to augment the independence of local allies by providing significant financial and military sustenance can end up fostering ever-deepening dependencies on external support. Over the years, foreign funding becomes institutionalized, which undermines local control instead of encouraging the development of self-sustaining local institutions.

What is asked of local allies has a greater impact on the success or failure of the counterinsurgency war than the rate of local compliance with requests from intervening forces. While somewhat obvious, this finding nevertheless discredits a primary assumption underpinning all counterinsurgency advisory missions, namely that following the prescriptions of intervening forces should increase the odds of victory. However, the findings here indicate that the ability of an intervening force to promote its agenda and coerce local allies may not be as critical in winning or losing as imagined, because intervening forces are not necessarily consistently promoting strategically sound policies or building local institutions that are likely to succeed in the long run.

Intervening forces can neglect certain important local political contexts that would better inform their strategic decisions and policy recommendations. Interestingly, to complicate the narrative of an ineffectual, lazy proxy being a primary cause of COIN defeat, there were examples of thoughtful local non-compliance helping, rather than hurting the collective war effort. Consider, for example, the 2007 US request that Afghans conduct year-round poppy elimination campaigns.[24] Kabul did not comply. As the governor of Badakhshan Province discovered, destroying poppy crops as sprouts gave local farmers time to replant different crops, while destroying poppy crops late in the season stressed farmers, fostering animosity against Kabul. Eliminating the crop once a year in the spring diminished poppy cultivation, yet enabled the governor to maintain popular support. When asked about his successful policies, the governor commented, "the reason I'm successful is because I've gained people's trust ... It certainly isn't because of an effective police force or the central government's efforts."[25] Where there are abundant examples of endemic corruption, nepotism, and ineptitude motivating local noncompliance, there are also examples of principled positions prompting local actors to refuse policy requests from intervening patrons.

Thus, noncompliance by local allies can cut both ways for a counterinsurgency. Over the course of an intervention, equal parts compliance, partial compliance, and noncompliance by local allies certainly decreases the efficacy of the intended political strategy, but it would be misleading to presume that the strategy of intervening forces motivating the policies recommended to local proxies was sound in the first place.

The Decisive Role of Local Partners

The observation that the particular rate of local compliance with requests from intervening forces is not correlated with winning or losing counter-insurgency wars should not be taken as a claim that the behavior of local allies is inconsequential to war outcome. On the contrary, the behavior of local allies is decisive for winning. Effective local COIN policies, how-ever, are a different issue than compliance with demands from foreign proxies.

Across large-scale counterinsurgency interventions, local allies readily asserted their influence by relying on established methods routinely employed by resource-deficient actors, such as selective defiance, partial compliance, or shirking. Certain combinations of interests and depend-encies provided the opportunity for local allies to be obstinate and assert their influence by foot-dragging. It is a mistake to imagine that defiant local allies are seeking to lose wars to insurgents, but it is also a mistake to assume that local allies will readily allow foreign allies to dictate their policies; instead they will reliably leverage their legal sovereign authority over state policy in defiance of intervening forces.

Writing on the influence of impoverished political actors, James Scott notes that traditionally weak actors subject to hegemonic narratives typically assert themselves by "passive noncompliance, subtle sabotage, evasion, and deception."[26] Although Scott is concerned with relations between peasants and elites, in the context of counterinsurgency alli-ances, too, persistent local evasion can whittle away at the goals of the dominant political force, thus asserting the power of the underprivileged in shaping the political environment. This is evident in the findings on compliance and noncompliance in counterinsurgency partnerships. For two-thirds of requests, local partners forced some compromise, whether unintentionally (because of deficient capacity to act) or willfully (through refusal to act). The effect was to bolster their influence and leverage the political dependencies of intervening partners on local action for policy implementation.

Finding that two-thirds of requests from intervening forces to local allies resulted in some degree of noncompliance or half-measure has significant implications for counterinsurgency interventions. Writing on Iraq, for example, Dexter Filkins vividly described the political conse-quences of the sort of "passive noncompliance, subtle sabotage, evasion, and deception" that was described by James Scott as typical pathways through which disadvantaged groups assert influence.[27] Reporting on a collaborative US–Iraqi reconstruction project, Filkins remarked that the US commander in charge was satisfied with the local partnerships, declaring, "We've made friends here." The Iraqis working alongside the commander played along with this narrative until US forces were out of earshot. In confidence an Iraqi official confessed to Filkins, "I take their money but I hate them."[28] Filkins noted how this Janus-faced behavior was pervasive and had grave consequences for the future of Iraq:

There were always two conversations in Iraq, the one the Iraqis were having with the Americans and the one they were having among themselves. The one the Iraqis were having with us – that was positive and predictable and boring, and it made the Americans happy because it made them think that they were winning. And the Iraqis kept it up because it kept the money flowing, or because it bought them a little peace. The conversation they were having with each other was the one that really mattered, of course. That conversation was the chatter of a whole other world, a parallel reality, which sometimes unfolded right next to the Americans, even right in front of them. And we almost never saw it.[29]

Filkins' observations demonstrate how negotiations between locals are foremost in shaping the most consequential political dynamics in the war. Foreign intervening forces play a critical role shaping the political environment and providing incentives for local actors; yet are ancillary to the political crux of the war being battled and negotiated for local political control. Filkins' thoughts as well as the findings here offer several insights regarding definitions of "weak," "strong," and "asymmetric" in counterinsurgency partnerships. As David Baldwin cautioned, it is best to model power as a relational process that is contingent on specific contexts and issues, which may or may not be affected by wealth and material resources.[30]

Despite significant asymmetries in the material means of local and foreign COIN allies, it was evident across the wars examined that each ally possessed political leverage vis-à-vis the other. Local proxies have material dependencies on foreign patrons. Meanwhile, intervening patrons have political dependencies on local proxies, due in large part to the norm of local legal sovereignty that dictates local regimes are the legitimate arbiters of local state policy. Local proxies regularly leverage this political position to defy intervening patrons. This finding confirms that local allies are not the puppets of foreign allies; they are highly influential partners with reliable sources of political leverage. Nevertheless, the extent of political influence at the disposal of resource-deficient local allies often surprises and frustrates intervening militaries. According to the Pentagon Papers, for example, US policymakers in Vietnam were frustrated with how much influence local partners like Ngo Dinh Diem exerted, a dynamic that put the Americans "in the unfortunate role of suitor to a fickle lover. Aware of our fundamental commitment to him, Diem could with relative impunity ignore our wishes. It reversed the real power relationship between the two countries."[31]

Expressing a pervasive, yet misguided logic, the comment conflates resources and power while failing to recognize that if Saigon was able to leverage US political dependencies effectively, the South Vietnamese were not weak in this particular context. Similar to other COIN alliances, in Vietnam resource asymmetries were often secondary to political dependencies in determining the influence of each partner in bargaining encounters. A more accurate description of the power dynamics between asymmetric counterinsurgency allies would instead recognize that neither the local nor the foreign ally is the prisoner of its partner. Each ally is bound to its own interests and political dependencies, which they will inevitably use to try to coerce the other partner. This alliance dynamic will inevitably lead to disjoined, compromised, and partially implemented counterinsurgency policy – yet another reason why success in counterinsurgency interventions is so often maddeningly elusive and why the politics of these wars is vexingly complex.

Notes

1 David A. Baldwin, "Power and International Relations," *Handbook of International Relations*, ed. Walter Carlsnaes, Thomas Risse, and Beth A. Simmons, 2nd ed. (Thousand Oaks, CA: SAGE), 273–97.

2 Ibid.

3 Douglas S. Blaufarb, *The Counterinsurgency Era: U.S. Doctrine and Performance, 1950 to the Present*, 1st ed. (New York: Free Press, 1977), 302.

4 The resource inequities between intervening and local allies are not always significant, including, for example, Vietnam in Cambodia or Cuba in Angola. However, due to the nature of intervention, the intervening state is in a position to dedicate significant resources to its ailing ally, a dynamic that signifies capacities beyond what locals are able to muster.

5 There are exceptions. Consider the wars examined in Chapter 9, where small states intervene, including Cuba in Angola, Egypt in Yemen, Vietnam in Cambodia, and Syria in Lebanon.

6 Partlow, *A Kingdom of Their Own*, 382.

7 Watts et al., *Countering Others' Insurgencies*, 178.

8 Nagl, *Learning to Eat Soup with a Knife*, xiv. Also note that the US operational COIN manual acknowledges the preponderant political role of local, not US forces, noting how "U.S. forces and agencies can help, but HN [host nation] elements must accept responsibilities to achieve real victory." US Army and Marine Corps, *Counterinsurgency, FM 3-24*, Section 1–147, 1–26.

9 US Marine Corps, "Counterinsurgency Measures B4S5499XQ – Student Handout," The Basic School, Marine Corps Training Command, Camp

Barrett, VA, 2016, 19, www.trngcmd.marines.mil/Portals/207/Docs/TBS/ B4S5499XQ%20CounterInsurgency%20Measures.pdf?ver=2016-02-10-114636-310.

10 Thomas C. Schelling, *The Strategy of Conflict* (New York: Oxford University Press, 1963), 11.

11 Galula, *Counterinsurgency Warfare*, xii.

12 This data is drawn only from wars where released documents enabled quantitative analysis, namely, the USA in Vietnam, Iraq, and Afghanistan; India in Sri Lanka; and the USSR in Afghanistan.

13 This conclusion, which suggests very modest, politically minimal agendas for counterinsurgency interventions, fits with other scholars who have examined these types of conflicts. See, for example, MacDonald, "'Retribution Must Succeed Rebellion'," 256, who summarized his findings on counterinsurgency by noting "the findings suggest that political conditions may make it difficult for incumbents to defeat guerilla opponents, no matter how sophisticated their particular strategies." See also Blaufarb, *The Counterinsurgency Era*, 302–3.

14 US Department of State, US Embassy Baghdad, "The New Joint Campaign Plan for Iraq."

15 US Department of State, US Embassy Baghdad, "Iraqi Anti-corruption Update for January 7," 10BAGHDAD0044, January 7, 2010. See also US Department of State, US Embassy Baghdad, "Anti-Corruption Update," 09BAGHDAD2454, September 11, 2009.

16 Ladwig, *The Forgotten Front*.

17 See Chapter 6 on Vietnam.

18 US Department of State, US Embassy Baghdad, "U.S. Sends Tough, but Necessary, Message during Detainee Meeting," 10BAGHDAD477, Secret, February 23, 2010.

19 Michael Weiss and Hassan Hassan, *ISIS: Inside the Army of Terror* (New York: Regan Arts, 2015), 88–9, citing reports by Anthony Shadid of the *Washington Post*. Some, but not all, of these prisoners were released by US or Iraqi officials, as many were also set loose by other insurgents via jailbreak operations.

20 Galula, *Counterinsurgency Warfare*, 61.

21 Hashim, *When Counterinsurgency Wins*.

22 Ronald E. Neumann, "Failed Relations between Hamid Karzai and the United States: What Can We Learn?" (United States Institute of Peace, May 2015), 3.

23 Gibbons, *The U.S. Government and the Vietnam War*, 628.

24 US Department of State, US Embassy Kabul, "Scenesetter: U.S.-Afghan Strategic Partnership Talks in Kabul – March 13." Poppy elimination programs in Afghanistan were later put on hold by the Obama administration.

25 Soraya Sarhaddi Nelson, "Teams Focus on Poppy Eradication in Afghanistan," *National Public Radio*, August 31, 2007, www.npr.org/templates/story/story.php?storyId=14088743. Similarly, in Vietnam, US requests that Saigon "aid anti-communist labor movements" backfired by "severely discrediting" the labor movements by being funded by "corrupt" officials in the government.

Edmund F. Wehrle, *Between a River and a Mountain* (Ann Arbor: University of Michigan Press, 2005), 3.

26 Scott, *Weapons of the Weak*, 31.

27 Ibid., 31.

28 Dexter Filkins, *The Forever War* (New York: Knopf, 2008), 115.

29 Ibid.

30 Baldwin, "Power and International Relations," 273–97.

31 The Pentagon Papers, "[Part IV. B. 5.] Evolution of the War. Counterinsurgency: The Overthrow of Ngo Dinh Diem, May–Nov. 1963," iii.

Appendix A   Comparing Requests across Wars

Table A.1 *Comparing types of requests made by intervening allies across counterinsurgency wars*

Category of request made by intervening ally	War and subject of policy request	
Expand local allied government	USA in Vietnam:	Form elected village executive councils
	USSR in Afghanistan:	Expand local government structures
	India in Sri Lanka:	Increase provincial autonomy
	USA in Iraq:	Pass provincial powers legislation
	USA in Afghanistan:	Increase district governance in Spin Boldak
Reconciliation with militants	USA in Vietnam:	Offer amnesty/reconciliation
	USSR in Afghanistan:	Offer political settlement
	India in Sri Lanka:	Amnesty to militants who surrender arms
		Detainees to be released if all groups enter negotiations
	USA in Iraq:	Amnesty and national reconciliation plan
	USA in Afghanistan:	Jirgas on peace and integration policy
Include opposition in government	USA in Vietnam:	Appoint opposition leaders to new government
	USSR in Afghanistan:	Negotiate with tribal and religious leaders to "attract" them to "the side of the party"
	India in Sri Lanka:	Announce openings for Tamils in government; prominent Tamil leaders to participate in elections
	USA in Iraq:	Cease sectarian appointments Urge prominent Sunnis to engage in the draft constitution

Table A.1 (*cont.*)

Category of request made by intervening ally	War and subject of policy request	
	USA in Afghanistan:	Retain internationals from the United Nations Assistance Mission in Afghanistan (UNAMA) and International Security Assistance Force (ISAF) on the Candidate Vetting Board
New constitution	USA in Vietnam:	Advisory council to draft constitution
	USSR in Afghanistan:	Enact a constitution
	India in Sri Lanka:	[No similar demands]
	USA in Iraq:	Form constitutional review committee
	USA in Afghanistan:	Preparation of new constitution
Decentralization	USA in Vietnam:	Route funds directly to province, make Province Chief the key figure in pacification efforts
	USSR in Afghanistan:	Institute local government structures
	India in Sri Lanka:	Increase provincial autonomy
	USA in Iraq:	Pass provincial powers legislation
	USA in Afghanistan:	[No similar demands – already highly decentralized]
Limit clientelism	USA in Vietnam:	Promotion based on merit and Army of the Republic of Vietnam (ARVN) efficiency
	USSR in Afghanistan:	Stop purging the military
	India in Sri Lanka:	[No similar demands]
	USA in Iraq:	Cease sectarian appointments Cease politically motivated prosecutions
	USA in Afghanistan:	Merit-based appointments of key ministers
Direct talks with enemies (for or against)	USA in Vietnam:	Offer to talk to the National Liberation Front (NLF)
	USSR in Afghanistan:	Bilateral talks with Pakistan and Iran
	India in Sri Lanka:	No bilateral talks between Colombo and Liberation Tigers of Tamil Eelam (LTTE)
	USA in Iraq:	Increased negotiated cease-fires
	USA in Afghanistan:	[No demands on similar issues recorded in available documents, but it has been widely reported that the USA encouraged Kabul to aid in negotiations]

Table A.1 (*cont.*)

Category of request made by intervening ally	War and subject of policy request	
Redistribute land	USA in Vietnam:	Land reform
	USSR in Afghanistan:	Land reform
	India in Sri Lanka:	[No similar demands]
	USA in Iraq:	Agricultural reform / increased privatization
	USA in Afghanistan:	Fair system for settlement of land disputes
Engage local youth	USA in Vietnam:	Increase funds for youth programs
	USSR in Afghanistan:	Organize youth programs – especially for students
	India in Sri Lanka:	Rehabilitate militant youths
	USA in Iraq:	Fund militia reintegration programs
		Young militia members to receive training
	USA in Afghanistan:	Programs to employ youth and ex-soldiers
Offer vocational training	USA in Vietnam:	Vocational training for graduating students
	USSR in Afghanistan:	[No similar demands]
	India in Sri Lanka:	[No similar demands]
	USA in Iraq:	Support vocational training
	USA in Afghanistan:	Services and vocational training for former child laborers
Domestic economic expansion	USA in Vietnam:	Simplify import procedures
		Require advanced deposit by importers
	USSR in Afghanistan:	Economic expansion – domestic and foreign trade
	India in Sri Lanka:	[No similar demands]
	USA in Iraq:	Expanding microfinance and SME lending
		Limit private sector subsidies
	USA in Afghanistan:	Transit trade with Pakistan
Regulation of local economy	USA in Vietnam:	Limit spending to prevent inflation
		Increase certain taxes
	USSR in Afghanistan:	[No similar demands]
	India in Sri Lanka:	Establish finance commission
	USA in Iraq:	State banking reform program
		Increase certain taxes
	USA in Afghanistan:	Registration and oversight for Hawalas
		Anti-money-laundering law

Table A.1 (*cont.*)

Category of request made by intervening ally	War and subject of policy request	
Focus on rural populations	USA in Vietnam:	Expand rural electrification
		Credit to farmers
		Distribute new seed varieties
	USSR in Afghanistan:	Organize poverty committees
	India in Sri Lanka:	[No similar demands]
	USA in Iraq:	[No similar demands]
	USA in Afghanistan:	Contribute to electricity development projects
Religious leaders (protect, divide, or co-opt)	USA in Vietnam:	Revitalize interreligious council
	USSR in Afghanistan:	Fractionalize mullahs – divide religious leaders
	India in Sri Lanka:	[No similar demands]
	USA in Iraq:	Protections for minority political and religious parties
	USA in Afghanistan:	Engage religious figures in public messaging
Military engagement	USA in Vietnam:	South Vietnam take strikes against North Vietnam
	USSR in Afghanistan:	[No similar demands]
	India in Sri Lanka:	Sri Lankan forces to stay confined to barracks
	USA in Iraq:	Increase Iraqi security forces operating independently
	USA in Afghanistan:	Engage in operations in Marja
Armed paramilitary organizations	USA in Vietnam:	[No similar demands]
	USSR in Afghanistan:	[No similar demands]
	India in Sri Lanka:	Civilian Volunteer Force (CVF) to be inducted into the reserve and regular police forces
	USA in Iraq:	Support Concerned Local Citizens (CLC), incorporating a percentage into Iraqi Security Forces (ISF)
	USA in Afghanistan:	Fund, expand the Community Defense Initiative (CDI)

Appendix B   Information Regarding US Department of State Cables from the US Interventions in Iraq and Afghanistan

Due to classification, much of the data for the chapters on the US wars in Iraq and Afghanistan relies on the US Department of State *Cablegate* document set released by *Wikileaks*. While not exhaustive compilations of all American policy requests to allies in Iraq and Afghanistan, the requests documented in the *Cablegate* set contain 6,677 cables from US officials in Baghdad and 2,961 from Kabul, providing important documents detailing alliance politics. The US embassies in Baghdad and Kabul were the center of high-level diplomatic engagement with local partners. The documents utilized contain instructions from Washington to embassy staff on policies handling allies, as well as reports from embassy personnel back to Washington, detailing interalliance bargaining and responses from local allies to policy requests.[1]

In this study I do not account for requests made to local allies through the US Department of Defense or military branches, due to a lack of availability of documents from those agencies and a focus on diplomatic engagement. However, because high-ranking US military personnel, such as US General David Petraeus, were often present in meetings and discussions documented in US Department of State cables, I would not expect alliance-related materials from Department of Defense sources to contradict the findings presented.[2]

It is important to compare the characteristics of the documents in the *Cablegate* set against the policy requests identified here in order to clarify potential bias from using this document set for data. For example, if the number of US requests to allies simply followed the trajectory of the available documents in *Cablegate*, the findings regarding the number of requests per year might be a by-product of the volume in SIPRNET, instead of a reliable indicator of trends in demands and interalliance relations. However, some correlation should be expected since an increase in activity at an embassy would likely result in an increase in both demands to allies and cable traffic, yet some variation between the two metrics would indicate independent processes and bolster claims

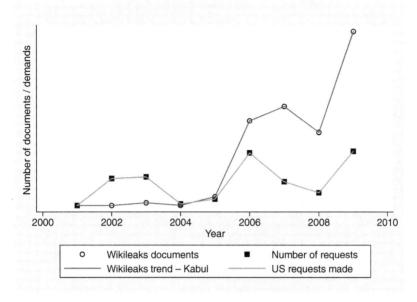

Figure B.1 Number of documents in the *Cablegate* set (released by *Wikileaks*) from the US Embassy Kabul and number of US requests to Kabul, 2001–10

about alliances offered here. Figure B.1 provides a graph of the number of documents in the *Cablegate* set from the US Embassy in Kabul per year, as well as the number of policy requests from Washington to Kabul that were identified over the same period.[3]

Illustrated by Figure B.1, there is correlation between the number of documents in the *Cablegate* database from the US Embassy in Kabul, and the number of US requests identified, in particular from 2004 and 2009. The number of documents contained in the *Cablegate* database and the size of the US mission in Afghanistan both expand after 2005, reflecting an increase in US diplomatic correspondence regarding Afghanistan. However, this does not necessarily indicate that characteristics of the *Cablegate* set are determining the observed patterns in requests and compliance. As violence in Afghanistan began to rise in Afghanistan after 2006, increasing attention was paid to US Operation Enduring Freedom, which included an increase in the number of requests to Kabul.

Graphing the trend in the number of documents available from the US Embassy in Baghdad in the *Cablegate* set with the number of requests from the USA to Baghdad identified is provided in Figure B.2.

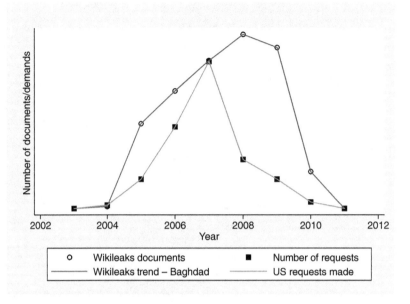

Figure B.2 Number of documents in the *Cablegate* set (released by *Wikileaks*) from the US Embassy Baghdad and number of US requests to Baghdad, 2003–11

According to Figure B.2, more documents are available in the *Cablegate* database from Iraq from 2004 to 2008. Yet the number of unique requests from the USA to Baghdad identified in those documents peaks in 2007, indicating that the number of requests identified does not rely on the number of cables in the *Cablegate* set. Furthermore, 2009 has significantly more documents available than requests issued.[4] Interestingly, trends in the number of requests made per year in Iraq and Afghanistan may correlate with the level of violence in these wars. Consider Figure B.3, estimating violence by improvised explosive device (IED) incidents.

Comparing Figures B.1, B.2, and B.3, it appears that increases and decreases in American demands may roughly correspond with levels of violence, in particular in Iraq.[5] As attacks against US forces in Iraq increased in 2007, so did American pressure on its Iraqi allies to reform. Note, however, that there isn't a corresponding spike in Iraqi compliance in this same period (see Chapter 4). There are several reasons why. As discussed in Chapters 4 and 5, violence in Iraq and Afghanistan created complications that made compliance with US requests increasingly difficult.

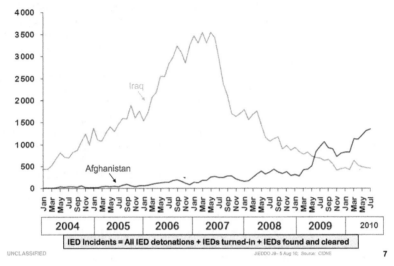

Figure B.3 IED attacks in Iraq and Afghanistan, 2004–10[6]

As a precaution against bias, most requests identified in this study through Department of State documents in the *Cablegate* set were correlated with declassified documents and public statements from US officials. In Iraq, for example, in May 2007 US President Bush publicly laid out eighteen "benchmarks" for Baghdad to achieve as a precondition for additional funding. These demands included policies to disarm militias, establish minority rights, hold provincial elections, perform a constitutional review, implement hydrocarbon legislation, and support the Baghdad Security Plan.[7] Each benchmark was the subject of debate in the cable traffic in the *Cablegate* database and was also captured as eighteen unique policy requests in this book. In Afghanistan, the 2006 "Afghanistan Compact" (also called "The London Compact") was similar. Benchmarks from the compact included policies to promote the employment of youth and demobilized soldiers, implementation of public administrative reform, increased judicial personnel in the provinces, providing assistance to refugees and the disabled, and programs to restructure state-owned banks.[8] Corroborating the information captured in the *Cablegate* database with other sources provides additional checks on the validity of the data.

Notes

1 Only a portion of US Department of State diplomatic cable traffic is contained in the *Cablegate* database, and therefore only a portion of existing diplomatic cable traffic is available for analysis for Chapter 4 and 5 that rely on these materials. Nevertheless, I do not expect this to be a source of bias in my findings, for several reasons. Only documents tagged for the Secret Internet Protocol Router Network (SIPRNET, SIPDIS) are in the *Cablegate* document set. The SIPRNET databank was established in the 1990s to facilitate interagency file sharing; see "SIPRNET: Where the Leaked Cables Came From," BBC, November 29, 2010, sec. US & Canada, www.bbc.co.uk/news/world-us-canada-11863618. Due to the counterinsurgency effort, Iraq and Afghanistan required substantial inter-agency coordination between the US Departments of State and Defense, and a significant portion of cable traffic regarding allies from the US Embassies in Baghdad and Kabul was routed through SIPRNET. The US Embassy in Baghdad, for example, which opened in June 2004, pro-duced the third most cables in *Cablegate*, namely 6,677 messages. This is second only to Department of State Headquarters (8,017 messages), which sends messages across the world, and the US Embassy in Ankara (7,918 messages), which sent 1,850 cables through SIPRNET before the US diplomatic post in Baghdad was established in June 28, 2004. From the time the US Embassy in Baghdad was functioning in 2004, it sent the most messages in SIPRNET of any US embassy; see Simon Rogers, "WikiLeaks Embassy Cables: Download the Key Data and See How It Breaks Down," *The Guardian*, November 28, 2010, sec. World News, www.guardian.co .uk/news/datablog/2010/nov/29/wikileaks-cables-data. Of the *Cablegate* set, 2,961 documents originated from the US Embassy in Kabul, less than half of the 6,677 messages from the Embassy in Baghdad, but Kabul was a fraction of the size of the US mission in Iraq.

2 See, for example, US Department of State, US Embassy Baghdad, "PM Maliki and President Talabani Discuss Elections and Terrorist Attacks with General Petraeus," 09BAGHDAD2998, November 15, 2009.

3 The graph ends in 2009 because the *Cablegate* database ends February 28, 2010, providing two months of data for 2010. Providing two months of figures for the year 2010 could provide a false impression of a decrease in requests, which may or may not be accurate. Therefore 2010 has been omitted.

4 The years 2004 and 2010 are unreliable, as the US Embassy in Baghdad did not exist for most of the year in 2004 to submit cables and there are only two months of data available for 2010.

5 IEDs were a preferred method of attack for insurgents in Iraq. The tactic took some time to catch on in Afghanistan, and was never as prevalent.

6 US Department of Defense, "IED Incidents Comparison – Iraq and Afghani-stan," Presentation Slide, Combined Information Data Network Exchange, Joint Improvised Explosive Device Defeat Organization – J9, August 5, 2010,

Unclassified. See also The Brookings Institute, "Iraq Index – Tracking Reconstruction and Security in Post Saddam Iraq," Saban Center for Middle East Policy, Washington, DC, January 31, 2011.
7 Beehner and Bruno, "What Are Iraq's Benchmarks?."
8 The London Conference on Afghanistan, "The Afghanistan Compact."

Appendix C Statistical Tables

Table C.1 *Local compliance with US demands – Iraq*

	Ordered probit – Baghdad and Erbil's compliance with US demands, 2004–11					
	(1)	(2)	(3)	(4)	(5)	(6)
US unilateral action possible	−0.437	0.092	0.088	−0.189	−0.130	0.030
	(0.253)	(0. 393)	(0.394)	(0.399)	(0.420)	(0.401)
GOI can be excluded	0.282					
	(0.286)					
Interaction term:						
GOI benefit and US unilateral action[a]				−0.383		
				(0.497)		
Interaction term:		−1.039*	−1.039*		−1.040*	−1.093*
US and GOI interest convergence-divergence and US unilateral action[b]		(0.514)	(0.514)		(0.517)	(0.524)
Local ally – potential benefit	**0.659****	**0.526***	**0.528***	**0.819***	**0.565***	
	(0.245)	(0.257)	(0.257)	(0.313)	(0.264)	
Local ally – potential threat	−0. 453	−0.318	−0.316	−0.390	−0.314	
	(0.268)	(0.281)	(0.282)	(0.263)	(0.284)	
Policy request – issue type						
Political reform		−1.039	−0.100	−0.150		
		(0.256)	(0.256)	(0.248)		
Development					0.324	
					(0.380)	
COIN strategy					0.512	
					(0.612)	
Economic reform					−0.278	
					(0.331)	
Military strategy and military reform					−0.466	
					(0.391)	
Capacity	**−0.911****	**−0.964****	**−0.938****	**−1.006****	**−0.941****	**−1.048****
	(0.240)	(0.252)	(0.285)	(0.249)	(0.366)	(0.248)
War or internal politics				−0.524		
				(0.272)		
Year						
2005						5.143
						(299.5)
2006						5.931
						(299.5)
2007						5.602
						(299.5)
2008						5.206
						(299.5)
2009						5.859
						(299.5)
2010						6.457
						(299.5)
N (Observations)	105	105	105	105	105	105

Notes: * $p < 0.05$, ** $p < 0.01$.
[a] Local interest is estimated via "Potential Benefit for Local Ally," a dichotomous variable measuring if local allies can utilize what is being requested by intervening allies in ways to benefit the short-term interests of local forces such as profit-making positions or political benefits. A more specific variable, "Interest Convergence/Divergence," is included in models 2, 3, 5, and 6, and in the section on the interaction term between interests and unilateral ability included in each case study chapter.
[b] Interest in this model is estimated by a composite variable, "Allied Interest Convergence/Divergence," which relies on the available documentary evidence to account for both private costs/threats and benefits of the request for the local allied regime.

Table C.2 *Local compliance with US demands – Afghanistan*

	Ordered probit – Kabul's compliance with US demands, 2002–10					
	(1)	(2)	(3)	(4)	(5)	(6)
US unilateral action possible	−0.040	1.185**	1.169**	0.782	1.119**	1.474**
	(0.226)	(0.351)	(0.358)	(0.415)	(0.356)	(0.373)
GIRoA can be excluded	0.469*					
	(0.219)					
Interaction term:						
GIRoA benefit and US unilateral action				−0.750		
				(0.469)		
Interaction term:						
US and GIRoA interest convergence-divergence and US unilateral action		−1.786**	−1.780**		−1.867**	−2.117**
		(0.448)	(0.449)		(0.455)	(0.470)
Local ally – potential benefit	0.423*	−0.141	0.161	0.701**	0.160	
	(0.219)	(0.237)	(0.251)	(0.274)	(0.239)	
Local ally – potential threat	−1.077**	−0.352	−0.342	−0.960**	−0.315	
	(0.222)	(0.245)	(0.248)	(0.213)	(0.246)	
Policy request – issue type						
Political reform		−0.036	0.031	−0.141		
		(0.217)	(0.218)	(0.201)		
Development					0.261	
					(0.290)	
COIN strategy					N/A	
Economic reform					−0.311	
					(0.298)	
Military strategy and military reform					−0.109	
					(0.489)	
Capacity	−0.525*	−1.028**	−1.021**	−0.646**	−1.005**	−1.152**
	(0.225)	(0.244)	(0.246)	(0.227)	(0.246)	(0.266)
War or internal politics			−0.057			
			(0.234)			
Year						
2002						−0.623
						(0.471)
2003						−0.316
						(0.448)
2004						−5.004
						(269.4)
2005						−0.642
						(0.777)
2006						−0.599
						(0.406)
2007						−1.533**
						(0.474)
2008						−0.855
						(0.551)
2009						−0.655
						(0.374)
N (Observations)	148	148	148	148	148	148

Notes: $^{*} p < 0.05$, $^{**} p < 0.01$.

Table C.3 *Local compliance with US demands – Vietnam*

	Ordered probit –Saigon's compliance with US demands, 1964–73						
	(1)	(2)	(3)	(4)	(5)	(6)	(7)
US unilateral action possible	−0.200 (0.307)	0.607 (0.358)	0.498 (0.364)	0.646 (0.405)	**0.719★** (0.370)	0.194 (0.460)	**0.775★** (0.368)
GVN can be excluded	0.519 (0.367)						
Interaction term: GVN benefit and US unilateral action					−1.024★ (0.491)		
Interaction term: US and GVN interest convergence-divergence and US unilateral action		−1.251★ (0.511)	−1.161★ (0.514)		−1.093★ (0.536)	−0.715 (0.576)	**−1.388★★** (0.518)
Local ally – potential benefit	**0.621★** (0.257)	0.449 (0.274)	0.444 (0.276)	**1.076★★** (0.370)	0.482 (0.284)		0.194 (0.291)
Local ally – potential threat	−0.423 (0.311)	−0.182 (0.339)	−0.188 (0.338)	−0.167 (0.294)	−0.375 (0.359)		−0.351 (0.304)
Policy request – issue type							
Political reform		−0.117 (0.304)	−0.001 (0.312)	−0.177 (0.275)			−0.168 (0.311)
Development					−0.244 (0.458)		
COIN strategy					−0.695 (0.535)		
Economic reform					0.450 (0.450)		
Military reform					0.185 (0.439)		
Military strategy					0.489 (0.673)		
Capacity issues	**−1.333★★** (0.292)	**−1.201★★** (0.294)	**−1.222★★** (0.296)	**−1.193★★** (0.290)	**−1.235★★** (0.301)	**−1.195★★** (0.304)	**−1.417★★** (0.310)
War or internal politics			0.322 (0.188)				
External threat							**1.195★★** (0.420)
Year							
1964						−5.417 (538.8)	
1965						−5.848 (538.8)	
1966						−4.888 (538.8)	
1967						−5.519 (538.8)	
1968						−5.912 (538.8)	
1969						−1.84e-09 (659.9)	
1970						−0.0824 (761.2)	
1971						−5.249 (538.8)	
1972						−5.706 (538.8)	
N (Observations)	105	105	105	105	105	105	105

Notes: ★ $p < 0.05$, ★★ $p < 0.01$.

Table C.4 *Local compliance with India's demands – Sri Lanka*

	Ordered probit – Sri Lanka's compliance with Indian demands, 1986–90					
	(1)	(2)	(3)	(4)	(5)	(6)
Indian unilateral action possible	−0.029 (0.292)	0.854* (0.390)	0.715 (0.401)	1.079* (0.465)	0.960* (0.399)	0.966* (0.408)
Sri Lanka can be excluded	0.175 (0.292)					
Interaction term: Sri Lankan benefit and Indian unilateral action				−1.481** (0.583)		
Interaction term: Sri Lankan and Indian interest convergence-divergence and US unilateral action		−1.338* (0.585)	−0.979 (0.615)		−1.507* (0.603)	−1.665** (0.598)
Local ally – potential benefit	1.026** (0.280)	0.254 (0.362)	0.324 (0.370)	1.568** (0.379)	0.131 (0.377)	
Local ally – potential threat	−0.796* (0.410)	−0.480 (0.473)	−0.166 (0.497)	−0.771 (0.441)	−0.200 (0.505)	
Political request – issue type						
Political reform		0.537 (0.303)	−0.316 (0.323)	0.446 (0.294)		
Development					0.068 (0.660)	
COIN strategy					−0.348 (0.419)	
Economic reform					4.046 (346.8)	
Military strategy and Military reform					−0.764 (0.384)	
Capacity	−0.552 (0.350)	−0.776 (0.368)	−0.523 (0.388)	−0.581 (0.356)	−0.878* (0.384)	−0.759* (0.382)
War or internal politics			−0.743* (0.354)			
Year						
1986						−3.690 (213.8)
1987						1.171* (0.466)
1988						0.907 (0.465)
N (Observations)	79	79	79	79	79	79

Notes: * $p < 0.05$, ** $p < 0.01$.

Bibliography

Abi-Habib, Maria. "Karzai Picks Ally to Fight Graft." *Wall Street Journal*, December 30, 2010. http://online.wsj.com/article/SB10001424052748 7045430045760516920923 14726.

"U.S. Blames Senior Afghan in Deaths." *Wall Street Journal*, April 1, 2012, sec. World News. www.wsj.com/articles/SB1000142405270230340470457731 1 522824172282.

"Afghanistan's Opium Poppies: No Quick Fixes." *The Economist*, June 19, 2008. www.economist.com/node/11591396.

Agence France-Presse (AFP). "U.S. Buys 'Concerned Citizens' in Iraq, but at What Price?" October 16, 2007. Accessed May 10, 2012. http://afp.google .com/article/ALeqM5iMzKGlyT_ahqRjtyXrAUrKIQLncA.

"Agreement between the United States of America and the Republic of Iraq on the Withdrawal of United States Forces from Iraq and the Organization of Their Activities during Their Temporary Presence in Iraq." US Department of State, November 27, 2008. www.state.gov/documents/organization/ 122074.pdf.

Ahmad, Sardar. "Karzai Appoints New Elections Chief for Afghanistan." *Agence France-Presse*, April 17, 2010.

"Aide Memoire Containing Points Conveyed by the Indian High Commissioner J.N. Dixit to the Sri Lankan President Recalling Certain Actions on the Part of the Sri Lankan Government," June 14, 1988. In *India–Sri Lanka: Relations and Sri Lanka's Ethnic Conflict Documents – 1947–2000*, ed. Avtar Singh Bhasin. Vol. IV, Document 832. New Delhi: Indian Research Press, 2001, 2257–9.

Al-Ali, Zaid. "How Maliki Ruined Iraq." *Foreign Policy* (blog), June 19, 2014. http://foreignpolicy.com/2014/06/19/how-maliki-ruined-iraq/.

Alden, Chris, and Mark Simpson. "Mozambique: A Delicate Peace." *The Journal of Modern African Studies* 31, no. 1 (March 1, 1993): 109–30.

Allawi, Ali A. *The Occupation of Iraq: Winning the War, Losing the Peace*. 1st ed. New Haven, CT: Yale University Press, 2007.

American Forces Press Service. "7th Iraqi Army Division Now Controlled by Iraqi Government," November 3, 2007. Accessed April 13, 2012. www .usf-iraq.com/news/press-releases/7th-iraqi-army-division-now-con trolled-by-iraqi-government.

"Iraqi Ground Forces Command Assumes Command and Control of 8th Iraqi Army Division," September 1, 2006. Accessed April 13, 2012. www.usf-iraq

.com/news/press-releases/iraqi-ground-forces-command-assumes-com
mand-and-control-of-8th-iraqi-army-division.

Andrew, Christopher, and Vasili Mitrokhin. *The Sword and the Shield: The
Mitrokhin Archive and the Secret History of the KGB*. New York: Basic Books,
2000.

Arango, Tim. "Transfer of Prison in Iraq Marks Another Milestone." *The New
York Times*, July 14, 2010, sec. World / Middle East. www.nytimes.com/
2010/07/15/world/middleeast/15iraq.html.

Arango, Tim, and Michael S. Schmidt. "Iraqis Say No to Immunity for
Remaining American Troops." *The New York Times*, October 4, 2011, sec.
World / Middle East. www.nytimes.com/2011/10/05/world/middleeast/
iraqis-say-no-to-immunity-for-remaining-american-troops.html.

Arnold, Anthony. *Afghanistan's Two-Party Communism: Parcham and Khalq*.
Stanford, CA: Hoover Press, 1983.

  The Fateful Pebble: Afghanistan's Role in the Fall of the Soviet Empire. Novato,
CA: Presidio, 1993.

Arnold, James R. *Americans at War: Eyewitness Accounts from the American Revolution to the 21st Century*. Santa Barbara, CA: ABC-CLIO, 2018.

Arreguin-Toft, Ivan M. *How the Weak Win Wars: A Theory of Asymmetric Conflict*.
Cambridge University Press, 2006.

Asselin, Pierre. *A Bitter Peace: Washington, Hanoi, and the Making of the Paris
Agreement*. Chapel Hill: University of North Carolina Press, 2002.

Associated Press. "10,000 Cubans Reported Killed in Angola War." *Los Angeles
Times*, June 16, 1987. http://articles.latimes.com/1987-06-16/news/mn-
7734_1_del-pino.

  "Afghanistan: U.N. Counts Cost of Bribes." *The New York Times*, January 20,
2010, sec. International / Asia Pacific. www.nytimes.com/2010/01/20/world/
asia/20briefs-Afghanistan.html.

  "New U.S. Embassy in Iraq Cloaked in Mystery." msnbc.com, April 14, 2006.
Accessed October 15, 2012. www.msnbc.msn.com/id/12319798/ns/world_
news-mideast_n_africa/t/new-us-embassy-iraq-cloaked-mystery/.

  "Roadside Bomb Kills 3 U.S. Soldiers in Iraq." *The New York Times*, June 29,
2004, sec. International. www.nytimes.com/2004/06/29/international/29WIRE-
BOMB.html.

  "U.S. Is Planning to Cut Its Staff at Iraq Embassy by as Much as Half."
The New York Times, February 7, 2012, sec. World / Middle East. www
.nytimes.com/2012/02/08/world/middleeast/united-states-planning-to-slash-
iraq-embassy-staff-by-half.html.

Auerswald, David P., and Stephen M. Saideman. *NATO in Afghanistan: Fighting
Together, Fighting Alone*. Princeton University Press, 2014.

Baldwin, David A. *Economic Statecraft*. Princeton University Press, 1985.

  "Power and International Relations." In *Handbook of International Relations*,
ed. Walter Carlsnaes, Thomas Risse, and Beth A. Simmons. 2nd ed. Thousand Oaks, CA: SAGE, 2012, 273–97.

Ball, George to President Lyndon B. Johnson, July 1, 1965. "A Compromise
Solution in South Vietnam." *The Pentagon Papers*, ed. Mike Gravel, Vol. 4.
Boston: Beacon Press, 1971, 615–19.

Bandarage, Asoka. *The Separatist Conflict in Sri Lanka: Terrorism, Ethnicity, Political Economy*. New York: Routledge, 2009.

Barakat, Sultan. "Reconstructing Post-Saddam Iraq: An Introduction." *Third World Quarterly* 26, no. 4 (2005): 565–70.

Barno, David. *2014 and Beyond: U.S. Policy towards Afghanistan and Pakistan*. Testimony before the House of Foreign Affairs Committee Subcommittee on the Middle East and South Asia. Center for a New American Security, November 3, 2011. www.cnas.org/files/documents/publications/CNAS%20Testimony%20Barno%20110311.pdf.

BBC. "Wikileaks Cables Say Afghan President Karzai 'Paranoid'," December 3, 2010. www.bbc.com/news/world-south-asia-11906216.

Beckett, Ian F. W. *Roots of Counterinsurgency: Armies and Guerrilla Warfare, 1900–1945*. London: Blandford Press, 1988.

Beehner, Lionel, and Greg Bruno. "What Are Iraq's Benchmarks?" *Council on Foreign Relations*. Updated March 11, 2008. Accessed April 20, 2012. www.cfr.org/iraq/iraqs-benchmarks/p13333.

Behr, Timo. "Germany and Regional Command – North." In *Statebuilding in Afghanistan: Multinational Contributions to Reconstruction*, ed. Nik Hynek and Péter Marton. London; New York: Routledge, 2012, 42–64.

Belasco, Amy. "Troop Levels in the Afghan and Iraq Wars, FY 2001–FY 2012: Cost and Other Potential Issues." Congressional Research Service, July 2, 2009.

Bender, Gerald J. "Angola, the Cubans, and American Anxieties." *Foreign Policy* no. 31 (Summer 1978): 3–30.

Bennett, Andrew. *Condemned to Repetition?: The Rise, Fall, and Reprise of Soviet-Russian Military Interventionism, 1973–1996*. Cambridge, MA: MIT Press, 1999.

Bennett, Andrew, Joseph Lepgold, and Danny Unger. "Burden-Sharing in the Persian Gulf War." *International Organization* 48, no. 1 (1994): 39–75.

Berman, Eli, and David A. Lake, eds. *Proxy Wars: Suppressing Transnational Violence through Local Agents*. Ithaca, NY: Cornell University Press, 2019.

Berman, Eli, David A. Lake, Gerard Padró i Miquel, and Pierre Yared, "Introduction: Principals, Agents, and Indirect Foreign Policies." In *Proxy Wars: Suppressing Transnational Violence through Local Agents*, ed. Eli Berman and David A. Lake. Ithaca, NY: Cornell University Press, 2019, 1–27.

Berman, Larry. *No Peace, No Honor: Nixon, Kissinger, and Betrayal in Vietnam*. New York: Touchstone, 2002.

Bhasin, Avtar Singh, ed. *India–Sri Lanka: Relations and Sri Lanka's Ethnic Conflict Documents – 1947–2000*. New Delhi: Indian Research Press, 2001, Volumes III, IV, V.

Biddle, Stephen. "Building Security Forces & Stabilizing Nations: The Problem of Agency." *Daedalus* 146, no. 4 (September 21, 2017): 126–38.

"Review of the New U.S. Army/Marine Corps Counterinsurgency Field Manual." *Perspectives on Politics* 6, no. 2 (2008): 347–50.

Biddle, Stephen, Jeffrey A. Friedman, and Jacob N. Shapiro. "Testing the Surge: Why Did Violence Decline in Iraq in 2007?" *International Security* 37, no. 1 (2012): 7–40.

Blaufarb, Douglas S. *The Counterinsurgency Era: U.S. Doctrine and Performance, 1950 to the Present.* 1st ed. New York: Free Press, 1977.

Boghani, Priyanka. "David Petraeus: ISIS's Rise in Iraq Isn't a Surprise." *PBS Frontline,* July 29, 2014. www.pbs.org/wgbh/frontline/article/david-petraeus-isiss-rise-in-iraq-isnt-a-surprise.

Bolduc, Donald C. *Bureaucracies at War: Organizing for Strategic Success in Afghanistan.* Carlisle Barracks, PA: US Army War College, 2009.

Boone, Jon. "Hamid Karzai Takes Control of Afghanistan Election Watchdog." *The Guardian,* February 22, 2010. www.guardian.co.uk/world/2010/feb/22/karzai-afghanistan-electoral-complaints-commission.

"Pakistan Border Closure Will Have Little Effect on NATO's Afghanistan Campaign." *The Guardian,* November 27, 2011, sec. World News. www.guardian.co.uk/world/2011/nov/27/pakistan-border-nato-afghanistan-supplies.

"U.S.-Afghan Relations Sink Further as Hamid Karzai Accused of Drug Abuse." *The Guardian,* April 7, 2010. www.guardian.co.uk/world/2010/apr/07/hamid-karzai-galbraith-substance-abuse.

Boot, Max. *The Savage Wars of Peace: Small Wars and the Rise of American Power.* New York: Basic Books, 2014.

Borer, Douglas A. *Superpowers Defeated: Vietnam and Afghanistan Compared.* London: Frank Cass, 1999.

Bowley, Graham. "Afghan Warlord Ismail Khan's Call to Arms Rattles Kabul." *The New York Times,* November 12, 2012, sec. World / Asia Pacific. www.nytimes.com/2012/11/13/world/asia/ismail-khan-powerful-afghan-stokes-concern-in-kabul.html.

Bradsher, Henry S. *Afghan Communism and Soviet Intervention.* Oxford University Press, 1999.

Bremer, L. Paul. "The Lost Year in Iraq." Interview. *PBS Frontline.* October 17, 2006. www.pbs.org/wgbh/pages/frontline/yeariniraq/interviews/bremer.html.

Bremer, L. Paul, and Malcolm McConnell. *My Year in Iraq: The Struggle to Build a Future of Hope.* 1st ed., 1st printing. New York: Simon & Schuster, 2006.

Brigham, Robert K. "Dreaming Different Dreams: The United States and the Army of the Republic of Vietnam." In *A Companion to the Vietnam War,* ed. Marilyn B. Young and Robert Buzzanco. Malden, MA: Wiley Blackwell, 2002, 146–61.

Brinkley, Joel. *Cambodia's Curse: The Modern History of a Troubled Land.* New York: PublicAffairs, 2012.

The Brookings Institution. "Iraq Index – Tracking Reconstruction and Security in Post Saddam Iraq," Saban Center for Middle East Policy, Washington, DC. January 31, 2011.

Brown, David. "The Development of Vietnam's Petroleum Resources." *Asian Survey* 16, no. 6 (June 1976): 553–70.

Brown, Robin. "Spinning the War: Political Communications, Information Operations and Public Diplomacy in the War on Terrorism." In *War and the Media: Reporting Conflict 24/7,* ed. Daya Kishan Thussu and Des Freedman. London: SAGE, 2003, 87–100.

Bunker, Ellsworth. *The Bunker Papers: Reports to the President from Vietnam, 1967–1973*, ed. Douglas Pike. Vol. 1. San Francisco: Asian Foundation, 1990.

Burrowes, Robert D. *The Yemen Arab Republic: The Politics of Development, 1962–1986*. Westview Special Studies on the Middle East. Boulder, CO: Westview Press, 1987.

Byman, Daniel. "Friends Like These: Counterinsurgency and the War on Terrorism." *International Security* 31, no. 2 (October 1, 2006): 79–115.

   Going to War with the Allies You Have: Allies, Counterinsurgency, and the War on Terrorism. Carlisle Barracks, PA: US Army War College, Strategic Studies Institute, 2005.

Byman, Daniel, and Matthew Waxman. *The Dynamics of Coercion: American Foreign Policy and the Limits of Military Might*. New York: Cambridge University Press, 2002.

Callwell, Charles E. *Small Wars: Their Principles and Practice*. 3rd new ed. of revised ed. Lincoln: University of Nebraska Press, 1996.

Camp, Dick. *Operation Phantom Fury: The Assault and Capture of Fallujah, Iraq*. Minneapolis, MN: Voyageur Press, 2009.

Can Iraq Pay for Its Own Reconstruction?: Joint Hearing before the Subcommittee on International Organizations, Human Rights and Oversights and the Subcommittee on the Middle East and South Asia of the Committee on Foreign Affairs, United States House of Representatives, 110th Congress, 1st Session, March 27, 2007.

Carlsnaes, Walter, Thomas Risse, and Beth A. Simmons, eds. *Handbook of International Relations*. Thousand Oaks, CA: SAGE, 2002.

Carney, Christopher P. "International Patron–Client Relationships: A Conceptual Framework." *Studies in Comparative International Development* 24, no. 2 (Summer 1989): 42–55.

Carter, James M. *Inventing Vietnam: The United States and State Building, 1954–1968*. 1st ed. New York: Cambridge University Press, 2008.

Castro, Fidel, Raul Castro, and Nelson Mandela. *Cuba and Angola: Fighting for Africa's Freedom and Our Own*, ed. Mary-Alice Waters. 1st ed. New York: Pathfinder Press, 2013.

Cavanaugh, Dan C. "Some Observations on the Current Plague Outbreak in the Republic of Vietnam." *American Journal of Public Health* (April 1968): 742–3.

Cavendish, Julius. "In Afghanistan War, Government Corruption Bigger Threat Than Taliban." *Christian Science Monitor*, April 12, 2010. www.csmonitor .com/World/2010/0412/In-Afghanistan-war-government-corruption-bigger-threat-than-Taliban.

Caverley, Jonathan D. "The Myth of Military Myopia: Democracy, Small Wars, and Vietnam." *International Security* 34, no. 3 (January 1, 2010): 119–57.

Central Intelligence Agency. "Afghanistan: The Revolution after Four Years." July 1982. NSEA 82-10341.

   "Memorandum for Director of Central Intelligence, From: Acting NIO for Latin America, Subject: Cuban Involvement in Angola, NI-1589-77," June 23, 1977. Declassified January 3, 2006. CREST Database, National Archives and Records Administration.

"National Intelligence Bulletin, Lebanon: Sarkis' Unanimous Election Is Set-back for Jumblatt," May 10, 1976. CREST Database, National Archives and Records Administration.

"Soviet Military Support to Angola: Intentions and Prospects," October 1985. Special National Intelligence Estimate, Director of Central Intelligence, SNIE 71/11-85. www.foia.cia.gov/sites/default/files/document_conversions/89801/DOC_0000261290.pdf.

Central Intelligence Agency, National Foreign Assessment Center. "Angola's Relations with the Soviet Union," November 15, 1978. RP M 78-10429. Declassified May 25, 2006. CREST Database, National Archives and Records Administration.

Central Intelligence Agency, National Intelligence Officer for Africa. "Memorandum for the Director of Central Intelligence, Assessment of Cuban and African Reactions to a U.S. Diplomatic Initiative Re Angola and Mozambique," July 6, 1977. Approved for Release 2004/03/11. CREST Database, National Archives and Records Administration.

Chadwick, Robert B. "Lebanon: The Uncertain Road to Reconstruction." Thesis, Naval Postgraduate School, Monterey, CA. Available from National Technical Information Service, Springfield, VA, 1997. http://archive.org/details/lebanonuncertain00chad.

Chanda, Nayan. *Brother Enemy: The War after the War.* New York: Free Press, 1988.

Chaudhuri, Rudra, and Theo Farrell. "Campaign Disconnect: Operational Progress and Strategic Obstacles in Afghanistan, 2009–2011." *International Affairs* 87, no. 2 (March 1, 2011): 271–96.

Chayes, Sarah. *Thieves of State: Why Corruption Threatens Global Security.* 1st ed. New York: W. W. Norton & Company, 2016.

Chiarelli, Peter W., and Patrick R. Michaelis. "Winning the Peace: The Requirement for Full-Spectrum Operations." *Military Review* 85, no. 4 (August 2005): 4–17.

Chin, Warren. "Examining the Application of British Counterinsurgency Doctrine by the American Army in Iraq." *Small Wars & Insurgencies* 18, no. 1 (March 1, 2007): 1–26.

Chulov, Martin. "Kurds and Shias Face off over Kirkuk in Vacuum Left by Iraqi Army." *The Guardian*, January 22, 2016. www.theguardian.com/world/2016/jan/22/kurds-and-shias-face-off-over-kirkuk-in-vacuum-left-by-iraqi-army.

Clarke, Jeffrey J. *Advice and Support: The Final Years, 1965–1973.* Washington, DC: US Government Printing Office, 1988.

Clausewitz, Carl von. *On War.* Translated by Michael Howard and Peter Paret. Reprint. Princeton University Press, 1989.

Clifford, Clark. "Annals of Government: Serving the President, the Vietnam Years." *The New Yorker*, May 13, 1991.

CNN News, "Pakistan Reopens NATO Supply Routes to Afghanistan." July 3, 2012. www.cnn.com/2012/07/03/world/asia/us-pakistan-border-routes/index.html.

The Coalition Provisional Authority. Archived Web Page, Regulations, Orders, Memoranda and Public Notices, Order Number 1. www.iraqcoalition.org/regulations.

Cockburn, Patrick. *The Rise of Islamic State: ISIS and the New Sunni Revolution*. London: Verso Books, 2015.

Cohen, Michael A. "The Myth of a Kinder, Gentler War." *World Policy Journal* 27, no. 1 (May 3, 2010): 75–86.

Cohen, Warren I., and Nancy Bernkopf Tucker, eds. *Lyndon Johnson Confronts the World: American Foreign Policy 1963–1968*. Cambridge University Press, 1994.

Cold War International History Project. "Documents on the Soviet Invasion of Afghanistan." Dossier No. 4. The Woodrow Wilson International Center for Scholars, Washington, DC. November 2001. www.wilsoncenter.org/sites/default/files/e-dossier_4.pdf.

Coleman, Fred. *The Decline and Fall of the Soviet Empire: Forty Years That Shook the World, from Stalin to Yeltsin*. New York: Macmillan, 1997.

Coll, Steve. *Directorate S: The C.I.A. and America's Secret Wars in Afghanistan and Pakistan*. New York: Penguin, 2019.

  Ghost Wars: The Secret History of the CIA, Afghanistan, and Bin Laden, from the Soviet Invasion to September 10, 2001. New York: Penguin, 2005.

Collelo, Thomas. *Angola: A Country Study*. United States Marine Corps, SSIC 03000 Operations and Readiness. Washington, DC: Federal Research Division, Library of Congress, 1989. www.marines.mil/News/Publications/ELECTRONICLIBRARY/ElectronicLibraryDisplay/tabid/13082/Article/125400/country-study-angola-click-here-fpts.aspx.

Collier, Craig A. "Now That We're Leaving Iraq, What Did We Learn?" *Military Review* 88 (October 2010): 91–2.

Combined Security Transition Command – Afghanistan. "Afghan National Police (ANP) Vetting and Recruiting Presentation," 2005. http://edocs.nps.edu/AR/topic/misc/09Dec_Haskell_appendix_II.pdf.

"Communiqué of the Sri Lankan Presidential Secretariat Underlining the Importance of Holding Elections to the North-East Provincial Council," Colombo, September 19, 1988. In *India–Sri Lanka: Relations and Sri Lanka's Ethnic Conflict Documents – 1947–2000*, ed. Avtar Singh Bhasin. Vol. IV, Document 857. New Delhi: Indian Research Press, 2001, 2312–13.

Conboy, Kenneth, and Paul Hannon. *Elite Forces of India and Pakistan*. Oxford: Osprey Publishing, 1992.

Congressional Record. "Afghanistan Stabilization and Reconstruction: A Status Report." Hearing before the Committee on Foreign Relations, United States Senate, 108th Congress, 2nd Session, January 27, 2004. Washington, DC: US Government Printing Office.

  "An Investigation of the U.S. Economic and Military Assistance Programs in Vietnam." 42nd Report by the Committee on Government Operations. House Report 2257, 89th Congress, 2nd Session, October 12, 1966. Washington, DC: US Government Printing Office.

  "Proceedings and Debates of the 109th Congress." 2nd Session, Vol. 152, No. 99. July 25, 2006, H5841. www.congress.gov/crec/2006/07/25/modified/CREC-2006-07-25-pt1-PgH5837.htm.

Cooper, Chester L. "Memorandum for Mr. Bundy, Subject: Status Report on Various Actions in Vietnam [The 41-Point Program]." LBJ Library, June 11, 1965. NLJ 84-130.

Cooper, Chester L., Judith E. Corson, Laurence J. Legere, David E. Lockwood, and Donald M. Weller. "The American Experience with Pacification in Vietnam – Volume I – An Overview of Pacification (Unclassified)." Institute for Defense Analyses (IDA), March 1972. IDA Log No. HQ 72-14046 CY # 94.

Cooper, Helene. "In Leaning on Karzai, U.S. Has Limited Leverage." *The New York Times*, November 11, 2009. sec. International / Asia Pacific. www .nytimes.com/2009/11/12/world/asia/12karzai.html.

Cordesman, Anthony H. *How America Corrupted Afghanistan: Time to Look in the Mirror*. Washington, DC: Center for Strategic and International Studies, September 9, 2010.

Cordesman, Anthony, and Adam Mausner. "How Soon Is Safe? Iraqi Force Development and Conditions-Based U.S. Withdrawals." Report. Center for Strategic and International Studies, December 16, 2009. www.ecoi.net/ en/file/local/1305282/1002_1236977391_csis-iraq.pdf.

Cordesman, Anthony H., and Emma R. Davies. *Iraq's Insurgency and the Road to Civil Conflict*. Vol. 2. Center for Strategic and International Studies. Westport, CT: Praeger Security International, 2008.

"Corruption Index 2011 from Transparency International: Find Out How Countries Compare." *The Guardian*, December 1, 2011, sec. News. www .guardian.co.uk/news/datablog/2011/dec/01/corruption-index-2011-transparency-international.

Costello, Charles 'Chuck'. Interviewed by Haven North. "Oral Histories: The Iraq Experience Project," United States Institute of Peace, October 14, 2004. Accessed April 4, 2012. www.usip.org/sites/default/files/file/ resources/collections/histories/iraq/costello.pdf.

"CPSU CC Politburo Decision," January 28, 1980, with Report by Gromyko-Andropov-Ustinov-Ponomarev, January 27, 1980, Top Secret, No. P181/ 34, to Comrades Brezhnev, Andropov, Gromyko, Suslov, Ustinov, Ponomarev, Rusakov. Cold War International History Project (CWIHP), The Woodrow Wilson International Center for Scholars, "documents on the Soviet Invasion of Afghanistan," CPSU Politburo session of January 28, 1980. E-Dossier No. 4. November 2001, 58–60.

"CPSU CC Politburo Decisions on Afghanistan (Excerpts) Calling on Puzanov to Meet with Taraki and Try to Mend the Rift between Taraki and Amin," September 13, 1979. Cold War International History Project (CWIHP), The Woodrow Wilson International Center for Scholars, Cold War in the Middle East, Soviet Invasion of Afghanistan, Document identifier: 111561. https://digitalarchive.wilsoncenter.org/document/111561.

Crane, Conrad. "Military Strategy in Afghanistan and Iraq: Learning and Adapting under Fire at Home and in the Field." In *Understanding the U.S. Wars in Iraq and Afghanistan*, ed. Beth Bailey and Richard H. Immerman. New York University Press, 2015, 124–46.

Crawford, Gordon. *Foreign Aid and Political Reform: A Comparative Analysis of Democracy Assistance and Political Conditionality*. Dordrecht: Springer, 2000.

Crawford, Neta C. "United States Budgetary Costs of the Post–9/11 Wars through FY2019: $5.9 Trillion Spent and Obligated." The Watson Institute

for International and Public Affairs, November 14, 2018. https://watson
.brown.edu/costsofwar/files/cow/imce/papers/2018/Crawford_Costs%20of%
20War%20Estimates%20Through%20FY2019.pdf.

"U.S. Costs of Wars through 2014: $4.4 Trillion and Counting." The Watson
Institute for International and Public Affairs, June 25, 2014. http://watson
.brown.edu/costsofwar/files/cow/imce/figures/2014/Costs%20of%20War%
20Summary%20Crawford%20June%202014.pdf.

Creighton, James L. "How Bureaucracy Impedes Victory in Afghanistan," World
Policy Blog. World Policy Institute, April 16, 2012. www.worldpolicy.org/
blog/2012/04/16/how-bureaucracy-impedes-victory-afghanistan.

Crilly, Rob. "Pakistan Permanently Closes Borders to NATO after Air Strike."
The Telegraph, November 28, 2011, sec. World News. www.telegraph.co.uk/
news/worldnews/asia/pakistan/8919960/Pakistan-permanently-closes-borders-
to-Nato-after-air-strike.html.

Cronin, Patrick. Civilian Surge: Key to Complex Operations. Washington, DC:
National Defense University Press, 2009.

Dacy, Douglas C. Foreign Aid, War, and Economic Development: South Vietnam,
1955–1975. Cambridge University Press, 1986.

Study S-337, The Fiscal System of Wartime Vietnam. Arlington, VA: Institute for
Defense Analyses Program Analysis Division, February 1969. www.dtic.mil/
docs/citations/AD0692611.

Dahlerup, Drude, and Anja Taarup Nordlund. "Gender Quotas: A Key to
Equality? A Case Study of Iraq and Afghanistan." European Political Science
3, no. 3 (2004): 91–8.

Dawisha, A. I. "Intervention in the Yemen: An Analysis of Egyptian Perceptions
and Policies." Middle East Journal 29, no. 1 (Winter 1975): 47–63.

"Decisions of the Security Co-ordination Group Regarding Security of the
Tamils, Colombo, October 8, 19; November 1, 3, 1989 – Minutes of the
Special Committee Appointed by the Security Co-ordination Group to Dis-
cuss the Details and the Numbers Required for the Citizens Volunteer Force,
Provincial Police and the Armed Services." In India–Sri Lanka: Relations and
Sri Lanka's Ethnic Conflict Documents – 1947–2000, ed. Avtar Singh Bhasin.
Vol. IV, Document 934. New Delhi: Indian Research Press, 2001, 2408–18.

Deni, John R. Alliance Management and Maintenance: Restructuring NATO for the
21st Century. London: Routledge, 2016.

DeRouen, Karl R., and U. K. Heo. Civil Wars of the World: Major Conflicts Since
World War II. Santa Barbara, CA: ABC-CLIO, 2007.

DeVotta, Neil. "Control Democracy, Institutional Decay, and the Quest for
Eelam: Explaining Ethnic Conflict in Sri Lanka." Pacific Affairs 73, no. 1
(2000): 55–76.

DeYoung, Karen. "Overworked U.S. Embassy in Kabul Straining to Meet
Administration's Demands." The Washington Post, March 11, 2010,
sec. Politics. www.washingtonpost.com/wp-dyn/content/article/2010/03/10/
AR2010031003975.html.

Dimitrakis, Panagiotis. Secret War in Afghanistan: The Soviet Union, China and
Anglo-American Intelligence in the Afghan War. 1st ed. London: I. B. Tauris,
2013.

Dobbins, James. *After the Taliban: Nation-Building in Afghanistan.* Washington, DC: Potomac Books, 2008.

Dobbins, James, Seth G. Jones, Benjamin Runkle, and Siddharth Mohandas. *Occupying Iraq: A History of the Coalition Provisional Authority.* Santa Monica, CA: RAND Corporation, 2009. www.rand.org/pubs/monographs/MG847 .html.

Donnelly, Jack. *Realism and International Relations.* New York: Cambridge University Press, 2000.

Dorman, Andrew M. "The United Kingdom: Innocence Lost in the War in Afghanistan?" In *Coalition Challenges in Afghanistan: The Politics of Alliance,* ed. Gale A. Mattox and Stephen M. Grenier. Stanford University Press, 2015, 108–22.

Dovkants, Keith. "Rebel Chief Begs: Don't Bomb Now; Taliban Will Be Gone in a Month." *Evening Standard,* October 5, 2001.

Efrat, Moshe, and Jacob Bercovitch, eds. *Superpowers and Client States in the Middle East: The Imbalance of Influence.* New York: Routledge, 1991.

Egnell, Robert. "Lessons from Helmand, Afghanistan: What Now for British Counterinsurgency?" *International Affairs* 87, no. 2 (March 1, 2011): 297–315.

Ehrenberg, John, J. Patrice McSherry, Jose Ramon Sanchez, and Caroleen Marji Sayej, eds. *The Iraq Papers.* 1st ed. New York: Oxford University Press, 2010.

Elias, Barbara. "The Likelihood of Local Allies Free-Riding: Testing Economic Theories of Alliances in US Counterinsurgency Interventions." *Cooperation and Conflict* 52, no. 3 (September 1, 2017): 309–31.

Elsey, George M. "Notes of Meeting." US Department of State, Office of the Historian, November 5, 1968. Document 195. *Foreign Relations of the United States, 1964–1968, Volume VII, Vietnam, September 1968–January 1969.* Johnson Presidential Library, George M. Elsey Papers, Van De Mark Transcripts [1 of 2]. https://history.state.gov/historicaldocuments/frus1964-68v07/d195.

Enterline, Andrew J., Emily Stull, and Joseph Magagnoli. "Reversal of Fortune? Strategy Change and Counterinsurgency Success by Foreign Powers in the Twentieth Century." *International Studies Perspectives* 14, no. 2 (May 1, 2013): 176–98.

"Exchange of Letters between the Prime Minister of India and the President of Sri Lanka, Prime Minister of India," New Delhi, July 29 1987. In *India–Sri Lanka: Relations and Sri Lanka's Ethnic Conflict Documents – 1947–2000,* ed. Avtar Singh Bhasin. Vol. IV, Document 723. New Delhi: Indian Research Press, 2001, 1946–51.

"Extract from the Statement of the Sri Lankan Minister of Foreign Affairs and Minister of State for Defence Ranjan Wijeratne in Parliament during the Debate on Foreign Affairs," March 31, 1989. In *India–Sri Lanka: Relations and Sri Lanka's Ethnic Conflict Documents – 1947–2000,* ed. Avtar Singh Bhasin. Vol. IV, Document 886. New Delhi: Indian Research Press, 2001, 2356–8.

Exum, Andrew. *Leverage: Designing a Political Campaign for Afghanistan.* Washington, DC: Center for a New American Security, 2010.

"Fallujah, Again." *The Economist*, May 28, 2016. www.economist.com/news/
    middle-east-and-africa/21699461-why-retaking-jihadist-stronghold-has-
    become-priority-fallujah-again.

Fearon, James D. "Domestic Political Audiences and the Escalation of
    International Disputes." *American Political Science Review* 88, no. 3 (Septem-
    ber 1994): 577–92.

Fearon, James D., and David D. Laitin. "Ethnicity, Insurgency, and Civil War."
    American Political Science Review 97, no. 1 (2003): 75–90.

Feifer, Gregory. *The Great Gamble: The Soviet War in Afghanistan*. 1st ed. New
    York: HarperCollins, 2009.

Felbab-Brown, Vanda. *Aspiration and Ambivalence: Strategies and Realities of
    Counterinsurgency and State-Building in Afghanistan*. Washington, DC:
    Brookings Institution Press, 2012.

"Counterinsurgency, Counternarcotics, and Illicit Economies in Afghanistan:
    Lessons for State-Building." In *Convergence: Illicit Networks and National
    Security in the Age of Globalization*, ed. Michael Miklaucic and Jacqueline
    Brewer. Washington, DC: National Defense University Press, 2013, 189–209.

Ferris, Jesse. "Egypt, the Cold War, and the Civil War in Yemen, 1962–1966."
    Ph.D. Dissertation, Princeton University, 2008.

Ferris-Rotman, Amie. "NATO Races to Secure Violent, Porous Afghanistan-
    Pakistan Border." *Reuters*, September 2, 2011. www.reuters.com/article/
    2011/09/02/us-afghanistan-pakistan-border-idUSTRE7814QY20110902.

Fields, Arnold. "Special Inspector General for Afghanistan Reconstruction
    (SIGAR): Quarterly Report to the United States Congress." SIGAR, 2010.

Filkins, Dexter. *The Forever War*. New York: Knopf, 2008.

Finnemore, Martha, and Judith Goldstein. *Back to Basics: State Power in a
    Contemporary World*. Oxford University Press, 2013.

Fisk, Robert. "Robert Fisk: 'Nobody Supports the Taliban, but People Hate the
    Government'." *The Independent*, November 26, 2008. www.independent.co
    .uk/voices/commentators/fisk/robert-fisk-nobody-supports-the-taliban-but-
    people-hate-the-government-1036905.html.

Fitzgerald, David. *Learning to Forget: US Army Counterinsurgency Doctrine and
    Practice from Vietnam to Iraq*. Stanford University Press, 2013.

FitzGerald, Frances. *Fire in the Lake: The Vietnamese and the Americans in Viet-
    nam*. Reprint edition. Boston: Back Bay Books, 2002.

Fox News. "White House Slams Karzai for Latest Anti-American Outburst."
    April 5, 2010. www.foxnews.com/politics/2010/04/05/white-house-slams-
    karzai-latest-anti-american-outburst/.

Galbraith, John Kenneth. *The Selected Letters of John Kenneth Galbraith*. Cam-
    bridge University Press, 2017.

Galula, David. *Counterinsurgency Warfare: Theory and Practice*. Westport, CT:
    Praeger, 2006.

Garcia-Navarro, Lourdes. "Bitterness Grows amid U.S.-Backed Sons of Iraq."
    National Public Radio, June 24, 2010. www.npr.org/templates/story/story
    .php?storyId=128084675.

Gardner, Lloyd C., and Ted Gittinger. *The Search for Peace in Vietnam,
    1964–1968*. College Station: Texas A&M University Press, 2004.

Garner, Lt. Gen. (Ret) Jay. "The Lost Year in Iraq." Interview. *PBS Frontline*, October 17, 2006. www.pbs.org/wgbh/pages/frontline/yeariniraq/interviews/garner.html.

Garrels, Anne. "Long-Awaited Fallujah Rebuilding Shows Promise." *National Public Radio*, January 23, 2008. www.npr.org/templates/story/story.php?storyId=18319948.

Garthoff, Raymond L. "On Estimating and Imputing Intentions." *International Security* 2, no. 3 (1978): 22–32.

Gentile, Gian. *Wrong Turn: America's Deadly Embrace of Counterinsurgency.* New York: The New Press, 2013.

George, Alexander L., and Andrew Bennett. *Case Studies and Theory Development in the Social Sciences.* Cambridge, MA: MIT Press, 2005.

George, Edward. *The Cuban Intervention in Angola, 1965–1991: From Che Guevara to Cuito Cuanavale.* Reprint. London; New York: Routledge, 2012.

Ghosh, P. A. *Ethnic Conflict in Sri Lanka and Role of Indian Peace Keeping Force (IPKF).* New Delhi: APH Publishing, 2000.

Gibbons, William Conrad. *The U.S. Government and the Vietnam War: Executive and Legislative Roles and Relationships, Part IV.* Princeton University Press, 1995.

Gibler, Douglas M. "The Costs of Reneging: Reputation and Alliance Formation." *The Journal of Conflict Resolution* 52, no. 3 (June 1, 2008): 426–54.

Gibler, Douglas M., and Scott Wolford. "Alliances, Then Democracy: An Examination of the Relationship between Regime Type and Alliance Formation." *The Journal of Conflict Resolution* 50, no. 1 (February 1, 2006): 129–53.

Girardet, Edward. *Afghanistan: The Soviet War.* London: Palgrave Macmillan, 1986.

Gladstone, Cary. *Afghanistan Revisited.* New York: Nova Publishers, 2001.

Glanz, James, and Eric Schmitt. "Iraq Attacks Lower, but Steady, New Figures Show." *The New York Times*, March 12, 2008, sec. International / Middle East. www.nytimes.com/2008/03/12/world/middleeast/12iraq.html.

Glaze, John. "Opium and Afghanistan: Reassessing U.S. Counternarcotics Strategy." US Army War College, Strategic Studies Institute, October 2007. www.strategicstudiesinstitute.army.mil/pdffiles/pub804.pdf.

Gleijeses, Piero. *Conflicting Missions: Havana, Washington, and Africa, 1959–1976.* Chapel Hill: University of North Carolina Press, 2002.

"H-Diplo Review Essay 135 on Cubans in Angola." H-Diplo, February 10, 2016. https://networks.h-net.org/system/files/contributed-files/re135.pdf.

Visions of Freedom: Havana, Washington, Pretoria and the Struggle for Southern Africa, 1976–1991. Chapel Hill: University of North Carolina Press, 2013.

Goldstein, Avery. *Deterrence and Security in the 21st Century: China, Britain, France, and the Enduring Legacy of the Nuclear Revolution.* 1st ed. Stanford University Press, 2007.

Gompert, David C., John Gordon, Adam Grissom et al. *War by Other Means – Building Complete and Balanced Capabilities for Counterinsurgency: RAND Counterinsurgency Study – Final Report.* Santa Monica, CA: RAND Corporation, 2008. www.rand.org/pubs/monographs/MG595z2.

Goodman, Allan E., and Lawrence M. Franks. "The Dynamics of Migration to Saigon, 1964–1972." *Pacific Affairs* 48, no. 2 (1975): 199–214.

Goodson, Larry P. "Afghanistan's Long Road to Reconstruction." *Journal of Democracy* 14, no. 1 (2003): 82–99.

Gorka, Sebastian, and David Kilcullen. "An Actor-centric Theory of War: Understanding the Difference between COIN and Counterinsurgency." *Joint Force Quarterly*, no. 60 (January 1, 2011): 14–19.

Gottesman, Evan R. *Cambodia after the Khmer Rouge: Inside the Politics of Nation Building*. New Haven, CT: Yale University Press, 2004.

Graff, Jonathan K. "United States Counterinsurgency Doctrine and Implementation in Iraq." ARMY Command and General Staff College – Fort Leavenworth, June 18, 2004. www.dtic.mil/docs/citations/ADA428901.

Grau, Lester W. *The Bear Went over the Mountain: Soviet Combat Tactics in Afghanistan*. New York: Psychology Press, 1996.

Grauer, Ryan, and Dominic Tierney. "The Arsenal of Insurrection: Explaining Rising Support for Rebels." *Security Studies* 27, no. 2 (April 3, 2018): 263–95.

Gravel, Mike. *The Pentagon Papers: The Defense Department History of United States Decisionmaking on Vietnam*. The Senator Gravel Edition. Boston: Beacon Press, 1971.

"Gromyko-Andropov-Ustinov-Ponomarev Report to CPSU CC on the Situation in Afghanistan," June 28, 1979, Cold War International History Project, Collection: Cold War in the Middle East, Soviet Invasion of Afghanistan, Document identifier: 5034D714–96B6–175C-9A356BA28915CAE5. www.wilsoncenter.org.

Gunaratna, Rohan. *Indian Intervention in Sri Lanka: The Role of India's Intelligence Agencies*. Colombo: South Asian Network on Conflict Research, 1993.

Gunewardene, Roshani M. "Indo-Sri Lanka Accord: Intervention by Invitation or Forced Intervention." *North Carolina Journal of International Law and Commercial Regulation* 16, no. 2 (1991): 211–34.

Gutcher, Lianne. "Afghanistan's Anti-corruption Efforts Thwarted at Every Turn." *The Guardian*, July 19, 2011. www.guardian.co.uk/world/2011/jul/19/afghanistan-anti-corruption-efforts-thwarted.

Hadley, Stephen. "Stephen Hadley: How Bush Started – and Ended – the Iraq War." Interview by Sarah Childress. *PBS Frontline*, July 29, 2014. www.pbs.org/wgbh/frontline/article/stephen-hadley-how-bush-started-and-ended-the-iraq-war/.

Halchin, L. Elaine. "The Coalition Provisional Authority (CPA): Origin, Characteristics, and Institutional Authorities." Text. UNT Digital Library, September 21, 2006. http://digital.library.unt.edu/ark:/67531/metacrs10420/.

Hallams, Ellen, and Benjamin Schreer. "Towards a 'Post-American' Alliance? NATO Burden-Sharing after Libya." *International Affairs* 88, no. 2 (March 1, 2012): 313–27.

Harris, William. "Syria in Lebanon." *MERIP Reports* no. 134 (July 1, 1985): 9–14.

Hasbullah, S. H., and Barrie M. Morrison. *Sri Lankan Society in an Era of Globalization: Struggling to Create a New Social Order*. New Delhi: SAGE, 2004.

Hashim, Ahmed. *When Counterinsurgency Wins: Sri Lanka's Defeat of the Tamil Tigers*. Philadelphia: University of Pennsylvania Press, 2013.

Hasrat, Zia-U-Rahman. "IS Runs Timber Smuggling in Business in Afghanistan, Officials Say." *Voice of America*, February 8, 2016. www.voanews.com/con tent/islamic-state-timber-smuggling-afghanistan/3182282.html.

Hatzky, Christine. *Cubans in Angola: South-South Cooperation and Transfer of Knowledge, 1976–1991*. Madison: University of Wisconsin Press, 2015.

Hazelton, Jacqueline L. "The 'Hearts and Minds' Fallacy: Violence, Coercion, and Success in Counterinsurgency Warfare." *International Security* 42, no. 1 (July 1, 2017): 80–113.

Heikkila, Pia. "Afghan Census Cancelled Due to Security Fears." *The Guardian*, June 11, 2008. www.guardian.co.uk/world/2008/jun/11/afghanistan .internationalaidanddevelopment.

Hellmann-Rajanayagam, Dagmar. *The Tamil Tigers: Armed Struggle for Identity*. Stuttgart: F. Steiner, 1994.

Hennessy, Michael A. *Strategy in Vietnam: The Marines and Revolutionary Warfare in I Corps, 1965–1972*. Westport, CT: Praeger, 1997.

Henriksen, Thomas. *WHAM: Winning Hearts and Minds in Afghanistan and Elsewhere*. Florida: Joint Special Operations University, MacDill Air Force Base, 2012.

Herd, Graeme P., and John Kriendler. *Understanding NATO in the 21st Century: Alliance Strategies, Security and Global Governance*. New York: Routledge, 2013.

Herring, George C. *America's Longest War: The United States and Vietnam 1950–1975*. 3rd ed. New York: McGraw-Hill Companies, 1995.

Hinnebusch, Raymond. "Pax-Syriana? The Origins, Causes and Consequences of Syria's Role in Lebanon." *Mediterranean Politics* 3, no. 1 (1998): 137–60.

Hodge, Nathan. "Afghans Probe Corruption at Borders." *Wall Street Journal*, December 10, 2012, sec. World News. http://online.wsj.com/article/ SB10001424127887324024004578171410335390372.html.

Holsti, Ole R., P. Terrence Hopmann, and John D. Sullivan. *Unity and Disintegration in International Alliances*. New York: John Wiley & Sons, 1973.

Holzgrefe, J. L., and Robert O. Keohane. *Humanitarian Intervention: Ethical, Legal and Political Dilemmas*. Cambridge University Press, 2003.

Hopmann, P. Terry. "International Conflict and Cohesion in the Communist System." *International Studies Quarterly* 11, no. 3 (1967): 212–36.

Hubbard, Andrew. "Plague and Paradox: Militias in Iraq." *Small Wars & Insurgencies* 18, no. 3 (September 2007): 345–62.

Hunt, Richard A. *Pacification: The American Struggle for Vietnam's Hearts and Minds*. Boulder, CO: Westview Press, 1998.

Immerman, Richard H. "'A Time in the Tide of Men's Affairs,' Lyndon Johnson and Vietnam." In *Lyndon Johnson Confronts the World: American Foreign Policy 1963–1968*, ed. Warren I. Cohen and Nancy Bernkopf Tucker. Cambridge University Press, 1994, 57–97.

"Information from CC CPSU to GDR leader E. Honecker," September 16, 1979. History and Public Policy Program Digital Archive, International History Declassified, The Woodrow Wilson International Center for Scholars. http://digitalarchive.wilsoncenter.org/document/111566.

International Development Association (IDA), The World Bank. "Improving Customs Performance for a Stronger Government," September 13, 2009. http://go.worldbank.org/LO2PTC0H21.

International Security Assistance Force Afghanistan. "Afghan Media Roundtable with General McChrystal." Transcript, April 28, 2010. Accessed April 19, 2013. www.isaf.nato.int/article/transcripts/transcript-afghan-media-roundta ble-with-gen.-mcchrystal.html.

"Iraq Study Estimates War-Related Deaths at 461,000." *BBC News*, October 16, 2013. www.bbc.com/news/world-middle-east-24547256.

"Iraq – U.S.A., Status of Forces Agreement (Nov 2008)." Accessed May 10, 2012. www.scribd.com/doc/70569755/Iraq-USA-Status-of-Forces-Agreement-Nov-2008.

Isaacs, Arnold R. *Vietnam Shadows: The War, Its Ghosts, and Its Legacy*. Baltimore, MD: Johns Hopkins University Press, 1997.

Isby, David. *Russia's War in Afghanistan*. London: Osprey Publishing, 1986.

James, W. Martin. *A Political History of the Civil War in Angola: 1974–1990*. New Brunswick, NJ: Transaction Publishers, 2011.

Jenne, Erik K. "Sri Lanka: A Fragmented State." In *State Failure and State Weakness in a Time of Terror*, ed. Robert I. Rotberg. Washington, DC: Brookings Institution Press, 2003, 219–44.

Jenzen-Jones, Nic. "Chasing the Dragon: Afghanistan's National Interdiction Unit." *Small Wars Journal*, September 5, 2011. http://smallwarsjournal .com/blog/chasing-the-dragon-afghanistan's-national-interdiction-unit.

Joes, Anthony James. *Victorious Insurgencies: Four Rebellions That Shaped Our World*. Lexington: The University Press of Kentucky, 2010.

Johnson, Lyndon B. "Our Objective in Vietnam." In *Public Papers of the Presidents of the United States: Lyndon B. Johnson: Containing the Public Messages, Speeches, and Statements of the President, 1963/64–1968/69*. Book 1. Washington, DC: US Government Printing Office, 1965.

Johnson, Thomas H., and Barry Scott Zellen. *Culture, Conflict, and Counterinsurgency*. Stanford Security Studies an imprint of Stanford University Press, 2014.

Johnston, Lauren. " 'Dirty Bomb' Depot Dispute." *CBS News*, July 9, 2004. www.cbsnews.com/news/dirty-bomb-depot-dispute/.

Jones, Seth G. *In the Graveyard of Empires: America's War in Afghanistan*. New York: W. W. Norton & Company, 2009.

Kahn, Owen Ellison. "Cuba's Impact in Southern Africa." *Journal of Interamerican Studies and World Affairs* 29, no. 3 (Autumn 1987): 33–54.

Kakar, Mohammed. *Afghanistan: The Soviet Invasion and the Afghan Response, 1979–1982*. Berkeley: University of California Press, 1997.

Karnow, Stanley. *Vietnam, a History*. Rev ed. New York: Viking, 1991.

Kattenburg, Paul M. *The Vietnam Trauma in American Foreign Policy: 1945–75*. New Brunswick, NJ: Transaction Publishers, 1980.

Katzman, Kenneth. "Afghanistan: Politics, Elections, and Government Performance." Congressional Research Service, RS21922, February 19, 2010.

"Afghanistan: Politics, Elections, and Government Performance." Congressional Research Service, RS21922, January 20, 2012.

"Afghanistan: Politics, Elections, and Government Performance." Congressional Research Service, RS21922, January 12, 2015.

"Afghanistan: Post-Taliban Governance, Security, and U.S. Policy." Congressional Research Service, RL30588, January 4, 2013.

"Afghanistan: Post-Taliban Governance, Security, and U.S. Policy." Congressional Research Service, RL30588, February 17, 2016.

"Iraq: Politics, Governance, and Human Rights." Congressional Research Service, RS21968, November 10, 2011.

"Iraq: Reconciliation and Benchmarks." Congressional Research Service, RS21968, May 12, 2008.

"Iraq: Reconciliation and Benchmarks." Congressional Research Service, RS21968, June 5, 2008.

"Iraq: Reconciliation and Benchmarks." Congressional Research Service, RS21968, August 4, 2008.

"Iraq: Reconciliation and Benchmarks." Congressional Research Service, RS21968, September 3, 2008.

Keane, Conor. *US Nation-Building in Afghanistan.* London: Routledge, 2016.

Keesing's. *South Vietnam: A Political History, 1954–1970.* New York: Scribner, 1970.

Kelegama, Saman. "Economic Costs of Conflict in Sri Lanka." In *Creating Peace in Sri Lanka: Civil War and Reconciliation,* ed. Robert I. Rotberg. Washington, DC: Brookings Institution Press, 1999, 71–87.

Kelley, Patrick, and Scott Sweetser. "The Spaces in between: Operating on the Afghan Border (or Not)." *Small Wars Journal,* April 1, 2010. smallwarsjournal.com/blog/journal/docs-temp/404-kelley.pdf.

Kelly, Thomas P. "The Northern Distribution Network and the Baltic Nexus." Remarks at the Commonwealth Club, Washington, DC. January 20, 2012. US Department of State. https://2009-2017.state.gov/t/pm/rls/rm/182317.htm.

Keohane, Robert O. "The Big Influence of Small Allies." *Foreign Policy* no. 2 (Spring 1971): 161–82.

Khan, Tahir. "Karzai's Anti-US Comments Win Little Support." *The Express Tribune,* March 16, 2013. http://tribune.com.pk/story/521471/karzais-anti-us-comments-win-little-support/.

Kilcullen, David. *Counterinsurgency.* Oxford: Oxford University Press, 2010.

"Counter-insurgency Redux." *Survival: Global Politics and Strategy* 48, no. 4 (2006): 111–30.

"Counterinsurgency Seminar 07." Small Wars Center of Excellence, September 26, 2007. http://smallwarsjournal.com/blog/coin-seminar-dr-david-kilcullen.

Kimball, Anessa L. "Alliance Formation and Conflict Initiation: The Missing Link." *Journal of Peace Research* 43, no. 4 (July 1, 2006): 371–89.

King, Gary, Robert O. Keohane, and Sidney Verba. *Designing Social Inquiry.* Princeton University Press, 1994.

Kirk-Greene, A. H. M. "The Thin White Line: The Size of the British Colonial Service in Africa." *African Affairs* 79, no. 314 (January 1, 1980): 25–44.

Kissinger, Henry. "Message from the President's Assistant for National Security Affairs (Kissinger) to President Nixon." US Department of State, Office of the Historian, January 11, 1973. Document 266. *Foreign Relations of the United States, 1969–1976, Volume IX, Vietnam, October 1972–January 1973.* https://history.state.gov/historicaldocuments/frus1969-76v09/d266.

Knights, Michael. "Kirkuk May Be Key to National Reconciliation in Iraq." *Al Jazeera*, September 23, 2015. www.aljazeera.com/indepth/opinion/2015/09/kirkuk-key-national-reconciliation-iraq-150923084554413.html.

Komer, Robert. "Bureaucracy Does Its Thing: Institutional Constraints on U.S.–GVN Performance in Vietnam." Santa Monica, CA: RAND Corporation, August 1972. www.rand.org/pubs/reports/R967/.

Koskinas, Ioannis. "President Karzai Is the One to Blame." *Foreign Policy*, June 20, 2014. https://foreignpolicy.com/2014/06/20/president-karzai-is-the-one-to-blame/.

Kramer, Andrew E. "U.S. Advised Iraqi Ministry on Oil Deals." *The New York Times*, June 30, 2008, sec. International / Middle East. www.nytimes.com/2008/06/30/world/middleeast/30contract.html.

Krasner, Stephen D. *Sovereignty: Organized Hypocrisy.* Princeton University Press, 1999.

Krepinevich Jr., Andrew F. *The Army and Vietnam.* Baltimore, MD: Johns Hopkins University Press, 1988.

Krishna, Sankaran. *Postcolonial Insecurities: India, Sri Lanka, and the Question of Nationhood.* New ed. Minneapolis: University of Minnesota Press, 1999.

Kuchins, Daniel, and Thomas M. Sanderson. "The Northern Distribution Network and Afghanistan." Center for Strategic and International Studies (CSIS), January 6, 2010. http://csis.org/publication/northern-distribution-network-and-afghanistan.

Ladwig III, Walter C. *The Forgotten Front: Patron–Client Relationships in Counter Insurgency.* Cambridge University Press, 2017.

"Influencing Clients in Counterinsurgency: U.S. Involvement in El Salvador's Civil War, 1979–92." *International Security* 41, no. 1 (July 1, 2016): 99–146.

Laffont, Jean-Jacques, and David Martimort. *The Theory of Incentives: The Principal–Agent Model.* Princeton University Press, 2009.

Lake, David A. *The Statebuilder's Dilemma: On the Limits of Foreign Intervention.* Ithaca, NY: Cornell University Press, 2016.

Lamberson, Anna. "A Capital Law for Baghdad: A Governance Framework for Iraq's Ancient Capital." *State & Local Government Review* 43, no. 2 (2011): 151–8. www.jstor.org/stable/41303186.

Lawrance, Benjamin N., Emily Lynn Osborn, and Richard L. Roberts. *Intermediaries, Interpreters, and Clerks: African Employees in the Making of Colonial Africa.* Madison: University of Wisconsin Press, 2006.

Leeds, Brett Ashley. "Alliance Reliability in Times of War: Explaining State Decisions to Violate Treaties." *International Organization* 57, no. 4 (October 1, 2003): 801–27.

"Do Alliances Deter Aggression? The Influence of Military Alliances on the Initiation of Militarized Interstate Disputes." *American Journal of Political Science* 47, no. 3 (July 1, 2003): 427–39.

"Letter from Ahmad Shah Masoud to the Soviet Chief Military Adviser," December 26, 1998, Cold War International History Project, Record ID 112471. http://digitalarchive.wilsoncenter.org/document/112471.

"Letter of the Indian High Commissioner in Colombo to the Sri Lankan President J. R. Jayewardene Conveying the Agreement Arrived at with the LTTE on the Interim Administrative Council in the Northern and Eastern Province," Colombo, September 28, 1987. In *India–Sri Lanka: Relations and Sri Lanka's Ethnic Conflict Documents – 1947–2000*, ed. Avtar Singh Bhasin. Vol. IV, Document 753. New Delhi: Indian Research Press, 2001, 2083–4.

Livingston, Ian S., Heather L. Messera, and Michael O'Hanlon. "Afghanistan Index: Tracking Variables of Reconstruction & Security in Post 9/11 Afghanistan." Washington, DC: The Brookings Institution, February 28, 2011. www.brookings.edu/about/programs/foreign-policy/afghanistan-index.

The London Conference on Afghanistan. "The Afghanistan Compact," February 2006. www.nato.int/isaf/docu/epub/pdf/afghanistan_compact.pdf.

Lyall, Jason, and Isaiah Wilson III. "Rage against the Machines: Explaining Outcomes in Counterinsurgency Wars." *International Organization* 63, no. 1 (January 1, 2009): 67–106.

Lynch, Marc. "How Can the U.S. Help Maliki When Maliki's the Problem?" *Washington Post*, June 12, 2014, sec. Monkey Cage. www.washingtonpost.com/news/monkey-cage/wp/2014/06/12/iraq-trapped-between-isis-and-maliki/.

Macdonald, Douglas J. *Adventures in Chaos: American Intervention for Reform in the Third World*. Cambridge, MA: Harvard University Press, 1992.

MacDonald, Paul K. " 'Retribution Must Succeed Rebellion': The Colonial Origins of Counterinsurgency Failure." *International Organization* 67, no. 2 (2013): 253–86.

MacVicar, Sheila. "SIGAR: We Built an Afghanistan They Can't Afford." *Al Jazeera America*, June 17, 2015. http://america.aljazeera.com/watch/shows/compass/articles/2015/6/17/john-sopko-sigar.html.

Mahoney, James, and Dietrich Rueschemeyer. *Comparative Historical Analysis in the Social Sciences*. Cambridge University Press, 2003.

Maley, William, and Susanne Schmeidl, eds. *Reconstructing Afghanistan: Civil-Military Experiences in Comparative Perspective*. Contemporary Security Studies. New York: Routledge, 2014.

Mandelbaum, Michael. *The Fate of Nations: The Search for National Security in the Nineteenth and Twentieth Centuries*. Cambridge University Press, 1988.

Manea, Octavian. Interview with Dr. David Kilcullen. *Small Wars Journal*, November 7, 2010. http://smallwarsjournal.com/jrnl/art/interview-with-dr-david-kilcullen.

Mankin, Justin. "Rotten to the Core." *Foreign Policy*, May 10, 2011. www.foreignpolicy.com/articles/2011/05/10/rotten_to_the_core.

Mardini, Ramzy. "Iraqi Leaders React to the U.S. Withdrawal." Institute for the Study of War, November 10, 2011. www.understandingwar.org/sites/default/files/Backgrounder_IraqLeadersReacttoWithdrawal.pdf.

Marlowe, Ann. "The Picture Awaits: The Birth of Modern Counterinsurgency." *World Affairs* 172, no. 1 (2009): 64–73.

Marr, David. "The Rise and Fall of 'Counterinsurgency': 1961–1964." In *Vietnam and America: The Most Comprehensive Documented History of the Vietnam War*, ed. Marvin Gettleman et al. 2nd ed. New York: Grove Press, 1995, 2015–15.

Marten, Kimberly. *Warlords: Strong-Arm Brokers in Weak States*. Ithaca, NY: Cornell University Press, 2012.

Matonse, Antonio. "Mozambique: A Painful Reconciliation." *Africa Today* 39, no. 1/2 (March 1, 1992): 29–34.

McConnell, John A. "The British in Kenya (1952–1960): Analysis of a Successful Counterinsurgency Campaign." Thesis, Naval Postgraduate School, June 2005. www.dtic.mil/docs/citations/ADA435532.

McKinlay, R. D., and R. Little. "A Foreign Policy Model of U.S. Bilateral Aid Allocation." *World Politics* 30, no. 1 (October 1977): 58–86.

McMahon, Robert J. *Major Problems in the History of the Vietnam War: Documents and Essays*. 2nd ed. Lexington, MA: D.C. Heath, 1995.

McNamara, Robert S., James G. Blight, Robert K. Brigham, Thomas J. Biersteker, and Herbert Schandler. *Argument without End: In Search of Answers to the Vietnam Tragedy*. 1st ed. New York: PublicAffairs, 1999.

Mehta, Ashok K. "India's Counterinsurgency in Sri Lanka." In *India and Counterinsurgency: Lessons Learned*, ed. Sumit Ganguly and David P. Fidler. New York: Routledge, 2009, 155–72.

"Memorandum for Director of Central Intelligence Helms." September 12, 1968, Washington, DC. *Foreign Relations of the United States, 1964–1968, Volume VII, Vietnam, September 1968–January 1969*. Document 11. https://history.state.gov/historicaldocuments/frus1964-68v07/d11.

Merom, Gil. *How Democracies Lose Small Wars: State, Society, and the Failures of France in Algeria, Israel in Lebanon, and the United States in Vietnam*. Cambridge University Press, 2003.

"Message of the Indian Government to the Sri Lankan Government on the Urgent Need to Send Relief Supplies to Jaffna through the Indian Red Cross," New Delhi, June 1, 1987. In *India–Sri Lanka: Relations and Sri Lanka's Ethnic Conflict Documents – 1947–2000*, ed. Avtar Singh Bhasin. Vol. III, Document 680. New Delhi: Indian Research Press, 2001, 1902.

"Message of the Indian Prime Minister Rajiv Gandhi to the Sri Lankan President J. R. Jayewardene Delivered through the Indian High Commissioner in Colombo J.N. Dixit," New Delhi, August 28, 1988. In *India–Sri Lanka: Relations and Sri Lanka's Ethnic Conflict Documents – 1947–2000*, ed. Avtar Singh Bhasin. Vol. IV, Document 851. New Delhi: Indian Research Press, 2001, 2292.

Metz, Steven. "Learning From Iraq: Counterinsurgency in American Strategy." US Army War College, Strategic Studies Institute, January 2007. www.dtic.mil/docs/citations/ADA459931.

"Unruly Clients: The Trouble with Allies." *World Affairs* 172, no. 4 (2010): 49–59.

Middle East Contemporary Survey. The Moshe Dayan Center, 1993.

Miles, Donna. "Petraeus Notes Differences between Iraq, Afghanistan Strategies." American Forces Press Service, April 22, 2009. www.defense.gov/news/newsarticle.aspx?id=54036.

Miller, Edward. *Misalliance: Ngo Dinh Diem, the United States, and the Fate of South Vietnam.* Cambridge, MA: Harvard University Press, 2013.

Miller, Gary J. "The Political Evolution of Principal–Agent Models." *Annual Review of Political Science* 8, no. 1 (2005): 203–25.

Mills, Nick. *Karzai: The Failing American Intervention and the Struggle for Afghanistan.* Hoboken, NJ: John Wiley & Sons, 2007.

Minter, William. *Apartheid's Contras: An Inquiry into the Roots of War in Angola and Mozambique.* Johannesburg: Witwatersrand University Press, 1994.

Misdaq, Nabi. *Afghanistan: Political Frailty and External Interference.* London; New York: Taylor & Francis, 2006.

Mitrokhin, Vasiliy. "The KGB in Afghanistan, English Edition." Cold War International History Project – Working Paper Series, February 2002. www.wilsoncenter.org/sites/default/files/WP40-english.pdf.

Mold, Andrew. "Policy Ownership and Aid Conditionality in the Light of the Financial Crisis." Paris: Organization for Economic Cooperation and Development, September 1, 2009. www.oecd-ilibrary.org/fr/development/policy-ownership-and-aid-conditionality-in-the-light-of-the-financial-crisis_9789264075528-en.

Morris, Stephen. *Why Vietnam Invaded Cambodia: Political Culture and the Causes of War.* 1st ed. Stanford University Press, 1999.

Morrow, James D. "Alliances and Asymmetry: An Alternative to the Capability Aggregation Model of Alliances." *American Journal of Political Science* 35, no. 4 (1991): 904–33.

Moyar, Mark. *A Question of Command: Counterinsurgency from the Civil War to Iraq.* New Haven, CT: Yale University Press, 2009.

Murdoch, James C., and Todd Sandler. "NATO Burden Sharing and the Forces of Change: Further Observations." *International Studies Quarterly* 35, no. 1 (March 1, 1991): 109–14.

Murphy, Dan. "Karzai Says Taliban No Threat to Women, NATO Created 'No Gains' for Afghanistan." *The Christian Science Monitor,* October 7, 2013.

Murray, Martin J. "The Post-colonial State: Investment and Intervention in Vietnam." *Politics & Society* 3, no. 4 (1973): 437–61.

Nagl, John A. "Foreword to the University of Chicago Press Edition: The Evolution and Importance of Field Manual 3-24, Counterinsurgency." In *The U.S. Army/Marine Corps Counterinsurgency Field Manual,* US Army and Marine Corps. University of Chicago Press, 2007.

  Learning to Eat Soup with a Knife: Counterinsurgency Lessons from Malaya and Vietnam. 1st ed. University of Chicago Press, 2005.

  "A Responsibility to Learn." In *Iraq Uncensored: Perspectives,* ed. Jim Ludes. Golden, CO: Fulcrum Publishing, 2009, 47–52.

Nanda, Ved P. "The 'Good Governance' Concept Revisited." *The Annals of the American Academy of Political and Social Science* 603 (January 2006): 269–83.

National Commission on Terrorist Attacks. *The 9/11 Commission Report: Final Report of the National Commission on Terrorist Attacks Upon the United States.* 2004.

National Security Archive. "Afghanistan Déjà vu? Lessons from the Soviet Experience," October 30, 2009. Ed. Svetlana Savranskaya. www.gwu.edu/%7Ensarchiv/NSAEBB/NSAEBB292/index.htm.

"Afghanistan: Lessons from the Last War," October 9. 2001. Ed. John Prados and Svetlana Savranskaya.

"Afghanistan and the Soviet Withdrawal 1989: 20 Years Later: Tribute to Alexander Lyakhovsky Includes Previously Secret Soviet Documents," February 15, 2009. Ed. Svetlana Savranskaya and Thomas Blanton. www .gwu.edu/%7Ensarchiv/NSAEBB/NSAEBB272/index.htm.

"Conflicting Missions: Secret Cuban Documents on the History of Africa Involvement," April 1, 2002. Ed. Peter Kornbluh. www.gwu.edu/ ~nsarchiv/NSAEBB/NSAEBB67/.

"The Diary of Anatoly Chernyaev: Former Top Soviet Adviser's Journal Chronicles Final Years of the Cold War," May 25, 2006. Ed. Svetlana Savranskaya. www.gwu.edu/%7Ensarchiv/NSAEBB/NSAEBB192/index.htm.

"U.S. Policy in the Vietnam War, Part I: 1954–1968." Ed. John Prados, Digital National Security Archive, Washington, DC. Chadwyck-Healey, 1992.

"U.S. Policy in the Vietnam War, Part II: 1969–1975." Ed. John Prados, Digital National Security Archive, Washington, DC. Chadwyck-Healey, 2004.

NATO Training Mission – Afghanistan, Combined Security Transition Command – Afghanistan (NTM-A/CSTC-A). "Afghan Ministry of Interior (MoI) Advisor Guide," May 2011. Version 1.0. http://publicintelligence .net/isaf-afghan-ministry-of-interior-advisor-guide/.

Navaratna-Bandara, A. M. "Ethnic Relations and State Crafting in Post-independence Sri Lanka." In *Sri Lanka: Current Issues and Historical Background*, ed. Walter Nubin. New York: Nova Publishers, 2002, 57–75.

Neef, Christian. "Return of the Lion: Former Warlord Preps for Western Withdrawal." *Spiegel Online*, September 23, 2013, sec. International. www .spiegel.de/international/world/afghan-warlords-like-ismail-khan-prepare-for-western-withdrawal-a-924019.html.

Nelson, Soraya Sarhaddi. "Teams Focus on Poppy Eradication in Afghanistan." *National Public Radio*, August 31, 2007. www.npr.org/templates/story/story .php?storyId=14088743.

Neumann, Ronald E. "Failed Relations between Hamid Karzai and the United States: What Can We Learn?" United States Institute of Peace, May 2015.

Nguyên, Anh Tuân. *South Vietnam, Trial and Experience: A Challenge for Development*. Ohio University, Monographs in International Studies, Southeast Asia Series, no. 80. Athens: Ohio University Press, 1986.

Nguyen, Lien-Hang T. "Cold War Contradictions: Toward an International History of the Second Indochina War, 1969–1973." In *Making Sense of the Vietnam Wars: Local, National, and Transnational Perspectives*, ed. Mark Philip Bradley and Marilyn B. Young. Oxford University Press, 2008, 219–50.

Nguyen-Vo, Thu-Huong. *Khmer-Viet Relations and the Third Indochina Conflict*. Jefferson, NC: Mcfarland, 1992.

Nichols, Michelle. "Afghanistan Plans National Electronic ID Cards." *Reuters*, December 12, 2010. www.reuters.com/article/2010/12/12/us-afghanistan-identification-cards-idUSTRE6BB0P720101212.

Nicoletti, Michael. "Opium Production and Distribution: Poppies, Profits and Power in Afghanistan." Theses and Dissertations. Paper 74. DePaul University, March 2011. http://via.library.depaul.edu/etd/74.

Nordland, Rod. "Afghan Tax on Foreign Contractors Hits Resistance." *The New York Times*, January 17, 2011, sec. World / Asia Pacific. www.nytimes.com/2011/01/18/world/asia/18afghan.html.

"U.S. Turns a Blind Eye to Opium in Afghan Town." *The New York Times*, March 20, 2010. www.nytimes.com/2010/03/21/world/asia/21marja.html.

Obama, Barack. "Remarks by the President and First Lady on the End of the War in Iraq." whitehouse.gov, December 14, 2011. www.whitehouse.gov/the-press-office/2011/12/14/remarks-president-and-first-lady-end-war-iraq.

Oberst, Robert. "A War without Winners in Sri Lanka." *Current History* 91, no. 563 (March 1992): 128–31.

Office of the Secretary of Defense Vietnam Task Force. "United States–Vietnam Relations, 1945–1967," January 1971.

Office of the Special Inspector General for Afghanistan Reconstruction (SIGAR). "Corruption in Conflict: Lessons from the U.S. Experience in Afghanistan," September 2016. www.dtic.mil/dtic/tr/fulltext/u2/1016886.pdf.

"Inspectors General Fiscal Year 2013 Joint Strategic Oversight Plan for Afghanistan Reconstruction," July 2012. www.sigar.mil/pdf/strategicoversightplans/fy-2013.pdf.

"Limited Interagency Coordination and Insufficient Controls over U.S. Funds in Afghanistan Hamper U.S. Efforts to Develop the Afghan Financial Sector and Safeguard U.S. Cash," July 20, 2011. www.sigar.mil/pdf/audits/2011-07-20audit-11-13.pdf.

"Quarterly Report to Congress," April 30, 2010. www.sigar.mil/pdf/quarterlyreports/2010-04-30qr.pdf.

"Quarterly Report to Congress," October 30, 2010. www.sigar.mil/quarterlyreports/2010-10-30qr.pdf.

"Quarterly Report to Congress," July 30, 2012. www.sigar.mil/pdf/quarterlyreports/2012-07-30qr.pdf.

"Quarterly Report – Special Inspector General for Afghanistan Reconstruction," July 30, 2012. www.sigar.mil/pdf/quarterlyreports/2012-07-30qr.pdf.

Office of the Special Inspector General for Iraq Reconstruction (SIGIR). "Falluja Waste Water Treatment System. Falluja, Iraq." SIGIR PA-08-144-08-148, October 27, 2008. www.dtic.mil/docs/citations/ADA529001.

Hard Lessons: The Iraq Reconstruction Experience. Stuart J. Bowen. Washington, DC: US Government Printing Office, 2009. https://permanent.access.gpo.gov/lps108462/Hard_Lessons_Report.pdf.

"Information on Government of Iraq Contributions to Reconstruction Costs." SIGIR 09-018, April 29, 2009. www.dtic.mil/docs/citations/ADA508864.

"Sons of Iraq Program: Results Are Uncertain and Financial Controls Were Weak." SIGIR 11-010, January 28, 2011.

"Quarterly Report to Congress," July 2008. www.globalsecurity.org/military/library/report/sigir/sigir-report-2008-07.html.

"Quarterly Report to Congress," October 30, 2008. www.globalsecurity.org/military/library/report/sigir/sigir-report-2008-10.htm.

O'Hanlon, Michael. "America's History of Counterinsurgency." Washington, DC: Brookings Institution, June 6, 2016. www.brookings.edu/wp-content/uploads/2016/06/06_counterinsurgency_ohanlon.pdf.

Oka, Takashi. *Newsletters about the Vietnamese War.* Washington, DC: Institute of Current World Affairs, 1968.

Olson, Mancur. *The Logic of Collective Action: Public Goods and the Theory of Groups, Second Printing with New Preface and Appendix.* Revised ed. Cambridge, MA: Harvard University Press, 1971.

Olson, Mancur, and Richard Zeckhauser. "An Economic Theory of Alliances." *The Review of Economics and Statistics* 48, no. 3 (August 1966): 266–79.

Oneal, John R. "Testing the Theory of Collective Action: NATO Defense Burdens, 1950–1984." *The Journal of Conflict Resolution* 34, no. 3 (September 1990): 426–48.

"The Theory of Collective Action and Burden Sharing in NATO." *International Organization* 44, no. 3 (July 1, 1990): 379–402.

Oneal, John R., and Paul F. Diehl. "The Theory of Collective Action and NATO Defense Burdens: New Empirical Tests." *Political Research Quarterly* 47, no. 2 (June 1, 1994): 373–96.

Otterman, Sharon. "Iraq: Interim Authority." Council on Foreign Relations, February 16, 2005. www.cfr.org/backgrounder/iraq-interim-authority.

"Iraq: Iraq's Governing Council." Council on Foreign Relations, May 17, 2004. www.cfr.org/iraq/iraq-iraqs-governing-council/p7665.

"Iraq: The June 28 Transfer of Power." Council on Foreign Relations. Accessed April 11, 2012. www.cfr.org/iraq/iraq-june-28-transfer-power/p7805.

Packer, George. "Statecraft as Psychiatry." *The New Yorker*, May 11, 2010. www.newyorker.com/news/george-packer/statecraft-as-psychiatry.

Paget, Julian. *Counter-insurgency Operations: Techniques of Guerrilla Warfare.* New York: Walker, 1967.

Parker, Christine, and Vibeke Nielsen. "The Challenge of Empirical Research on Business Compliance in Regulatory Capitalism." *Annual Review of Law and Social Science* 4 (October 23, 2009): 45–70.

Parker, Michelle. "Programming Development Funds to Support a Counterinsurgency: Nangarhar, Afghanistan 2006." National Defense University, DTP-053, September 2008. http://ndupress.ndu.edu/Media/News/News-Article-View/Article/1227862/dtp-053-programming-development-funds-to-support-a-counterinsurgency-nangarhar/.

Partlow, Joshua. *A Kingdom of Their Own: The Family Karzai and the Afghan Disaster.* New York: Knopf Doubleday Publishing Group, 2016.

Paul, T. V. *Asymmetric Conflicts: War Initiation by Weaker Powers.* Cambridge University Press, 1994.

PBS NewsHour. "Iraq's Transfer of Power." Transcript. June 28, 2004. Accessed April 11, 2012. www.pbs.org/newshour/bb/middle_east/jan-june04/sovereignty_6-28.html.

The Pentagon Papers. "[Part IV. B. 5.] Evolution of the War. Counterinsurgency: The Overthrow of Ngo Dinh Diem, May–Nov. 1963," 1/1969 1967. Series: Report of the Office of the Secretary of Defense Vietnam Task Force,

Record Group 330: Records of the Office of the Secretary of Defense, 1921–2008. US National Archives. www.archives.gov/research/pentagon-papers.

"[Part IV. C. 1.] Evolution of the War. U.S. Programs in South Vietnam, November 1963–April 1965," NASM 273 – NSAM 288, ARC Identifier 5890498. Series: Report of the Office of the Secretary of Defense Vietnam Task Force, 6/1967–1/1969, Record Group 330: Records of the Office of the Secretary of Defense, 1921–2008. US National Archives. https://catalog.archives.gov/id/5890498.

"[Part IV. C. 2. c.] Evolution of the War. Military Pressures against NVN. November – December 1964," 1/1969 1967. Series: Report of the Office of the Secretary of Defense Vietnam Task Force, 6/1967 – 1/1969, Record Group 330: Records of the Office of the Secretary of Defense, 1921–2008. US National Archives. https://catalog.archives.gov/id/5890501.

"[Part IV. C. 8.] Evolution of the War. Re-emphasis on Pacification: 1965–1967," ARC Identifier 5890510. Series: Report of the Office of the Secretary of Defense Vietnam Task Force, 6/1967–1/1969, Record Group 330: Records of the Office of the Secretary of Defense, 1921–2008. US National Archives. https://catalog.archives.gov/id/5890510.

"[Part IV. C. 9. a.] Evolution of the War. U.S. GVN Relations. Volume 1: December 1963– June 1965," ARC Identifier 5890511. Series: Report of the Office of the Secretary of Defense Vietnam Task Force, 6/1967–1/1969, Record Group 330: Records of the Office of the Secretary of Defense, 1921–2008. US National Archives. https://catalog.archives.gov/id/5890511.

Peter, Tom A. "Afghanistan War: Gap Grows between US Efforts, Afghan Expectations." *Christian Science Monitor*, January 19, 2010. www.csmonitor.com/World/Asia-South-Central/2010/0119/Afghanistan-war-gap-grows-between-US-efforts-Afghan-expectations.

Peterson, John E. "The Yemen Arab Republic and the Politics of Balance." *Asian Affairs* 12, no. 3 (November 1981): 254–66.

Pierson, Paul. *Politics in Time: History, Institutions, and Social Analysis.* Princeton University Press, 2004.

Pipes, Daniel. *Syria beyond the Peace Process.* Washington, DC: Washington Institute for Near East Policy, 1996.

"Points of Verbal Message of the Indian Prime Minister Rajiv Gandhi and the Reaction of the Sri Lankan President J. R. Jayewardene Conveyed through the Indian High Commissioner J. N. Dixit," Colombo, January 13, 1988. In *India–Sri Lanka: Relations and Sri Lanka's Ethnic Conflict Documents – 1947–2000*, ed. Avtar Singh Bhasin. Vol. IV, Document 793. New Delhi: Indian Research Press, 2001, 2184–5.

Porch, Douglas. *Counterinsurgency: Exposing the Myths of the New Way of War.* Cambridge University Press, 2013.

Prados, John. "The Shape of the Table." In *The Search for Peace in Vietnam, 1964–1968*, ed. Lloyd C. Gardner and Ted Gittinger. College Station: Texas A&M University Press, 2004, 355–70.

Pressman, Jeremy. *Warring Friends: Alliance Restraint in International Politics.* 2nd ed. Ithaca, NY: Cornell University Press, 2008.

"Press Briefing of the Indian High Commissioner in Colombo J. N. Dixit on the Peace Efforts to Bring about a Negotiated Settlement between the LTTE and the Sri Lankan Government," Colombo, February 27, 1988. In *India–Sri Lanka: Relations and Sri Lanka's Ethnic Conflict Documents – 1947–2000*, ed. Avtar Singh Bhasin. Vol. IV, Document 806. New Delhi: Indian Research Press, 2001, 2206–7.

"Press Release of the Indian High Commissioner in Colombo," July 6, 1987. In *India–Sri Lanka: Relations and Sri Lanka's Ethnic Conflict Documents – 1947–2000*, ed. Avtar Singh Bhasin. Vol. III, Document 714. New Delhi: Indian Research Press, 2001, 1933.

Pritchard, James, and M. L. R. Smith. "Thompson in Helmand: Comparing Theory to Practice in British Counter-Insurgency Operations in Afghanistan." *Civil Wars* 12, no. 1–2 (January 1, 2010): 65–90.

"Question in the Sri Lankan Parliament Regarding the Implementation of the Indo-Sri-Lanka Peace Accord of July 1987," June 20, 1989. In *India–Sri Lanka: Relations and Sri Lanka's Ethnic Conflict Documents – 1947–2000*, ed. Avtar Singh Bhasin. Vol. IV, Document 914. New Delhi: Indian Research Press, 2001, 2385–8.

Rabinovich, Itamar. *The War for Lebanon, 1970–1985*. Ithaca, NY: Cornell University Press, 1985.

Rasanayagam, Angelo. *Afghanistan: A Modern History*. London: I. B. Tauris, 2005.

Rashid, Ahmed. *Descent into Chaos: The U.S. and the Disaster in Pakistan, Afghanistan, and Central Asia*. Revised ed. London: Penguin Books, 2009.

Rathmell, Andrew. "Planning Post-conflict Reconstruction in Iraq: What Can We Learn?" *International Affairs* 81, no. 5 (2005): 1013–38.

Rayburn, Joel. *Iraq after America: Strongmen, Sectarians, Resistance*. Stanford, CA: Hoover Institution Press, 2014.

Refugees International. "Iraq: Fix the Public Distribution System to Meet Needs of the Displaced." April 10, 2007. www.unhcr.org/refworld/docid/47a6eef311.html.

Reis, Bruno C. "The Myth of British Minimum Force in Counterinsurgency Campaigns during Decolonisation (1945–1970)." *Journal of Strategic Studies* 34, no. 2 (April 1, 2011): 245–79.

Reiter, Dan, and Allan C. Stam. *Democracies at War*. Princeton University Press, 2002.

Remarks to the Press at Conclusion of Rehearsal of Concept (ROC) Drill. US Department of State, Special Representative for Afghanistan and Pakistan. Transcript, April 11, 2010. https://2009-2017.state.gov/s/special_rep_afghanistan_pakistan/2010/140010.htm.

"Reply Message of the Sri Lankan Government to the Indian Government," Colombo, June 1, 1987. In *India–Sri Lanka: Relations and Sri Lanka's Ethnic Conflict Documents – 1947–2000*, ed. Avtar Singh Bhasin. Vol. III, Document 681. New Delhi: Indian Research Press, 2001, 1902.

Resnick, Evan N. "Hang Together or Hang Separately? Evaluating Rival Theories of Wartime Alliance Cohesion." *Security Studies* 22, no. 4 (October 1, 2013): 672–706.

Roberts, Michael. "Language and National Identity: The Sinhalese and Others over the Centuries." *Nationalism and Ethnic Politics* 9, no. 2 (2003): 75–102.

Rogers, Simon. "WikiLeaks Embassy Cables: Download the Key Data and See How It Breaks Down." *The Guardian*, November 28, 2010, sec. World News. www.guardian.co.uk/news/datablog/2010/nov/29/wikileaks-cables-data.

Roggio, Bill, and Lisa Lundquist. "Green-on-Blue Attacks in Afghanistan: The Data." *The Long War Journal*, August 23, 2012. www.longwarjournal.org/archives/2012/08/green-on-blue_attack.php.

Rohde, David, and David E. Sanger. "How a 'Good War' in Afghanistan Went Bad." *The New York Times*, August 12, 2007, sec. International / Asia Pacific. www.nytimes.com/2007/08/12/world/asia/12afghan.html.

Root, Hilton L. *Alliance Curse: How America Lost the Third World*. 1st ed. Washington, DC: Brookings Institution Press, 2008.

Rose, Charlie. A Conversation with General Petraeus, *The Charlie Rose Show*, April 26, 2007. www.nytimes.com/2007/04/30/world/americas/30iht-30petraeus-charlie-rose.5499787.html.

Rosen, Nir. "The Flight from Iraq." *The New York Times*, May 13, 2007, sec. Magazine. www.nytimes.com/2007/05/13/magazine/13refugees-t.html.

Rosenberg, Matthew. "False Claims in Afghan Accusations on U.S. Raid Add to Doubts on Karzai." *The New York Times*, January 25, 2014.

Rubin, Alissa J. "Karzai's Words Leave Few Choices for the West." *The New York Times*, April 4, 2010, sec. Asia Pacific. www.nytimes.com/2010/04/05/world/asia/05karzai.html.

Rubin, Alissa J., and Campbell Robertson. "U.S. Helps Remove Uranium from Iraq." *The New York Times*, July 7, 2008. www.nytimes.com/2008/07/07/world/middleeast/07iraq.html?_r=0.

Rubin, Alissa J., and Mark Fineman. "Council Moves to Further 'De-Baathify' Iraq." *Los Angeles Times*, September 17, 2003. http://articles.latimes.com/2003/sep/17/world/fg-council17.

Rubin, Alissa J., and Rod Nordland. "U.S. General Puts Troops on Security Alert after Karzai Remarks." *The New York Times*, March 13, 2013. www.nytimes.com/2013/03/14/world/asia/karzais-remarks-draw-us-troop-alert.html.

Rubin, Barnett R. *Afghanistan in the Post-Cold War Era*. New York: Oxford University Press, 2013.

"Crafting a Constitution for Afghanistan." *Journal of Democracy* 15, no. 3 (2004): 5–19.

Rubin, Elizabeth. "Karzai in His Labyrinth." *New York Times Magazine*, August 4, 2009. www.nytimes.com/2009/08/09/magazine/09Karzai-t.html.

Rumsfeld, Donald. "Risk in the Way ahead in Iraq," October 28, 2003. Secret. Declassified September 2007. http://papers.rumsfeld.com/. http://library.rumsfeld.com/doclib/sp/353/re%20Risk%20in%20the%20Way%20Ahead%20in%20Iraq%2010-28-2003.pdf.

"Rumsfeld's Memo of Options for Iraq War." *The New York Times*, December 3, 2006, sec. International / Middle East. www.nytimes.com/2006/12/03/world/middleeast/03mtext.html.

Russett, Bruce M., and John D. Sullivan. "Collective Goods and International Organization." *International Organization* 25, no. 4 (October 1, 1971): 845–65.

Sabha, India. Parliament. Lok. *Lok Sabha Debates*. Lok Sabha Secretariat, 1987.

Saikal, Amin. *Modern Afghanistan: A History of Struggle and Survival*. London; New York: I. B. Tauris, 2012.

Sandler, Todd, and Keith Hartley. "Economics of Alliances: The Lessons for Collective Action." *Journal of Economic Literature* 39, no. 3 (September 2001): 869–96.

Sappington, David E. M. "Incentives in Principal–Agent Relationships." *Journal of Economic Perspectives* 5, no. 2 (1991): 45–66.

Sardeshpande, S. C. *Assignment Jaffna*. New Delhi: Lancer Publishers, 1992.

Savage, Kevin, Lorenzo Delesgues, Ellen Martin, and Gul Pacha Ulfat. "Corruption Perceptions and Risks in Humanitarian Assistance: An Afghanistan Case Study." HPG Working Paper. Humanitarian Policy Group, July 2007.

Schaffer, Howard B. *Ellsworth Bunker: Global Troubleshooter, Vietnam Hawk*. Chapel Hill: University of North Carolina Press, 2003.

Schelling, Thomas C. *The Strategy of Conflict*. New York: Oxford University Press, 1963.

Schmemann, Serge. "Soviet Archives: Half-Open, Dirty Window on Past." *The New York Times*, April 26, 1995. www.nytimes.com/1995/04/26/world/soviet-archives-half-open-dirty-window-on-past.html.

Schmidt, Dana Adams. *Yemen: The Unknown War*. London: Bodley Head, 1968.

Schmidt, Michael S. "Immunity Gone, Contractor in Iraq Sentenced to Prison." *The New York Times*, February 28, 2011, sec. World / Middle East. www.nytimes.com/2011/03/01/world/middleeast/01iraq.html.

Schmitt, Eric. "U.S. Envoy's Cables Show Worries on Afghan Plans." *The New York Times*, January 26, 2010, sec. International / Asia Pacific. www.nytimes.com/2010/01/26/world/asia/26strategy.html.

Schroeder, Paul W. "Alliances, 1815–1945: Weapons of Power and Tools of Management." In *Historical Dimensions of National Security Problems*, ed. Klaus Knorr. Lawrence: University of Kansas Press, 1976, 247–86.

Schulman, Daniel. "Corruption in Afghanistan: It's Even Worse Than You Think." *Mother Jones*, January 21, 2010. http://motherjones.com/politics/2010/01/corruption-afghanistan-its-even-worse-you-think.

Schwarz, Benjamin. "American Counterinsurgency Doctrine and El Salvador." Santa Monica, CA: Rand Corporation, 1991. www.rand.org/pubs/reports/R4042.html.

Schwartz, Michael. *War without End: The Iraq War in Context*. Chicago: Haymarket Books, 2008.

Scott, James C. *Domination and the Arts of Resistance: Hidden Transcripts*. New Haven, CT: Yale University Press, 1990.

   Weapons of the Weak: Everyday Forms of Peasant Resistance. New Haven, CT: Yale University Press, 1987.

Sellström, Tor. *Sweden and National Liberation in South Africa: Solidarity and Assistance, 1970–1994*. Uppsala: Nordic Africa Institute, 1999.

Sewall, Sarah. "Introduction to the University of Chicago Press Edition." In *The U.S. Army/Marine Corps Counterinsurgency Field Manual*, US Army and Marine Corps. University of Chicago Press, 2007.

Sewall, Sarah, John A. Nagl, David H. Petraeus, and James F. Amos. *The U.S. Army/Marine Corps Counterinsurgency Field Manual*, US Army and Marine Corps. University of Chicago Press, 2007.

Shafer, D. Michael. *Deadly Paradigms: The Failure of U.S. Counterinsurgency Policy*. Princeton University Press, 1988.

Shapiro, Jeremy. "The Latest Battle in Fallujah Is a Symbol of the Futility of US Efforts in Iraq." *Vox*, May 25, 2016. www.vox.com/2016/5/25/11750054/battle-fallujah-iraq.

Shaplen, Robert. *The Lost Revolution: The U.S. in Vietnam, 1946–1966*. New York: Harper & Row, 1966.

Sharp, Jeremy M. "The Iraqi Security Forces: The Challenge of Sectarian and Ethnic Influences." Congressional Research Service, RS21968, January 18, 2007. Accessed May 15, 2012. www.cfr.org/iraq/crs-iraqi-security-forces-challenge-sectarian-ethnic-influences/p12616.

Sheehan, Neil. *A Bright Shining Lie: John Paul Vann and America in Vietnam*. New York: Random House, 1988.

Silva, K. M. *Regional Powers and Small State Security: India and Sri Lanka, 1977–1990*. Washington, DC: Woodrow Wilson Center Press, 1995.

Simpson, Mark. "Foreign and Domestic Factors in the Transformation of Frelimo." *The Journal of Modern African Studies* 31, no. 2 (June 1, 1993): 309–37.

Singh, Harkirat. *Indian Intervention in Sri Lanka: The IPKF Experience Retold*. Colombo: Vijitha Yapa Publications, 2006.

"SIPRNET: Where America Stores Its Secret Cables." *The Guardian*, November 28, 2010, sec. World News. www.guardian.co.uk/world/2010/nov/28/siprnet-america-stores-secret-cables.

"SIPRNET: Where the Leaked Cables Came From." BBC, November 29, 2010, sec. US & Canada. www.bbc.co.uk/news/world-us-canada-11863618.

Sloan, Stanley R. *Permanent Alliance?: NATO and the Transatlantic Bargain from Truman to Obama*. New York: Bloomsbury Publishing USA, 2010.

Smith, Alastair. "Alliance Formation and War." *International Studies Quarterly* 39, no. 4 (December 1, 1995): 405–25.

"International Crises and Domestic Politics." *American Political Science Review* 92, no. 3 (September 1998): 623–38.

Smith, Chris. "South Asia's Enduring War." In *Creating Peace in Sri Lanka: Civil War and Reconciliation*, ed. Robert I. Rotberg. Washington, DC: Brookings Institution Press, 1999, 17–40.

Smith, M. L. R., and David Martin Jones. *The Political Impossibility of Modern Counterinsurgency: Strategic Problems, Puzzles, and Paradoxes*. New York: Columbia University Press, 2015.

Snyder, Glenn H. *Alliance Politics*. Ithaca, NY: Cornell University Press, 2007.

"Alliance Theory: A Neorealist First Cut." *Journal of International Affairs* 44, no. 1 (Spring 1990): 103–23.

Snyder, Glenn Herald, and Paul Diesing. *Conflict among Nations: Bargaining, Decision Making, and System Structure in International Crises*. Princeton University Press, 2015.

Sorley, Lewis. *A Better War: The Unexamined Victories and Final Tragedy of America's Last Years in Vietnam*. New York: Mariner Books, 2007.

"Soviet Briefing on the Talks between Brezhnev and B. Karmal in Moscow," October 29, 1980. Top Secret, Central Committee Foreign Department Bulletin, Budapest, Hungary. Cold War International History Project, The Woodrow Wilson International Center for Scholars. http://digitalarchive.wilsoncenter.org/document/112500. Original source, National Archives of Hungary (MOL), M-KS 288 f. 11/4391.o.e.

"Special File Record of Conversation of L.I. Brezhnev with N.M. Taraki." March 20, 1979. History and Public Policy Program Digital Archive. http://digitalarchive.wilsoncenter.org/document/111282.

Spector, Ronald H. *After Tet: The Bloodiest Year in Vietnam*. New York: Vintage, 1994.

Sprecher, Christopher. "Alliances, Armed Conflict, and Cooperation: Theoretical Approaches and Empirical Evidence." *Journal of Peace Research* 43, no. 4 (July 1, 2006): 363–9.

"Statement of the Indian Minister for External Affairs P.V. Narasimha Rao in the Lok Sabha: 'North-Eastern Provincial Council Elections in Sri Lanka'," New Delhi, November 22, 1988. In *India–Sri Lanka: Relations and Sri Lanka's Ethnic Conflict Documents – 1947–2000*, ed. Avtar Singh Bhasin. Vol. IV, Document 870. New Delhi: Indian Research Press, 2001, 2335–7.

"Statement of the Leader of the Opposition in the Sri Lanka Parliament Anura Bandaranaike Regarding Statements Made by the Commander of the IPKF," Colombo, December 8, 1987. In *India–Sri Lanka: Relations and Sri Lanka's Ethnic Conflict Documents – 1947–2000*, ed. Avtar Singh Bhasin. Vol. IV, Document 789. New Delhi: Indian Research Press, 2001, 2178–80.

Stephens, Hampton. "Analysts, U.S. Officials Differ on Maliki's Plans for Sons of Iraq." *World Politics Review*, September 11, 2008. www.worldpoliticsreview.com/articles/2651/analysts-u-s-officials-differ-on-malikis-plans-for-sons-of-iraq.

Stewart, Geoffrey C. *Vietnam's Lost Revolution: Ngô Đình Diệm's Failure to Build an Independent Nation, 1955–1963*. Cambridge University Press, 2017.

Stewart, Scott. "U.S. Diplomatic Security in Iraq after the Withdrawal." *STRATFOR, Security Weekly*, December 22, 2011. www.stratfor.com/weekly/us-diplomatic-security-iraq-after-withdrawal.

Strayer, Robert W. *Why Did the Soviet Union Collapse? Understanding Historical Change*. New York: M. E. Sharpe, 1998.

Sturcke, James. "Syria Confirms Full Troop Withdrawal from Lebanon." *The Guardian*, March 30, 2005. www.guardian.co.uk/world/2005/mar/30/syria.unitednations.

Taber, Robert. *War of the Flea: The Classic Study of Guerrilla Warfare*. Washington, DC: Potomac Books Inc., 2002.

Takahashi, Takamichi. "Japan: A New Self-Defense Force Role ... or Not?" In *Coalition Challenges in Afghanistan: The Politics of Alliance*, ed. Gale Mattox and Stephen Grenier. Stanford University Press, 2015, 214–24.

Talbot, Stephen. "Syria/Lebanon: The Occupier and the Occupied." *PBS Frontline World*, August 3, 2004. www.pbs.org/frontlineworld/elections/syria .lebanon/.

Tankel, Stephen. *With Us and against Us: How America's Partners Help and Hinder the War on Terror*. New York: Columbia University Press, 2018.

Taylor, John B. *Global Financial Warriors: The Untold Story of International Finance in the Post–9/11 World*. 1st ed. New York: W. W. Norton & Company, 2007.

"Text of U.S. Security Adviser's Iraq Memo." *The New York Times*, November 29, 2006, sec. International / Middle East. www.nytimes.com/2006/11/29/ world/middleeast/29mtext.html.

Thayer, Thomas. "How to Analyze a War without Fronts: Vietnam, 1965–72." *Journal of Defense Research* 7B, no. 3 (1975): 767–943.

Thayer, Thomas C., ed. "A Systems Analysis View of the Vietnam War: 1965–1972. Volume 2. Forces and Manpower." Washington, DC: Office of the Assistant Secretary of Defense Southeast Asia Intelligence Division, February 18, 1975. https://apps.dtic.mil/dtic/tr/fulltext/u2/a051609.pdf.

Thies, Wallace J. *Friendly Rivals: Bargaining and Burden-Shifting in NATO*. New York: Routledge, 2003.

Thompson, Robert. *Defeating Communist Insurgency: The Lessons of Malaya and Vietnam*. New York: F. A. Praeger, 1966.

Thompson, William R., and David P. Rapkin. "Collaboration, Consensus, and Détente: The External Threat-Bloc Cohesion Hypothesis." *The Journal of Conflict Resolution* 25, no. 4 (1981): 615–37.

Tomsen, Peter. *The Wars of Afghanistan: Messianic Terrorism, Tribal Conflicts, and the Failures of Great Powers*. New York: PublicAffairs, 2011.

Tomz, Michael. "Domestic Audience Costs in International Relations: An Experimental Approach." *International Organization* 61, no. 4 (October 2007): 821–40.

"TOP SECRET Note of the RAW Agent to the Chief of the Sri Lankan Intelligence and Security RE: LTTE's Surrender of Weapons," Colombo, June 20, 1988. In *India–Sri Lanka: Relations and Sri Lanka's Ethnic Conflict Documents – 1947–2000*, ed. Avtar Singh Bhasin. Vol. IV, Document 835. New Delhi: Indian Research Press, 2001, 2263.

Tse-Tung, Mao. *On Guerrilla Warfare*. Translated by Samuel B. Griffith. Santiago: BN Publishing, 2007.

Tucker, Spencer C., ed. *Encyclopedia of Insurgency and Counterinsurgency: A New Era of Modern Warfare*. Santa Barbara, CA: ABC-CLIO, 2013.

United Nations Convention against Corruption Signature and Ratification Status. www.unodc.org/unodc/en/treaties/CAC/signatories.html.

United Nations Mission in Afghanistan. *Second Six Month Report: Independent Joint Anti-Corruption Monitoring and Evaluation Committee*. July 25, 2012. unama.unmissions.org.

United Nations Office on Drugs and Crime (UNDOC). "Corruption in Afghanistan: Bribery as Reported by Victims," January 19, 2010. www.unodc.org/documents/data-and-analysis/Afghanistan/Afghanistan-corruption-survey 2010-Eng.pdf.

US Agency for International Development. Joseph A. Medenhall, "U.S. Government, USAID and U.S. CORDS Objectives and Organization in Vietnam," July 22, 1969.

"United States Economic Assistance to South Vietnam 1954–75." Terminal Report. Washington, DC: Asia Bureau, Office of Residual Indochina Affairs. December 31, 1975. Vol. II.

US Army, "Operational Report – Lessons Learned for Period 1 November 1966–31 January 1967," RCS CSFOR – 65 (U), Department of the Army, Headquarters, United States Army, Vietnam. https://archive.org/stream/DTIC_AD0386164/DTIC_AD0386164_djvu.txt.

US Army and Marine Corps. *Counterinsurgency, FM 3-24*. Washington, DC: Headquarters, Department of the Army, 2006.

Field Manual FM 3-24 MCWP 3-33.5 Insurgencies and Countering Insurgencies. Washington, DC: Headquarters, Department of the Army, 2014. www.hqmc.marines.mil/Portals/135/JAO/FM%203_24%20May%202014 .pdf.

US Congress. "110th Congress Public Law 28, Conditioning of Future United States Strategy in Iraq on the Iraqi Government's Record of Performance on Its Benchmarks," May 25, 2007. www.congress.gov/110/plaws/publ28/PLAW-110publ28.pdf.

Public Law 110–28, "U.S. Troop Readiness, Veterans' Care, Katrina Recovery, and Iraq Accountability Appropriations Act," May 25, 2007. www .congress.gov/110/plaws/publ28/PLAW-110publ28.pdf.

US Department of Defense. "Backgrounder on Reconstruction and Humanitarian Assistance in Post-War." News Transcript, Office of the Assistant Secretary of Defense (Public Affairs), March 11, 2003. www.defense.gov/transcripts/transcript.aspx?transcriptid=2037.

"A Bold Shift in Iraq Policy – Accelerate the Transition, Sustain the Partnership, and Stabilize the Region." December 4, 2006. Declassified July 19, 2010.

"IED Incidents Comparison – Iraq and Afghanistan." Presentation Slide, Combined Information Data Network Exchange, Joint Improvised Explosive Device Defeat Organization – J9, August 5, 2010. Unclassified.

"Measuring Stability and Security in Iraq." Report to Congress, December 2009. www.defense.gov/Portals/1/Documents/pubs/Master_9204_29Jan10_FINAL_SIGNED.pdf.

"Memorandum from the Director, Far East Region, Office of the Assistant Secretary of Defense for International Security Affairs (Blouin) to the Assistant Secretary of Defense for International Security Affairs (McNaughton)," Washington, March 30, 1964. *Foreign Relations of the United States, 1964–1968, Volume 1, Vietnam,* 1964, Document 101. http://history.state .gov/historicaldocuments/frus1964–68v01.

Military Personnel Historical Reports, 2010, http://siadapp.dmdc.osd.mil/person nel/MILITARY/history/309hist.htm.

"Report for the President," June 28, 1966. Secret, Special Handing Required. Declassified June 28, 1978. Johnson Library, White House Central File, Confidential File, Subject Reports, Department of Defense, June 1966. Declassified Documents Reference System.

"Report on Progress toward Security and Stability in Afghanistan," April 2012. www.defense.gov/pubs/pdfs/Report_Final_SecDef_04_27_12.pdf.

"South Vietnam's Internal Security Capabilities." National Security Study. Secret. May 1, 1969. Declassified March 17, 2004.

"Status of Actions Approved in NSAM No. 288." Memorandum from the Director, Far East Region, Office of the Assistant Secretary of Defense for International Security Affairs (Blouin) to the Assistant Secretary of Defense for International Security Affairs (McNaughton), March 30, 1964, Washington National Records Center, RG 330, OASD/ISA Files: FRC 68 A 4023, 092 Vietnam. Top Secret. Document 101. *Foreign Relations of the United States, 1964–1968, Volume I, Vietnam, 1964.*

"Strategy," Attachment, "U.S. Strategy in Afghanistan," Donald Rumsfeld to Douglas Feith, National Security Council, October 16, 2001, 7:42 a.m., Secret/Close Hold/ Draft for Discussion. http://nsarchive.gwu.edu/ NSAEBB/NSAEBB358a/doc18.pdf.

"Subject: Iraq – Illustrative New Courses of Action." November 6, 2006. 10-M-1231. Special Collections. Accessed January 7, 2020. http://library .rumsfeld.com/.

United States Plan for Sustaining the Afghanistan National Security Forces. June 2008. Report to Congress in Accordance with the 2008 National Defense Authorization Act for Fiscal Year 2008 (Public Law 110-181), June 2008. www.defense.gov/pubs/United_States_Plan_for_Sustaining_the_Afghani stan_National_Security_Forces_1231.pdf.

"U.S. Department of Defense Casualty Status." January 19, 2016. www .defense.gov/casualty.pdf.

US Department of Defense, Defense Human Resources Activity (DHRA). "Using the SIPRNET." Accessed April 20, 2012. www.dhra.mil/perserec/ csg/s1class/siprnet.htm.

US Department of Defense, DoD/OGC, "Fact Sheet: Tax Exceptions Accorded U.S. Contractors and U.S. Contractor Personnel under the Agreement Regarding the Status of United States Military and Civilian Personnel of the U.S. Department of Defense Present in Afghanistan in Connection with Cooperative Efforts in Response to Terrorism, Humanitarian and Civic Assistance, Military Training and Exercises, and Other Activities (U.S.-Afghanistan Status of Forces Agreement (SOFA))," March 28, 2011. www .acq.osd.mil/dpap/ops/docs/TAB_A_-_Incoming%5B1%5D.pdf.

US Department of Defense, Donald Rumsfeld Files. "Iraq Policy: Proposal for the New Phase – Memo from Secretary Rumsfeld to President George W. Bush," December 8, 2006. 08-M-1641. Special Collections. www.dod.mil/pubs/foi/ specialCollections/Rumsfeld/DocumentsReleased ToSecretaryRumsfeldUnderMDR.pdf.

"Memorandum for the President, Subject: Iraqi Interim Authority," April 1, 2003. 09-M-2634. Declassified July 7, 2009.

"Subject: Afghanistan, Memorandum to Douglas Feith," April 17, 2002.

"U.S. Strategy in Afghanistan, Memo from Secretary of Defense Donald Rumsfeld to Under Secretary of Defense Douglas Feith," October 30, 2001.

US Department of State. "Ambassador Bunker Discusses Post Election Priorities in South Vietnam." Cable. September 5, 1967. Declassified December 15, 1994.

"Ambassador Bunker's Seventy-Second Weekly Message from Saigon Briefing President Johnson on the Present Situation in Vietnam." Cable. Secret. October 30, 1968. Declassified October 27, 1994.

"Ambassador Bunker's Weekly Report to President Johnson Regarding the Situation in Vietnam," January 24, 1968. Declassified June 12, 1997.

"Authorization to Sign Implementing Agreement for Immediate-Impact Counternarcotics Project," 2003STATE068522, March 14, 2003. The National Security Archive, Freedom of Information Act Request Number 2001105588.

"Blueprint for Vietnam," National Archives Records Administration, RG 59, S/S–S Files: Lot 70 D 48, Misc. VN Rpts. & Briefing Books. Chapter IV – National Development, 10–11. Referenced in "Editorial Note," *Foreign Relations of the United States, 1964–1968, Volume V, Vietnam, 1967*, Document 296.

"The Formation of Afghan Military and Police – U.S. President Bush to Hamid Karzai," January 10, 2002.

"Letter of Agreement on Police and Justice Projects," 02STATE244042, November 29, 2002. Washington, DC: The National Security Archive. Freedom of Information Act Request Number 2001105588.

"Memorandum for Walt Rostow, from William Leonhart, 'Blueprint for Vietnam'," September 11, 1967. NLJ 87-50. Lyndon B. Johnson Library.

"More for Our Effort: U.S. Leverage in Vietnam." First Draft/ HH &VW. August 7, 1967. Secret. Declassified February 20, 1992. US National Archives, 89–231.

"Negotiations Committee," Top Secret, Memorandum, October 14, 1966, 1 pp. Collection U.S. Policy in the Vietnam War, 1954–1968. Item VI01714. Library of Congress. W. Averell Harriman Papers. Special Files: Public Service. Box 520. Vietnam, General, Oct–Dec 1966.

"Next Steps in the Political Process, President Bush to President Karzai." Washington, DC: The National Security Archive, January 18, 2002.

"The Northern Distribution Network and the Baltic Nexus." Remarks. January 20, 2012. www.state.gov/t/pm/rls/rm/182317.htm.

"Objectives for Certification to the Government of Afghanistan," June 6, 2003.

"Political Stability and Security in South Vietnam." Samuel P. Huntington. Miscellaneous. Secret. December 1, 1967. Declassified September 17, 1990.

"For the President from Bunker," April 19, 1968. Declassified August 9, 1996. Douglas Pike (ed.) *The Bunker Papers, Reports to the President from Vietnam,*

1967–1973, Volume 2. Institute of East Asian Studies University of California at Berkeley, 411.

"For the President from Bunker, Saigon 16850," January 24, 1968, NLJ 96–207, LBJ Presidential Library, 14–15.

"Prime Minister Nguyen Van Loc Announces an Action Program to Attack Social Problems and Governmental Deficiencies in South Vietnam," November 18, 1967.

"Project TAKEOFF – Action Program." Undated. Central File. Pol27 Viet S., Douglas Pike (ed.), *The Bunker Papers, Reports to the President from Vietnam, 1967–1973, Volume I.* Institute of East Asian Studies University of California at Berkeley, 52.

"Response to DOS Telegram 19056 on Refugees," Secret. August 17, 1966. Declassified July 7, 1993.

"Secretary Rusk's Cable to Prepare Gaud for Response at Press Conference on Senator Edward Kennedy's Report on His Recent Trip to Vietnam Which Emphasized Corruption," January 27, 1968.

"Sen. Edward Kennedy's Report on His Recent Trip to Vietnam. For Komer and MacDonald from Grant," January 27, 1968. Declassified May 5, 1994.

"State Department Negotiations Committee Meeting on Manila Summit Top Secret," Minutes, October 14, 1966, 8 pp. Collection U.S. Policy in the Vietnam War, 1954–1968 Item VI01715. Ambassador-at-Large Library of Congress. W. Averell Harriman Papers. Public Service, Box 520. Vietnam, General, Oct–Dec 1966.

"Status of Activities Social, Political, and Economic Pertaining to the Hono- lulu Conference," Memorandum. Secret. April 7, 1966. Declassified Feb- ruary 25, 1980.

"Status Report on Political and Economic Reform in Vietnam," Report. Secret May 26, 1966. Declassified February 25, 1980.

"Transmittal Memorandum, Leonard Unger, Dep. Asst. Secy of State for Far Eastern Affairs and Chairman, Vietnam Coordinating Committee, to Chet [Chester Cooper]." Secret, March 20, 1965. Declassified March 17, 1980. Sanitized. Johnson Library, NSF, Countries, Vietnam, Vol. 31.

"Viet-Nam Political Situation Report," November 18, 1967, NLJ 94-480, LBJ Presidential Library.

US Department of State and the Broadcasting Board of Governors, Office of Inspector General. PAE Operations and Maintenance Support at Embassy Kabul, Afghanistan – Performance Evaluation, Report Number MERO-I-11-05, December 2010. http://oig.state.gov/documents/organization/156020 .pdf.

US Department of State, Bureau for International Narcotics and Law Enforce- ment Affairs. "International Narcotics Control Strategy Report, Volume I, Drug and Chemical Control," March 2010. www.state.gov/documents/ organization/137411.pdf.

US Department of State, Bureau of Public Affairs. "2012 Investment Climate Statement – Afghanistan." Report, June 7, 2012. www.state.gov/e/eb/rls/ othr/ics/2012/191093.htm.

US Department of State, Department of Defense Inspectors General. "Inter-agency Assessment of Afghan Police Training and Readiness, Department of State Report No. ISP-IQO-07-07, Department of Defense Report No. IE-2007-001," November 2006. http://oig.state.gov/documents/organization/76103.pdf.

US Department of State, Office of the Historian. *Foreign Relations of the United States, 1964–1968, Volume I, Vietnam, 1964*, ed. Edward C. Keefer and Charles S. Sampson. Washington, DC: US Government Printing Office, 1992. http://history.state.gov/historicaldocuments/frus1964–68v01.

Foreign Relations of the United States, 1964–1968, Volume II, Vietnam January–June 1965, ed. David C. Humphrey, Ronald D. Landa, and Louis J. Smith. Washington, DC: US Government Printing Office, 1996. http://history.state.gov/historicaldocuments/frus1964–68v02.

Foreign Relations of the United States, 1964–1968, Volume III, Vietnam, June–December 1965, ed. David C. Humphrey, Edward C. Keefer, and Louis J. Smith. Washington, DC: US Government Printing Office, 1996. http://history.state.gov/historicaldocuments/frus1964–68v03.

Foreign Relations of the United States, 1964–1968, Volume IV, Vietnam, 1966, ed. David C. Humphrey. Washington, DC: US Government Printing Office, 1998. http://history.state.gov/historicaldocuments/frus1964–68v04.

Foreign Relations of the United States, 1964–1968, Volume V, Vietnam, 1967, ed. Kent Sieg. Washington, DC: US Government Printing Office, 1998. http://history.state.gov/historicaldocuments/frus1964–68v05.

Foreign Relations of the United States, 1964–1968, Volume VI, Vietnam, January–August 1968, ed. Kent Sieg. Washington, DC: US Government Printing Office, 2002. http://history.state.gov/historicaldocuments/frus1964–68v06.

Foreign Relations of the United States, 1964–1968, Volume VII, Vietnam, September 1968– January 1969, ed. Kent Sieg. Washington, DC: US Government Printing Office, 2003. http://history.state.gov/historicaldocuments/frus1964–68v07.

Foreign Relations of the United States, 1969–1976, Volume VI, Vietnam, January 1969–July 1970, ed. Edward C. Keefer and Carolyn Yee. Washington, DC: US Government Printing Office, 2006. http://history.state.gov/historicaldocuments/frus1969–76v06.

Foreign Relations of the United States, 1969–1976, Volume VII, Vietnam, July 1970–January 1972, ed. David Goldman and Erin Mahan. Washington, DC: US Government Printing Office, 2010. http://history.state.gov/historicaldocuments/frus1969–76v07.

Foreign Relations of the United States, 1969–1976, Volume VIII, Vietnam, January–October 1972, ed. John M. Carland. Washington, DC: US Government Printing Office, 2010. http://history.state.gov/historicaldocuments/frus1969–76v08.

Foreign Relations of the United States, 1969–1976, Volume IX, Vietnam, October 1972–January 1973, ed. John M. Carland. Washington, DC: US Government Printing Office, 2010. http://history.state.gov/historicaldocuments/frus1969–76v09.

Foreign Relations of the United States, 1969–1976, Volume X, Vietnam, January 1973–July 1975, ed. Bradley Lynn Coleman. Washington, DC: US Government Printing Office, 2010. http://history.state.gov/historicaldocuments/frus1969–76v10.

US Department of State, US Embassy Baghdad. "The 2006 Iraq Budget," 06BAGHDAD955, March 23, 2006.

"Allawi Back in Iraq; Seeks to Build Centrist Coalition," 07BAGHDAD612, February 20, 2007.

"AMB, CG and PM Discuss SOFA, SOI, Ambassador's Trip to Erbil, GOI/KRG Relations and Election Law," 08BAGHDAD3031, September 21, 2008.

"Anti-corruption Update," 09BAGHDAD2454, September 11, 2009.

"Are the Iraqi Prisons Working Yet? – An Assessment of Ministry of Justice/Iraqi Corrections Service (ICS) Operations," 09BAGHDAD2384, September 4, 2009.

"Badr Leader Agrees the Militia Should Demobilize," 06BAGHDAD930, March 21, 2006.

"Barzani Agrees to Push for New GOI Article 140 Committee Chair, Invigorate Committee," 07BAGHDAD2466, July 25, 2007.

"Building Capacity at the Ministry of Oil," 07BAGHDAD3837, November 25, 2007.

"CG and CDA Discuss Foreign Fighters and Syria, Turkey and the PKK, and UNSCR with PM," 07BAGHDAD3911, December 2, 2007.

"CODEL Inhofe Meets with TNA Members," 05BAGHDAD5051, December 19, 2005.

"Concerned Local Citizens Program: Securing Communities," 08BAGHDAD164, January 22, 2008.

"Delayed Gratification: Election Law Adopted," 09BAGHDAD3157, December 7, 2009.

"Demarche to Iraqi Interior Minister on Site 4," 06BAGHDAD2842, August 7, 2006.

"Embassy Delivers U.S. Proposal on Out of Country Voting DIP Note," 09BAGHDAD3310, December 22, 2009.

"Fallujans Mobilized for Election amid Increased Tension in City," 05BAGHDAD4971, December 13, 2005.

"Human Rights in the Interior Ministry; Director Complains of Marginalization, Threats," 07BAGHDAD1377, April 23, 2007.

"Implementation of Recommendations on Personal Protective Services: Status Report Update #1," 07BAGHDAD4001, December 10, 2007.

"Iraqi Anti-corruption Update for January 7," 10BAGHDAD0044, January 7, 2010.

"Iraqi Politics: Shifting Alliances and the Emergence of Issue-Based Coalitions," 08BAGHDAD3791, December 3, 2008.

"Iraq's Council of Representatives," 08BAGHDAD495, February 21, 2008.

"JCRED – Progress Report," 05BAGHDAD4084, October 3, 2005.

"Joint Reconstruction Efforts Support Baghdad Security Plan," 07BAGHDAD3045, September 11, 2007.

"Justice Pass to ODAG- JJOnes, OPDAT, IITAP, State Pass to INL, NEA-I," 09BAGHDAD2384, September 4, 2009.

"KRG Officials on Article 140 and Kirkuk," 10BAGHDAD64, January 11, 2010.

"Legal Ambiguity in Baghdad Governance Structures and Political Violence," 07BAGHDAD2040, June 20, 2007.

"Maliki on Cabinet Shake-Up, Return of Tawafaq, and Major Legislative Challenges," 08BAGHDAD166, January 22, 2008.

"Maliki on Concerned Local Citizens, Strategic Partnership Declaration, and Large-Scale Detainee Amnesty," 07BAGHDAD3721, November 13, 2007.

"Meeting between Deputy Secretary Negroponte and Deputy Prime Minister Barham Salih," 07BAGHDAD1991, June 17, 2007.

"Meeting with the Minister of Science and Technology Regarding the Tuwaitha Site," 07BAGHDAD2028, June 20, 2007.

"Meetings with GOI Officials Regarding the Tuwaitha Site," 07BAGH-DAD1960, June 14, 2007.

"Ministerial Capacity Surge Assessment," 08BAGHDAD1008, April 1, 2008.

"The New Joint Campaign Plan for Iraq," 07BAGHDAD2464, July 25, 2007.

"PDS Inefficient, but Potential Electoral and Confidence Building Tool," 08BAGHDAD713, March 10, 2008.

"PDS Issues to Watch Closely, Tariff and WTO Developments," 06BAGH-DAD3005, August 18, 2006.

"PM Maliki and President Talabani Discuss Elections and Terrorist Attacks with General Petraeus," 09BAGHDAD2998, November 15, 2009.

"Post Proposes Shoulder-Fired Missile Abatement Program for Iraq (MAN-PADS Reduction)," 09BAGHDAD2736, October 11, 2009.

"Reforming the Public Distribution System – Easier Said Than Done," 09BAGHDAD2621, September 29, 2009.

"Riding High after the Conference: Iraq's National Investment Commission's Next Projects," 09BAGHDAD3042, November 18, 2009.

"Summing Up Iraq's Year of Anti-corruption," 08BAGHDAD4058, December 29, 2008.

"Tuwaitha Request Letter," 07BAGHDAD2924, August 31, 2007.

"Tuwaitha Update: Meeting at Most – GOI Letter to IAEA Signed," 08BAGHDAD36, January 8, 2008.

"UK Ambassador Proposes Mediation for Iraq's Constitutional Review, Other Key Issues," 07BAGHDAD1756, May 27, 2007.

"USEB 018: Iraqi Government Signals Tougher Sentences for Criminals and Terrorists," 04BAGHDAD17, July 2, 2004.

"U.S. Sends Tough, but Necessary, Message during Detainee Meeting," 10BAGHDAD477, Secret, February 23, 2010.

US Department of State, US Embassy Kabul. "Afghan Commerce Minister Visit – Opportunity to Stress Our Priorities," 06KABUL5238, October 29, 2006.

"Afghanistan – A Spate of Good Economic News," 09KABUL872, April 6, 2009.

"Afghanistan Reports Better Than Expected Annual Revenues," 10KABUL488, February 9, 2010.

"Afghanistan Terror Finance – Disrupting External Financing to the Taliban," 07KABUL1555, May 8, 2007.

"Brokering Eradication Consensus in Helmand," 07KABUL1045, March 29, 2007.

"Charge's Initial Call on New Afghan Finance Minister," 09KABUL558, March 12, 2009.

"Codel Hayes Meets Karzai, Wardak," 06KABUL2723, June 15, 2006.

"Complaints to GIRoA on Pre-trial Releases and Pardons of Narco-traffickers," 09KABUL2246, August 6, 2009, "WikiLeaks Archive – A Selection from the Cache of Diplomatic Dispatches." *The New York Times*, June 19, 2011. Accessed January 8, 2013. www.nytimes.com/interactive/2010/11/28/world/20101128-cables-viewer.html#report/corruption-09KABUL2246.

"Corrected Copy: IDLG Director Popal Meets with Ambassador: Moving Ahead on the District Delivery Program," 10KABUL570_a, February 15, 2010.

"Corruption Threatens Mobile Money Pilot Program for Police," 09KABUL3863, December 3, 2009.

"Economic Agenda Items for Af-Pak Trilateral Commission," 09KABUL943, April 15, 2009.

"Election Preparations in RC-North: A Message to Governor Atta," 09KABUL2425, August 19, 2009.

"EXBS Afghanistan Advisor Monthly Border Management Initiative Reporting Cable – April 2007," 07KABUL1731, May 24, 2007.

"Finance Ministry Deflects Customs Reform," 05KABUL5052, December 14, 2005.

"GIRoA Appears to Retreat on Electoral Reform," 10KABUL577, February 15, 2010.

"Helmand Province – Poppy Eradication Force's Pre-planting Campaign Has Good Start," 07KABUL3135, September 18, 2007.

"IMF and Afghanistan Agree on Terms for Completing the Fifth Review; Ball in Afghans' Court," 09KABUL317, February 11, 2009.

"Karzai on ANSF, Cabinet, and 2010 Elections," 09KABUL4027, December 16, 2009.

"Karzai on Elections and the Future: September 1 Meeting at the Palace," 09KABUL2681, September 3, 2009.

"Karzai on the State of U.S.-Afghan Relations," 09KABUL1767, July 7, 2009.

"Karzai Urges Codel McCain to Support Zardari and Welcome Increase in U.S. Forces, 08KABUL3237," December 21, 2008.

"Kunar's Timber Industry and Smuggling: Solutions Await a New Cabinet," 09KABUL3792, November 28, 2009.

"Letter of Agreement on Police and Justice Projects," November 29, 2002.

"The New Cabinet: Better but Not Best," 09KABUL4070, December 19, 2009.

"Operation Moshtarak Moving to Governance Phase," 10KABUL695, February 25, 2010.

"Powerbroker and Governance Issues in Spin Boldak," 10KABUL467, February 7, 2010.

"Precariously Perched between Crisis and Recovery, Ghor Residents Hope for a Plentiful 2009 Wheat Harvest," 09KABUL1000, April 21, 2009.

"PRT/Badghis: Provincial Officials Steal Humanitarian Aid," 07KABUL1861, June 5, 2007.

"PRT/Lashkar GAH-Naw Zad Prosecutor Says Official Corruption Causes Alienation," 06KABUL874, March 1, 2006.

"Scenesetter: U.S.-Afghan Strategic Partnership Talks in Kabul – March 13," 07KABUL804, March 8, 2007.

"Urgent Resource Request to Support Afghan Border Management Initiative (BMI)," 05KABUL5185, December 20, 2005.

US Department of State, US Embassy Saigon, "Memorandum of Conversation," Saigon, December 17, 1970, 6 p.m. *Foreign Relations of the United States, 1969–1976, Volume VII, Vietnam, July 1970–January 1972*, Document 91.

"Memorandum of Conversation," Saigon, July 4, 1971, 10:40 a.m.–12:20 p. m. *Foreign Relations of the United States, 1969–1976, Volume VII, Vietnam, July 1970–January 1972*, Document 231.

"Post-election Priorities in South Vietnam Detailed," Cable. Secret September 2, 1967. Declassified December 15, 1994. Unsanitized. Complete.

"For the President's Files – Lord, Vietnam Negotiations," Sensitive, Camp David, Cables, 10/69–12/31/71.

"Telegram 10019 from Saigon," June 24, 1971. National Archives, RG 59, Central Files 1970–73, POL 14 VIET S, Telegram 6169 from Saigon.

"Telegram from the Embassy in Vietnam to the Department of State," Saigon, November 9, 1964 – 7 p.m. *Foreign Relations of the United States, Volume 1, Vietnam, 1964*, Document 408. http://history.state.gov/historicaldocuments/frus1964–68v01/d408.

"Telegram from the Embassy in Vietnam to the Department of State," Saigon, August 20, 1971, see Documents 250 and 225 – Editorial Note, *Foreign Relations of the United States, 1969–1976, Vietnam, Volume VII, 1971*.

US Department of State, US Regional Embassy Office Hillah. "Taking Steps to Uproot the Statist Legacy in Babil's Agriculture," 09HILLAH24, April 5, 2009.

US Government Accountability Office. "Securing, Stabilizing, and Rebuilding Iraq, Iraqi Government Has Not Met Most Legislative, Security, and Economic Benchmarks," GAO-07-1195, September 4, 2007.

"Stabilizing and Rebuilding Iraq – U.S. Ministry Capacity Development Efforts Need an Overall Integrated Strategy to Guide Efforts and Manage Risk," GAO 08-117, October 2007. Accessed April 11, 2012. www.gao.gov/new.items/d08117.pdf.

US House of Representatives. *Making Emergency Supplemental Appropriations for the Fiscal Year Ending September 30, 2005, and for Other Purposes, Conference Report* (To Accompany H.R. 1268). (109 H. Rpt. 72). www.congress.gov/congressional-report/109th-congress/house-report/72/1.

US House of Representatives, Committee on Foreign Affairs. "Foreign Policy and Mutual Security, Draft Report Submitted to the Committee on Foreign Affairs." Committee Print. Washington, DC: US Government Printing

Office, December 24, 1956. https://hdl.handle.net/2027/mdp.390150771
82163.

US Marine Corps. "Counterinsurgency Measures B4S5499XQ – Student Handout." The Basic School, Marine Corps Training Command, Camp Barrett,
VA, 2016. www.trngcmd.marines.mil/Portals/207/Docs/TBS/B4S5499XQ
%20CounterInsurgency%20Measures.pdf?ver=2016-02-10-114636-310.

US Marine Corps. *Small Wars Manual*. Philadelphia, PA: Pavilion Press, 2004.

US Navy. "Danang U.S. Naval Support Activity/Facility Danang – Command
History – 1970." OPNAV Report 5750 –1, May 10, 1971.

US Senate Armed Services Committee. *Iraq, Afghanistan, and the Global War on
Terrorism: Hearings before the Committee on Armed Services*, 109th Congress,
2nd Session, § Armed Services Committee, 2006.

US Senate Foreign Relations. "Evaluating U.S. Foreign Assistance to Afghanistan:
A Majority Staff Report Prepared for the Use of the Committee on Foreign
Relations." 120th Congress, 1st Session, 2011, S. Prt. 112-21, 19. www.gpo
.gov/fdsys/pkg/CPRT-112SPRT66591/pdf/CPRT-112SPRT66591.pdf.

US Senate Foreign Relations. *Vietnam: Policy and Prospects, 1970: Hearings before
the Committee on Foreign Relations, U.S. Senate, on Civil Operations and Rural
Development Support Program, Feb. 17, 18, 19, and 20, and March 3, 14, 17,
and 19, 1970.* Washington, DC: US Government Printing Office, 1970.

Van Buren, Peter. "Inside the World's Largest Embassy." *Mother Jones*. September 28, 2011. Accessed October 15, 2012. http://motherjones.com/politics/
2011/09/baghdad-peter-van-buren-we-meant-well.

"Vietnam Conflict – U.S. Military Forces in Vietnam and Casualties Incurred:
1961 to 1971," No. 402. US Bureau of the Census, Statistical Abstract of the
United States 1971 (92nd Annual Edition). US Department of Commerce.
Library of Congress. Washington, DC. No. 4-18089. 253.

Viggo Jakobsen, Peter. "Right Strategy, Wrong Place – Why NATO's Comprehensive Approach Will Fail in Afghanistan." *UNISCI Discussion Papers*, no.
22 (2010). www.redalyc.org/resumen.oa?id=76712438006.

Voice of America. "Iraqis Take Control of Last US Prison in Iraq," July 14, 2010.
www.voanews.com/english/news/middle-east/Iraqis-Take-Control-of-Last-
US-Prison-in-Iraq-98504894.html.

Vonnegut, Kurt. *Armageddon in Retrospect*. New York: Putnam, 2008.

Walt, Stephen M. *The Origins of Alliances*. Ithaca, NY: Cornell University Press,
1990.

Watts, Stephen, Jason H. Campbell, Patrick B. Johnston, Sameer Lalwani, and
Sarah H. Bana. *Countering Others' Insurgencies: Understanding U.S. Small-
Footprint Interventions in Local Context*. Santa Monica, CA: RAND Corpor-
ation, 2014.

Wehrle, Edmund F. *Between a River and a Mountain*. Ann Arbor: University of
Michigan Press, 2005.

Weigert, Stephen L. *Angola: A Modern Military History, 1961–2002*. 1st ed. New
York: Palgrave Macmillan, 2011.

Weiss, Michael, and Hassan Hassan. *ISIS: Inside the Army of Terror*. New York:
Regan Arts, 2015.

Weitsman, Patricia A. *Dangerous Alliances: Proponents of Peace, Weapons of War.* Stanford University Press, 2004.

Waging War: Alliances, Coalitions, and Institutions of Interstate Violence. Stanford University Press, 2013.

Wenner, Manfred W. *The Yemen Arab Republic: Development and Change in an Ancient Land.* Boulder, CO: Westview Press, 1991.

West, Bing. *No True Glory: A Frontline Account of the Battle for Fallujah.* New York: Random House Publishing Group, 2011.

Westmoreland, William C. "Appendix B – Republic of Vietnam Armed Forces," in "Report on Operations in South Vietnam, January 1964–June 1968," in *Vietnam War: After Action Reports.* Beverly Hills, CA: BACM Research, 2009.

"Chapter IV – The Year of Decision (1968)," in "Report on Operations in South Vietnam, January 1964–June 1968," in *Vietnam War: After Action Reports.* Beverly Hills, CA: BACM Research, 2009.

The White House. "Henry A. Kissinger Cable to Ambassador Bunker on President Thieu's Reply to President Nixon Regarding Peace Negotiations with Hanoi and the Withdrawal of North Vietnamese Troops from South Vietnam Meeting with Alexander Haig, Äù," Cable. Top Secret. November 12, 1972. Gerald R. Ford Presidential Library. Declassified May 6, 1997.

The White House, The National Security Council. "Instructions from the President to the Ambassador to Vietnam (Taylor)," Washington, DC, December 3, 1964. Johnson Library, National Security File, Aides File, McGeorge Bundy, Memos to the President. Top Secret. *Foreign Relations of the United States, 1964–1968, Volume I, Vietnam 1964,* Document 435.

"The Situation in Vietnam, Memorandum for the President," February 7, 1965. Declassified November 10, 1976. In the possession of the author. Musgrove Paper Files 6-103 (4). Copies likely available at the Johnson Presidential Library.

The White House, Office of the Press Secretary. "Remarks by the President in Address to the Nation on the Way Forward in Afghanistan and Pakistan." Eisenhower Hall Theatre, United States Military Academy at West Point, West Point, New York, December 1, 2009. Accessed October 4, 2012. www .whitehouse.gov/the-press-office/remarks-president-address-nation-way-for ward-afghanistan-and-pakistan.

Williams-Bridgers, Jacquelyn. *Iraq and Afghanistan: Security, Economic, and Governance Challenges to Rebuilding Efforts Should Be Addressed in U. S. Strategies.* GAO-09-476T Congressional Testimony. Washington, DC: US Government Accounting Office, 2009.

Winters, Paul A. *The Collapse of the Soviet Union.* San Diego, CA: Greenhaven Press, 1999.

Woods, Randall. *LBJ: Architect of American Ambition.* New York: Simon & Schuster, 2007.

Wright, Donald P., and Timothy R. Reese. *On Point II: Transition to the New Campaign: The United States Army in Operation Iraqi Freedom, May 2003–January 2005.* US Department of the Army, 1st ed. Fort Leavenworth, KS:

Combat Studies Institute Press, 2008. https://history.army.mil/html/book shelves/resmat/GWOT/OnPointII.pdf.

Yarhi-Milo, Keren. *Knowing the Adversary.* Princeton University Press, 2014.

Young, Marilyn Blatt. *The Vietnam Wars, 1945–1990.* New York: HarperCollins, 1991.

Young, Marilyn Blatt, and Robert Buzzanco. *A Companion to the Vietnam War.* Malden, MA: Wiley-Blackwell, 2006.

Index

(Page numbers in Bold refer to Tables.)

CPSIA information can be obtained
at www.ICGtesting.com
Printed in the USA
LVHW082022280720
661756LV00014B/286